PHASE SEPARATION IN CUPRATE SUPERCONDUCTORS

Proceedings of the Workshop on

REGIONAL OFFICE FOR SCIENCE
AND TECHNOLOGY FOR EUROPE

PHASE SEPARATION IN CUPRATE SUPERCONDUCTORS

Erice, Italy 6 – 12 May 1992

Edited by

K. A. Müller and G. Benedek

World Scientific
Singapore • New Jersey • London • Hong Kong

Published by

World Scientific Publishing Co. Pte. Ltd.
5 Toh Tuck Link, Singapore 596224
USA office: 27 Warren Street, Suite 401-402, Hackensack, NJ 07601
UK office: 57 Shelton Street, Covent Garden, London WC2H 9HE

British Library Cataloguing-in-Publication Data
A catalogue record for this book is available from the British Library.

PHASE SEPARATION IN CUPRATE SUPERCONDUCTORS

ISBN-13 978-981-02-1274-2
ISBN-10 981-02-1274-7

PREFACE AND SCOPE

All cuprate superconductors contain as an essential structural element CuO_2-planes with the Cu^{2+} and O^{2-} ions sited on a perovskite lattice and substantially hybridized: Undoped they form antiferromagnetic (AFM) insulators owing to the correlation. Upon doping with carriers — these essentially are holes for the highest T_c. — first a singlet (Zhang and Rice) is formed with the carriers in oxygen orbitals. A main question now arises: What happens if progressively more holes are present? Will one highly correlated phase form out of the singlet, or will a phase separation be induced with metallic (hole-rich) clusters and AFM areas?

As early as 1988, Hyzhnyakov and Sigmund of the University of Stuttgart had considered such a dynamic type of phase separation of the doped Mott-AFM insulator theoretically using molecular methods. With metallic two-dimensional droplets, one could understand the well-known phase diagrams of the cuprates as a function of hole concentration. These clusters would be slightly ferromagnetic and are called ferroic. Having close contacts to Stuttgart from earlier times, I followed their efforts continually, but there was not much experimental evidence or theoretical resonance in the field. Then in 1990, on the occasion of an honor at the University of La Sapienza in Rome, I learned of the results obtained by the group under Di Castro. With field-theoretical methods, he, Castellani and Grilli had arrived at the existence of phase separation in the normal state. With this knowledge, I visited the Brookhaven National Laboratory in spring of 1991, where I was introduced to the considerations by V. Emery on the subject, which he, Kivelson and Lin had published in 1990, regarding the dimensionality two as a boundary for dynamic phase separation: They discussed the gain in kinetic energy versus the loss of potential magnetic energy in the presence of metallic clusters. Thus I was confronted with the fact that three theoretical efforts pointed to the same possibility, unaware of each other despite being clearly sequential in time.

In the history of science, a number of famous cases are known in which a discovery occurred independently in different locations, if the time was ripe. Therefore I thought one should take the dynamic phase separation problem in the cuprates seriously, and decided to bring together in a small workshop not only all the above-mentioned groups, who agreed on phase separation, but also other interested theoreticians as well as an equal number of experimental physicists whose experiments appeared to be relevant to the subject. The total number of participants should be on the order of 30 to 40. Sufficient time for exposing recent results and discussions would be foreseen, and also some work related indirectly to the main subject could be included. With such a program in mind, I approached Giorgio Benedek of the University of Milano, with whom we had together with J.G. Bednorz successfully organized a school on "Earlier and Recent Aspects of Superconductivity" in the summer of 1989. My proposal met with an enthusiastic response, which, on his part, was translated not only into finding a convenient time in Erice for the workshop but also into raising the essential funding. Being myself an experimentalist, his well-known theoretical capacity would moreover ensure an equilibrated undertaking from the physics point of view as well. Even at this relatively short notice, we were able to set up a program for May 6 to 12 of this year.

May 1992 we felt was also the right time to come together. Furthermore, I had learned of another theoretical statistical mechanics effort by Dichtel using the Emery model that was in favor of phase separation. On the other hand, neutron-scattering experts and NMR researchers were rather skeptic. When they looked at their data, many of them were more in favor of a single phase — however, there is more than one notable exception among them. Therefore it was decided to invite representatives of such groups to bring a balanced workshop of "believers" and skeptics together. In retrospect, perhaps the number of NMR papers may have been too large for this purpose. Especially the NMR method reflects responses over long times despite being a local probe.

It became apparent during the workshop that local probes, if they are fast, such as Mössbauer spectroscopy, are quite relevant. Indeed, in the time since the workshop, it has turned out that electron paramagnetic resonance, three orders of magnitude faster than NMR, appears to contribute. This does not mean that the workshop was missing relevant experiments. On the contrary, the optical doping experiment by Alan Heeger and coworkers as well as the oxygen doping and thermal quenching experiment of the University of Stuttgart—Max Planck Institute collaboration clearly showed evidence of phase separation, probably driven electronically. However, also a chemical clustering of the negative countercharge balancing the created mobile holes would be conjointly present. Owing to the migration of the countercharge, these experiments may circumvent the difficulty of phase separation suppression induced by Coulomb repulsion as investigated by Emery. This is but one of the important issues for further research that became apparent during the workshop.

The sequence of papers printed in these proceedings is that of the workshop. From the titles of the various presentations it is apparent that discussions of the phase separation of structural or chemical origin were also emphasized. Another question addressed was, "If phase separation is present, what is the pairing mechanism in the metallic clusters?" Before the conclusion of the workshop, an extensive discussion took place and was recorded. Summaries of the various interventions of the participants as deduced from the recorded material have been compiled by Giorgio Benedek and are also included in this book. It is hoped that these proceedings will be a representative publication on the subject.

K. Alex Müller

ACKNOWLEDGMENTS

Phase Separation in Cuprate Superconductors was the Third Workshop of the International School of Solid State Physics, and was held at the Ettore Majorana Centre for Scientific Culture, Erice, Italy. The Editors are grateful to the Director of the Centre, Prof. A. Zichichi, and to the staff of the Centre for the excellent assistance and facilities. Mirjam Mann of the IBM Zurich Research Laboratory is gratefully acknowledged for her substantial efforts in preparing the workshop and her help in compiling the proceedings, and Charlotte Bolliger of the Publications Department for her assistance in assembling and editing the proceedings on our side. The Workshop was held under the auspices of the UNESCO Regional Office for Science and Technology in Europe (ROSTE) and the European Physical Society, and sponsored by the UNIDO International Centre for Science and High Technology (ICS), the Fondazione IBM Italia and INFM Italy.

TABLE OF CONTENTS

Frustrated Phase Separation and High Temperature Superconductivity

V.J. Emery
Dept. of Physics Brookhaven National Laboratory, Upton, NY 11973

S. A. Kivelson
Dept. of Physics UCLA, Los Angeles, CA 90024

ABSTRACT

A dilute system of neutral holes in an antiferromagnet separates into a hole-rich and a hole-poor phase. The phase separation is frustrated by long-range Coulomb interactions but, provided the dielectric constant is sufficiently large, there remain large-amplitude low-energy fluctuations in the hole density at intermediate length scales. The extensive experimental evidence showing that this behavior gives a reasonable picture of high temperature superconductors is surveyed. Further, it is shown that the scattering of mobile holes from the local density fluctuations may account for the anomalous normal-state properties of high temperature superconductors and also provide the mechanism of pairing.

1. Introduction

On the experimental level, high temperature superconductivity[1] is a robust phenomenon, occurring in a wide variety of materials containing CuO_2 planes. However, despite various analytical arguments suggesting that there is an attractive interaction between charge carriers, numerical experiments[2] on Hubbard, extended Hubbard, $t-J$, and related models have so far failed to produce any indication of a significantly enhanced pairing susceptibility, although the models are believed to incorporate, at least crudely, the most important physical interactions.[3] It is conceivable that more extensive numerical experiments may still succeed in finding evidence for superconductivity; nevertheless it is striking that such a gross property has failed to show up so far, and it seems likely that something is missing. We shall argue that the long-range Coulomb interaction is the essential piece of physics that has been ignored.

The central feature of high temperature superconductors is that they are doped insulators,[3] obtained by chemically adding charge carriers to a highly-correlated antiferromagnetic insulating state. We have argued previously,[4,5] on the basis of analytical and numerical studies of the two dimensional $t-J$ model that a low concentration of holes in an antiferromagnet is unstable to phase separation into a hole-rich and a hole-deficient phase. The $t-J$ model describes neutral holes (no long-range Coulomb interaction) with hopping amplitude t, and local moments with exchange interaction J. Phase separation certainly can be thought of as a strong attractive interaction between holes, although in a real sense the mechanism is more properly regarded as the ejection of holes from the antiferromagnet. The characteristic energy scale of this interaction is set by magnetic energies, so one expects to see phase separation only below temperatures of order the exchange energy J. In the presence of the long-range Coulomb repulsion between holes, the system cannot macroscopically phase separate unless the counterions are mobile. Thus, the system is frustrated. However, so long as the Coulomb interactions are not too strong, i.e. if the background dielectric constant is large enough, the local tendency toward phase separation will still have important consequences. Specifically, there will be large local density fluctuations so that a snapshot of the system at moderate doping will show metallic regions, in which the hole density is greater than average, and antiferromagnetic droplets, in which it is smaller than average. This structure may be frozen, as in a pinned charge density wave state, or fluctuating with very slow dynamics due to the metastability implied by the frustration.

In this paper, we propose to see what we can learn about the the properties of high temperature superconductors by looking at the materials from this new perspective. In particular, it will be shown that there is considerable evidence of a purely electronic tendency towards phase separation of the holes, and that the consequent large-amplitude local density fluctuations may provide an understanding of low-energy long-wavelength phenomena, such as the anomalous normal-state properties[6,7] and high temperature superconductivity itself. Indeed, it is widely recognized that the normal-state properties of high temperature superconductors imply the existence of a class of low-frequency collective excitations which strongly scatter the conduction electrons. Here we have, ready made, a rather obvious candidate. The presence of low-frequency collective modes typically indicates a nearby phase in which these fluctuations condense into a new ordered state. The only known ordered state which overlaps the metallic phases of the cuprate superconductors is the so-called spin-glass phase, in which there are frozen local magnetic moments but no long-range magnetic order.[8,9] We suggest that this phase should be regarded as a "cluster spin-glass", in which there is substantial local charge inhomogeneity, and the spins in the hole-deficient regions are locally Neel ordered, but with a random direction of the staggered magnetization. Thus the slow density fluctuations in the metal are related to the existence of the nearby cluster spin-glass phase.

Frustrated phase separation may also give rise to superconductivity, possibly with a high transition temperature. Indeed there are known examples of such behavior. One is the $t-J-V$ model in the weak hopping limit. Here, V is a repulsive interaction between holes on neighboring lattice sites. It has been shown[5,10] that a finite value of V frustrates the phase separation that occurs for $V = 0$ and that, for $0.168 < V/J < 0.75$, the system becomes a superconducting square liquid or dimer liquid. A second example, which is more physical but comes from another area of physics is provided by atomic nuclei, which have a gap in their energy spectrum analogous to that of a BCS superconductor.[11] Nuclei are droplets formed because the Coulomb interaction between protons frustrates the condensation of nuclear matter into a liquid state. The energy gap is quite small in nuclear matter[12] but relatively large in a finite nucleus. The reason is that pairing takes place in the low-density surface region of a nucleus, and the energy gap increases as the density decreases.[12] Thus the first effect of the nuclear force is condensation into a liquid state (phase separation), and the secondary effect is low temperature superconductivity. However the long-range Coulomb interaction frustrates the phase separation and leads to droplets (nuclei), which are high T_c superconductors. Part of this story is relevant for the cuprates, but, in that case, high temperature superconductivity is more subtle because, unlike nucleons, electrons have a short-range interaction which is repulsive.

This paper is organized as follows: In Sec. **2**, we summarize the theoretical arguments leading to the conclusion that dilute holes in an antiferromagnet are unstable to phase separation. These arguments are discussed in greater detail in our previously published work on the subject.[4,5] In Sec. **3**, we summarize arguments showing that collective modes with slow dynamics are a consequence of frustrated phase separation. We also introduce an effective model from which low-energy properties may be calculated. An important feature of this model is that the behavior is governed by a fixed point, which is consistent with the robustness and universality of the behavior of high temperature superconductors. In Sec. **4** we review experiments that demonstrate the existence of frustrated phase separation in the high temperature superconductors. First, we emphasize the fact, discussed extensively at this conference, that phase separation on laboratory time scales occurs in oxygen-doped La_2CuO_4 and in photo-doped materials. In both cases, the negative counter charges are mobile enough to allow phase separation to take place. We then consider the more usual situation in which the dopants are not mobile, and describe some of the experimental evidence to support the notion that the doped state is inhomogeneous at intermediate distances and times. We argue that the fact that it is difficult to produce electronically homogeneous materials is a reflection of the tendency of the system to phase separate locally in response to a fluctuation in the local potential. Whereas a normal metal would screen small local variations in the concentration of dopant ions, the doped antiferromagnet tends to overscreen, thus amplifying the effect of any small inhomogeneity. We also show that large-amplitude density fluctuations provide a natural explanation for a number of experimental anomalies in these materials and outline our understanding of the implications for the mechanism of high temperature superconductivity. Finally, in Sec. **5** summarize our results. Brief accounts of some of our ideas have previously been given in conference reports.[13] More detailed and extended accounts will be presented in papers currently under preparation.[14]

2. Neutral Holes in an Antiferromagnet

We have previously reported[4,5] on the results of extensive analytical and numerical studies of the $t-J$ model, which led us to suggest that phase separation always occurs at zero temperature, provided the hole concentrations x is less than a critical value $x_c(J/t)$. It is easy to see that two holes in an antiferromagnet attract each other. At the shortest distances, this attraction arises from the fact that two holes on nearest-neighbor sites break one less antiferromagnetic bond than two far-separated holes. At larger distances, two holes attract each other through the exchange of magnon pairs. In essence, phase separation is the best way to minimize the zero-point kinetic energy of the electrons and to reconcile the mobility of doped holes with the maintenance of local antiferromagnetic order (which optimizes the zero-point energy for the half-filled band).

A wide variety of models and materials exhibit phase separation, although the reasons may have nothing to do with antiferromagnetism. For instance, it has been shown in various three-band models of the copper oxide planes that, near to the charge-transfer instability, phase separation occurs on doping.[15] It also is possible that non-electronic effects, such as chemical interactions between the oxygen dopant atoms, could cause phase-separation in some range of doping. While these other interactions might augment the tendency toward phase separation in high temperature superconductors, we focus on the role of the antiferromagnetism because it is a prominent feature of the doped and undoped systems. Moreover, as we shall see, experiments indicate that one of the phases is essentially hole-free, suggesting a magnetic mechanism, rather than a charge-transfer instability, which would lead to hole-rich and hole-poor phases.[15]

As outlined above, studies of various models in various limits have *established* that phase separation is a common behavior of dilute holes in an antiferromagnet. We *believe*, but have not yet proven, that it is generic, essentially model-independent, behavior. However, if it were shown that a given model of holes in an antiferromagnet did not exhibit phase separation for a physically reasonable choice of parameters, we would turn to the experimental evidence (discussed below) that there is a strong local tendency toward phase separation in the high temperature superconductors, and conclude that that there is an important piece of physics missing from that model.

3. Implications of Frustrated Phase Separation

It is easy to demonstrate that a model with a short-range tendency for phase separation together with long-range Coulomb interactions is highly frustrated. In a coarse-grained sense, we can represent the hole-rich, hole-poor, and average-density phases as different orientations of a local "block spin". The local tendency toward phase separation is modelled as a short-range ferromagnetic interaction between spins while, the long-range Coulomb interaction corresponds to a long-range antiferromagnetic interaction. The latter is known to be highly frustrating; at the classical level it produces Devil's staircases[16] and a large degree of metastability. At the quantum level, the frustration is reflected in the existence of spatially-localized, large-amplitude density fluctuations, or collective modes, in which local regions of the hole-rich and hole-poor phases nucleate and disappear.[14] The large amplitude of the distortion implies that the collective modes are very "heavy" in the sense that they have little dispersion; even a modest disorder will localize them. Once again there is a large degree of metastability and, associated with it, extremely slow "glass-like" dynamics,[14] which is manifested via the spin fluctuations as spin freezing. The mobile holes create, annihilate and scatter from the collective modes. These processes, together with the distribution of energies of the collective modes determine the low-energy properties of the system. At present, it is not possible to give a fully deductive theory of all of these phenomena and their consequences, starting from e.g. an extended Hubbard model with long-range Coulomb interactions. Therefore we proceed in stages, first constructing a lower-level model from which the low-energy physics may be calculated. Such a theory will be described in detail in a future publication:[14] for the present, we describe the essential ideas.

Our approach is to calculate directly the consequences of the interactions between the mobile holes and the collective modes, which dominate the low-energy behavior of the system and determine the temperature-dependence of physical quantities. Because the collective modes correspond to local phase separation, they have significant internal structure; in particular, where the hole density is low, we expect behavior characteristic of the antiferromagnetic insulating state. For the internal degrees of freedom, we shall use the single-mode approximation, in which we ignore all internal excited states save one. This approximation is valid both for small clusters[17] and in the thermodynamic limit.[18] The single internal mode is a spin-1 excitation with momentum (π, π) and energy ω_g. An NMR experiment, or a neutron scattering experiment which measures the q-integrated intensity of the (π, π) peak, does not probe the spatial structure of the collective modes. Consequently, for many purposes, we may regard them as point-like objects and introduce operators $b_0^\dagger(\vec{r})$ and $b_{1,\sigma}^\dagger(\vec{r})$, which create a collective mode at position $\vec{r}$ in the spin-0 ground-state or in the spin-1, $S_z = \sigma = (-1, 0, +1)$ excited state respectively. These operators are well-defined in the adiabatic limit, where the collective-mode fluctuations are either frozen (due to the disorder) or slow compared to the equilibration times of the microscopic modes. Since phase separation is driven by the tendency of the antiferromagnetic ground-state to exclude holes, it follows that the existence of large-amplitude density fluctuations requires that the system be reasonably close to the adiabatic limit. By the same token, the maximum energy scale (ultraviolet cutoff) of the fluctuations must be of the order of the exchange integral, J. In general, we expect this approximation to be particularly good in low- or moderately-doped materials whereas, in sufficiently heavily-doped materials, corrections to the adiabatic approximation may be more significant. (A more complete discussion will be presented in a forthcoming publication[14]).

In order to calculate correlation functions, we note that the collective modes are local in space and that there cannot be two at the same point. Moreover, the fact the collective modes correspond to a local charge inhomogeneity implies that they have a dipolar character and that there is a local flow of current as they are created. Using these properties, we have shown[14] that the scattering of the mobile holes from the collective modes is equivalent to a two-channel Kondo problem. It has already been noted by Cox[19] that the behavior of the latter has much in common with the phenomenology of the normal state of high-temperature superconductors.[6,7,20] When the excitation energies of the collective modes may be neglected, the imaginary part of the susceptibility for $b_0^\dagger(\vec{r})$ has the form[21,22]:

$$\chi''(\omega,T) = \frac{1}{2} tanh(\omega/kT) \frac{\Gamma}{\omega^2+\Gamma^2} \tag{1}$$

where Γ is the Kondo energy scale.

We have found that this structure is impressed on many of the observed properties of high-temperature superconductors, in particular the optical conductivity[22] and spin fluctuations. As a specific example, we shall consider the dynamical structure factor $S(\omega,T)$ of the operator $b_{1,\sigma}^\dagger(\vec{r})b_0(\vec{r})$. For a material such as $YBa_2Cu_3O_{7-\delta}$ this function determines the relaxation rate T_1 of Cu spins measured by an NMR experiment,[23] and the q-integrated intensity of the (π,π) peak in neutron scattering. The point is that (π,π) is not a special nesting vector of the Fermi surface[24] in $YBa_2Cu_3O_{7-\delta}$ and thus the amplitude for spin excitations at this wave vector is significant only within the hole-poor part of a collective mode. We have shown[14] that, when ω is small compared to Γ, $S(\omega,T)$ is given by

$$S(\omega,T) = A(\frac{\omega_g}{kT})[e^{\frac{\omega_g}{kT}}F(\omega-\omega_g,T)+F(\omega+\omega_g,T)] \tag{2}$$

where

$$F(\omega,T) = \frac{\pi}{2}\delta(\omega) + \frac{1}{\Gamma(1+e^{-\frac{\omega}{kT}})} \tag{3}$$

and

$$A(x) = \frac{const.}{e^x+3} \tag{4}$$

If the excitation energy of the collective modes had been included it would have broadened the δ-function in Eq. (3). A comparison of these results with NMR and neutron scattering experiments will be made in Sec. **4**.

An additional feature of the fixed point of the two-channel Kondo problem is a resonance between the "impurity" and a pair of conduction electrons,[22,25] which may be responsible for high-temperature superconductivity. It is important to note that this is a fixed point theory and hence the conclusions are insensitive to the detailed structure of the materials. The dipolar nature of the collective modes is essential, otherwise we would have obtained a single-channel Kondo problem which does not have the same low-energy behavior as the normal state of high temperature superconductors.

The spatial structure of a collective mode may be probed by a q-resolved neutron scattering experiment. The major effect is a cutoff in the q-width of the (π,π) peak when the correlation length reaches the size of the cluster. The latter may be quite large for lightly-doped materials but is a few lattice spacings for superconducting materials, close to optimum doping, as we shall see later.

4. Experimental Evidence for Frustrated Phase Separation

The existence of phase separation in model calculations suggests an examination of experiments that might show if it is a feature of real materials. There are two quite different situations. If the dopant atoms are absolutely frozen, then clearly phase separation can occur only as a short-distance, fluctuation effect. This true of the majority of the cuprate perovskites. On the other hand, if the charge donors are mobile on laboratory time scales, they can be dragged along by the holes and compensate the long-range part of the Coulomb interactions; in this case, the holes can actually phase separate. This occurs in two special cases: In photo-doped materials and in oxygen-doped $La_2CuO_{4-\delta}$ where the oxygen ions remain fairly mobile to low temperatures. The evidence for true phase separation is discussed in other lectures at this workshop.[26]

Of course, there is an issue as to whether the observed phase separation is driven by the physics of the CuO_2 planes, or is induced by extraneous factors such as the oxygen chemistry. However we feel that the fact that phase separation occurs in *both* of the two known cases where the dopants are mobile provides strong experimental support for an electronically driven tendency for phase separation whenever the long-range Coulomb interactions do not prohibit it.

We shall now go on to consider the evidence for frustrated phase separation in materials in which the countercharges are immobile.

4.1 Lightly Hole-Doped Antiferromagnets

When the antiferromagnetic insulating "parent" materials of the high temperature superconductors are lightly doped with immobile charge-donors such as Sr, the antiferromagnetic long-range order is rapidly destroyed (at about 2% in Sr doped La_2CuO_4), although superconductivity requires somewhat larger dopant concentrations (of order 5% in Sr doped La_2CuO_4). We imagine that, in this entire range, if the dopant atoms were mobile, the system would phase separate below a characteristic temperature which is some fraction of the antiferromagnetic exchange energy J. Consequently, chemical or other kinds of inhomogeneity would be enhanced rather than screened, and would have an unusual effect on the electronic properties. Indeed, worries about sample homogeneity plague all studies of lightly doped materials.

4.1.1 Neutron Scattering

Some of the most interesting recent data on the lightly-doped material comes from neutron scattering studies of $La_{2-x}Sr_xCuO_{4-\delta}$ with x between 2% and 5%, where there is neither Neel order nor superconductivity. Recent data by Keimer et al[27] show a peak in the dynamical spin structure factor, $S(\vec{q},\omega)$, at $\vec{q} = (\pi,\pi)$, with an inverse correlation length $\kappa(x,T)$ which they fitted by the simple empirical relation

$$\kappa(x,T) = \kappa_o(x) + \kappa(0,T) \tag{5}$$

where κ_o is the zero temperature value of the inverse correlation length, a monotonically increasing function of the dopant concentration, and $\kappa(0,T)$ is the inverse correlation length of the undoped material.

There are other ways to fit this data, especially considering the large error bars.[28] Nevertheless, it is worth noting that Eq. (5) displays precisely the form we expect on the basis of frustrated phase separation and that we may deduce the size of the antiferromagnetic regions from the data: for x=2%, $\kappa_o^{-1} = 200\text{Å}$ and, for x=4%, $\kappa_o^{-1}(x) = 40\text{Å}$.

4.1.2 Optical Absorption

It has been found by Thomas et al[29] that the electronic contribution to the optical absorption is quite similar in lightly hole-doped and electron-doped materials. The salient features are two very broad characteristic absorption peaks in the insulating gap, one with an energy of order 500 meV, and the other with an energy of order 100 meV. They seem to be present in all the cuprate superconductors, although they may be more or less well-resolved, depending on their widths in the different materials. The transitions are certainly associated with bound charges, since in all cases the optical absorption vanishes as T and ω tend to 0. Later, when we consider more heavily-doped materials, it will be shown that there is a natural interpretation of the lower energy peak in terms of fluctuating phase separation.

4.2 Local Droplet Fluctuations in Superconducting Materials

We now examine the evidence that fluctuating local phase separation occurs in more heavily hole-doped, superconducting materials.

4.2.1 Photoemission

Thermodynamically, the *definition* of phase separation is that, in a region of the phase diagram, the chemical potential remains constant over a range of concentrations. From this perspective, the most direct evidence of phase separation, at least at short length scales, comes from photoemission studies[30] of both $La_{2-x}Sr_xCuO_{4-\delta}$ and $Nd_{2-x}Ce_xCuO_4$, which show that, as the hole concentration varies, the chemical potential remains within the insulating gap, while the density of states in the gap grows in proportion to the hole concentration. This is precisely the expected behavior for an inhomogeneous system in which insulating and metallic regions coexist in chemical equilibrium: The density of states in the gap arises from the metallic region, whereas the gap features are associated with the insulating regions. The conditions of chemical equilibrium require that the chemical potential be equal in the two regions, and hence it must lie within the insulating gap. This is quite different from the behavior of a semiconducting model or the *one-dimensional* Hubbard model.[31] Of course, photoemission is basically a high-energy, short-wavelength probe, so it is sufficient to invoke *dynamical* phase separation in order to account for these experiments.

4.2.2 Motion of the Apex Oxygens

Phase separation of the holes also has structural implications. The apical oxygens, which sit directly above or below the Cu ions, are a likely probe.[32] In the first place, it is generally true that only cuprates possessing apical oxygens can be hole-doped; the so called electron-doped materials, which do not have apical oxygens, cannot be hole doped.[33] A possible exception is the infinite-layer compound[34] $(Sr_{1-x}Ca_x)_{1-y}CuO_2$. Nevertheless, it is clear that the apical oxygens may play a significant role in compensating the charge of the holes in the CuO_2 planes. Moreover, measurements of the crystal structure as a function of hole concentration[35] in $YBa_2Cu_3O_{7-\delta}$ show that the average distance of an apical oxygen from the CuO_2 planes increases by about 0.15Å as δ increases from 0 to 1. We therefore conclude that there must be a significant coupling between the *local* hole concentration and the apical oxygen distance. In effect the oxygen motion modulates the local chemical potential of the holes and compensates the local charge (i.e. it reduces the effective strength of the long-range Coulomb repulsion between holes). Thus, the polarizability of the apical oxygens enhances phase separation and, at the same time, implies slow (ionic) dynamics for large amplitude density fluctuations. However this is not the only way in which the Coulomb interaction may be compensated: the essential ingredient is a large dielectric constant, which is a common feature of oxide superconductors. We emphasize that the important role of the apex oxygens is to contribute to the dielectric constant, *not* to provide an electron-phonon interaction.

Evidence for these effects has been obtained by Egami et al[36] in neutron scattering experiments on $Tl_2Ba_2CaCu_2O_8$. They found that there are dynamical fluctuations of 0.3Å in the position of the apical oxygens. This is a rather large displacement and it signifies large local differences in the local hole concentration.

4.2.3 Frozen Moments

Muon spin rotation experiments (μSR) on $La_{2-x}Sr_xCuO_{4-\delta}$ observe a so-called "spin-glass" phase, characterized by frozen magnetic moments, extending from the point at which long-range antiferromagnetic order is destroyed to higher Sr concentrations where the low temperature phase is superconducting.[8] The fraction of frozen moments and the spin freezing temperature are found to be continuously decreasing functions of the Sr concentration and to extrapolate to zero at x=15%. Similarly, in $YBa_2Cu_3O_{7-\delta}$ the fraction of frozen moments and the spin-freezing temperature extrapolate to zero for $\delta \approx 0.3$. At higher doping the hole-deficient droplets are fluctuating more rapidly than μSR time scales, which are quite long (of the order of a microsecond). No information about the droplet sizes or shapes can be extracted from the μSR data. Measurements using the Mossbauer effect give a similar picture and are reviewed by P. Imbert at this workshop.[9]

In the context of frustrated phase separation, the existence of frozen moments is quite natural. The system should be regarded as a cluster spin glass, in which the charge is inhomogeneous and the spins in the hole-poor regions locally Neel ordered, but with a random orientation of the staggered magnetization.

4.2.4 Magnetic Correlations

Nuclear magnetic resonance experiments on $YBa_2Cu_3O_{7-\delta}$ show that the relaxation rate of the nuclear spins is an order of magnitude larger on Cu than on oxygen[37]. The current interpretation is that the Cu nuclear spin is relaxed by antiferromagnetic fluctuations, which are strongly suppressed at an oxygen nucleus by geometric form factors[23]. The latter reflect the position of an oxygen atom between two Cu atoms and are given by $(1 + cosq_a)$ or $(1 + cosq_b)$, both of which vanish when $(q_a, q_b) = (\pi, \pi)$, the wave vector for antiferromagnetic order in the insulating phase[23]. Thus the experiments require that the wave vector at which spin fluctuations peak in the metallic phase is not too different from (π, π) and, indeed, this is confirmed by neutron scattering experiments[38,39] on $YBa_2Cu_3O_{7-\delta}$. From a microscopic point of view, this behavior is not easy to understand. The wave vector (π, π) does not appear to be special for the Fermi surface[24] of $YBa_2Cu_3O_{7-\delta}$ and indeed nesting vectors are, if anything, along the (1,0) and (0,1) directions. However, it is quite natural to have spin fluctuations at (π, π) if hole-poor regions occur as low-energy fluctuations. In fact the neutron scattering experiments provide our best estimate of the size of the hole-poor domains (about four lattice spacings for $YBa_2Cu_3O_{7-\delta}$ with $\delta = 0$).

According to the discussion of Sec. **3**, the relaxation rate of a Cu nucleus ${}^{63}T_1^{-1}$ is proportional to $S(\omega = 0, T)$, where $S(\omega, T)$ is given by Eqs. (2)-(5). It follows that:

$$^{63}T_1 = B(4 + e^{\frac{\omega_g}{kT}} + 3e^{-\frac{\omega_g}{kT}}) \tag{6}$$

where B is a constant. This expression provides a quite good description of the temperature-dependence of ${}^{63}T_1$ in $YBa_2Cu_3O_{6.63}$ (ref.40), with $\omega_g = 240K$, and in $YBa_2Cu_4O_8$ (ref.41), with $\omega_g = 285K$. The calculation of the Knight shift and T_1 for other nuclei will be presented in a future publication.[14]

Recent neutron-scattering experiments[42] on $YBa_2Cu_3O_{7-\delta}$ for $\delta = 0.4$ and $T_c = 53K$, were fitted to a phenomenological expression for $S(\omega, T)$, similar to that quoted in Eqs. (2) and (3), except that the $\delta(\omega)$ term in Eq. (3) was omitted. It was found that $\omega_g = 9meV$, which is about half as large as as the value required to fit the NMR experiments. As pointed out by Tranquada et al,[42] this may not be a serious discrepancy, since ω_g varies rapidly with small changes in oxygen content. Moreover, the value of ω_g was determined above T_c by NMR but below T_c by neutron scattering experiments which, at least initially, were designed to find the superconducting gap. Clearly it is desirable to fit the the results of neutron scattering experiments carried out above T_c to the expressions given in Sec. **3**.

$La_{2-x}Sr_xCuO_{4-\delta}$ is different from $YBa_2Cu_3O_{7-\delta}$ because there are incommensurate peaks near to (π, π) which are related to a Fermi surface instability.[43] Consequently, NMR and neutron scattering experiments on this material do not give *direct* information about the "doped-insulator" effects that we have explored here and it is inappropriate to invoke a spin gap. In our picture, any specific frequency- and temperature dependence is obtained because the dynamics of the collective modes are impressed on the motion of the mobile holes.[14] Experimentally,[44,45] the frequency-dependence of the q-integrated susceptibility was found to be consistent with $arctan(\omega/2kT)$. However, within experimental error, an equally good fit can be obtained using $tanh(\omega/2kT)$, as suggested by Eq. (1).

4.2.5 Optical Absorption Spectrum

While there is by now considerable agreement on the nature of the optical absorption spectrum in the cuprate superconductors from a few hundreths of an eV to a few eV, the interpretation of this data is still very controversial. There are two schools of thought: the one-component, or Drude school,[46] infers a single species of (mobile) charge carrier for which the scattering rate and effective mass depend strongly on frequency. On the other hand, the two-component or Drude-Lorentz school[29,47-50] proposes that there are two types of charge carrier: mobile charges which dominate the absorption at low frequency and which, at low temperatures, form the superconducting condensate; and bound charges which dominate the mid-infrared absorption but depend weakly on temperature.

There is no doubt that a two-component theory is required by the data in reduced T_c materials, where a well-defined finite-frequency peak in the absorption is observed in the normal state, as we saw earlier. However, in the highest T_c materials, the normal-state absorption is monotonic and is dominated by its Drude part (regardless of interpretation). It is not at all clear that the current data are sufficient to distinguish these two models unambiguously; the whole controversy turns on how seriously one is to take the continuity of behavior as a function of doping. There is, microscopically, only one type of charge carrier, hence the appeal of the one-component theory. On the other hand, continuity of phenomena at finite frequency is expected on general grounds, as stressed by the two-component enthusiasts. From the point of view of frustrated phase separation, it is natural to expect the spontaneous generation of a two component absorption spectrum, one associated with the metallic regions, the other with the insulating (hole-poor) regions. Moreover, it is clear that the separation into two components should begin to break down when the Drude response of the metallic regions extends out to frequencies characteristic of the density fluctuations. This interpretation should be testable, especially in reduced T_c materials.

Following the argument of Sec. **3**, we may obtain an expression for the contribution $\sigma_L(\omega)$ to the optical conductivity from the local collective modes. It is found that:[14]

$$\sigma_L(\omega) = const.\omega\chi''(\omega) \tag{7}$$

where $\chi''(\omega)$ is given in Eq. (1). This expression has a peak at Γ and, as argued in Sec. **3**, Γ should be of the order of the antiferromagnetic exchange integral J. It has been emphasized by Thomas[29] that there is a mid-infrared peak at this energy in the cuprates at all levels of doping. We have also found that, as a consequence of the fluctuating trapping of charge in the local collective modes, the major contribution to the optical conductivity at low frequency has the form:

$$\sigma_c(\omega) = const.Re\frac{\chi(\omega)}{i\omega} \tag{8}$$

where the imaginary part of $\chi(\omega)$ is given in Eq. (1). This expression is proportional to ω^{-1} when ω is larger than T, and to T^{-1} when T is larger than ω, as observed in optimally-doped materials[46,51]. Note that the frequency- and temperature dependence is attributed to the $tanh(\omega/2kT)$ prefactor in Eq. (1), and not to the scattering rate Γ.

5. Summary

We conclude by summarizing the gist of our approach and our most important conclusions.

1) There is a strong tendency for an antiferromagnet to expel holes, which leads to phase separation in a wide class of models of dilute, neutral holes in an antiferromagnet.

2) This tendency is present in the high temperature superconductors, as evidenced by the fact that whenever the countercharges are mobile on laboratory time scales, the holes do, in fact, phase separate into a hole-rich metallic phase and a hole-poor antiferromagnetic phase.

3) If the background charge is fixed, the Coulomb interactions between the holes prohibit true phase separation but, so long as the dielectric constant is large enough, the frustrated tendency toward phase separation is still important, and gives rise to large-amplitude, collective density fluctuations which have little dispersion and are readily pinned by even moderate disorder, so in practice they are spatially localized.

4) There are strong indications of local, fluctuating phase separation in all the superconducting cuprates. Specifically, the most striking evidence is: a) The chemical potential changes relatively little as a function of doping concentration, suggesting a tendency toward local phase separation. b) In most of the high temperature superconductors the dopant concentration is relatively high ($\sim 0.1 - 0.3$), and in the case of $YBa_2Cu_3O_{7-\delta}$ the Fermi surface is rotated by 90° relative to that of a simple tight-binding model on a square lattice. As a result, $\vec{Q} = (\pi, \pi)$ is not close to a nesting vector, and may not even be a spanning vector of the Fermi surface. Nevertheless, the dominant structure in the spin susceptibility is peaked in the neighborhood of $\vec{Q}$, as evidenced by neutron scattering and the difference between the O and Cu NMR. This surely implies that memory of the antiferromagnetic insulating state is retained even in the heavily doped metallic regime. Frustrated phase separation naturally accounts for the persistence of pronounced structure at $\vec{Q}$, whereas theories based on homogeneous states have greater difficulty. c) Less direct, but still compelling evidence comes from the optical response, in which spectral features characteristic of the lightly-doped insulator are found to persist in superconducting materials. d) Neutron scattering studies of apical oxygen distances provide support for the notion that the system is inhomogeneous on some intermediate time scale.

In our opinion, these relatively gross features of high temperature superconductors confirm the relevance of phase separation in these materials. But they also *suggest* an approach to a more-detailed understanding of low-energy phenomena; specifically the temperature- and frequency dependences of the optical conductivity and the dynamical spin susceptibility, the occurence of high temperature superconductivity, and the properties of the superconducting state. We have quoted some preliminary results which are sufficiently promising that we feel they are also worth summarizing here.

1) The dynamics of the localized collective density fluctuations are governed by the same fixed-point Hamiltonian as the two-channel Kondo problem. Since this model is critical, it implies that there are singnificant fluctuations on all energy scales below some high energy cutoff Γ, the Kondo energy scale. Despite the formal similarity, the present theory should be distinguished from the one-dimensional electron gas where the low-energy behavior is determined by the fixed *line* of the Tomonaga-Luttinger model,[51] rather than a fixed *point*. The difference is important because the behavior along a fixed line is non-universal, and typically depends on coupling constants, the carrier density, and the details of short-distance cutoffs. It is an essential feature of our theory that a fixed point determines the behavior of the system in the temperature range between Γ and T_c (where new coherent phenomena come into play). This behavior accounts for the robustness and universality of many of the normal-state features of the high temperature superconductors.

2) The existence of two-channel Kondo behavior is apparent in the physical properties of the normal state and provides a plausible explanation for the optical conductivity and the spin dynamics.

3) There is an enhanced local superconducting susceptibility as a result of a pairing resonance involving the local collective mode and the metallic electrons. Superconductivity appears when coherence develops between nearby regions of the solid. Elsewhere, we and others[22,25] have examined the nature and symmetry of this resonance. While it is clear that the resulting superconducting order parameter is spin singlet, we have not yet fully explored the remaining symmetry, although it appears[22] that it may be odd in frequency.[52]

Clearly, what we have sketched in the last three points is not a fully microscopic theory; it starts from an intermediate scale model based on localized collective modes of a specific variety. In that sense it is logically independent of the central discussion of the present paper. The connection with frustrated phase separation is that this concept provides an appealing rationale for such a model.

Finally, we wish to stress that there are many significant areas of overlap between the present theory and several other existing approaches to the theory of high temperature superconductivity.

1) The spin-bag theory[53] is based on the observations that it is energetically costly to place a hole in an antiferromagnet, and that the consequent tendency of holes to share regions of supressed antiferromagnetism amounts to an effective attraction between holes. We diverge from the spin-bag picture in our belief that the consequence is phase separation, not superconductivity. Long-range Coulomb interactions are essential, in our view, to frustrate the phase separation.

2) As was first noted by Cox,[19] the behavior of the two-channel Kondo problem has important similarities to the phenomenology emphasized in the marginal Fermi liquid theory.[7,54] As a minimum, our theory of the specific low-energy localized collective modes outlined above can be viewed as a higher level model, one step closer to the

microscopic, which rationalizes the more successful aspects of that phenomenology. Important features of these collective modes are that they are dipolar and that they generate their own (two-channel Kondo) dynamics through their interaction with the mobile holes. Also they have an internal structure, which is especially important in understanding the magnetic properties of the system. Their energy scale is found to be of order J, as required to explain the experiments. On the other hand, the collective modes previously invoked in the marginal Fermi liquid theory[7] were thought to be featureless charge-density excitations, and their dynamical properties were ***assumed*** at the outset. We also note that, although the behavior of the two-channel Kondo problem is marginal in some respects, it is truly *non*-Fermi liquid in character.[25] In particular, the electron self energy is proportional to $(\omega/\Gamma)^{1/2}$ at T=0.

3) The present theory satisfies, for the most part, the experimental constraints on any theory of the normal state of high temperature superconductors, outlined recently by Anderson.[20] In particular, the fact that the essential dynamical structure is determined by the two-channel Kondo *fixed point* accounts for the universality and robustness of the anomalous normal-state properties (i.e. those features inconsistent with Fermi liquid theory). However, we find enhanced pairing within a single CuO_2 plane and do not need to invoke a fundamental role for the tunnelling of electrons between planes.

4) At this workshop, quite distinct approaches to the implications of phase separation, that did not invoke frustration by long-range Coulomb interactions, are considered. C. Di Castro describes models in which superconductivity either occurs in a region of the phase diagram close to phase separation, or suppresses phase separation altogether. E. Sigmund attributes the metal-to-insulator transition, which occurs with increased doping, to the percolation of ferromagnetic polarons in the otherwise antiferromagnetic material, although it has yet to be established that ferromagnetic polarons occur for the strong antiferromagnetic exchange found in the cuprate superconductors.

6. Acknowledgments

We benefitted greatly from discussions with more of our colleagues then we can possibly list. However, we must particularly acknowledge useful discussions with G. Aeppli, A. Auerbach, R. Birgeneau, J. Budnick, D.L. Cox, A. Heeger, H.Q. Lin, G. Shirane, J. Tranquada, and A.Tsvelik. Innumerable discussions with Sudip Chakravarty have been central to the development of these ideas, and to our understanding of this problem. SK acknowledges the hospitality of Brookhaven National Laboratory and IBM Almaden Research Center. Part of this work was carried out while VJE was Kramers Professor at The University of Utrecht, The Netherlands, and he thanks Professor J.E. van Himbergen for his hospitality. SK was supported in part by NSF grant #DMR -90-11803. This work also was supported by the Division of Materials Sciences, U.S. Department of Energy, under contract DE-AC02-76CH00016.

7. References

1. J. G. Bednorz and K. A. Mueller, *Z. Phys.* **B64** (1986) 89.
2. For a review, see D. J. Scalapino in *High Temperature Superconductivity*, eds. K. S. Bedell, D. Coffey, D.E. Meltzer, D. Pines, and J. R. Schrieffer (Addison-Wesley, Redwood City, 1990) p. 314.
3. P. W. Anderson, *Science*, **235** (1987) 1196; V. J. Emery, *Phys. Rev. Lett.* **58** (1987) 2794; C. M. Varma, S. Schmitt-Rink, and E. Abrahams in *Novel Superconductivity*, eds. S. A. Wolf and V. Z. Kresin (Plenum, New York 1987).
4. V. J. Emery, S. A. Kivelson, and H-Q. Lin, *Phys. Rev. Lett.* **64** (1990) 475.
5. S. A. Kivelson, V. J. Emery, and H-Q. Lin, *Phys. Rev.* **B42** (1990) 6523.
6. P. W. Anderson, in *Frontiers and Borderlines in Many Particle Physics*, eds. J. R. Schrieffer and R. Broglia (North Holland, Amsterdam, 1988).
7. C. M. Varma, P. B. Littlewood, S. Schmitt-Rink, E. Abrahams, and A. E. Ruckenstein, *Phys. Rev. Lett.* **63** (1989) 1996.
8. A. Weidinger et al, *Hyp. Int.* **63** (1990) 147; B. J. Sternlieb et al., *Phys.Rev.* **B41** (1990) 8866; E. Torikai et al, *Hyp. Int.*, **63** (1990) 271.
9. P. Imbert, lecture at this workshop.
10. E. Dagatto and J. Riera, preprint.
11. A. Bohr, B. Mottelson, and D. Pines, *Phys. Rev.* **110** (1958) 936.
12. V. J. Emery and A. M. Sessler, *Phys. Rev.* **119** (1960) 248.
13. V. J. Emery, *Hyp. Int.* **63** (1990) 13; *Physica* **B169** (1991) 17.
14. V. J. Emery and S. A. Kivelson, to be published.
15. C. Di Castro, lecture at this workshop.
16. P. Bak and R. Bruinsma, *Phys. Rev. Lett.* **49** (1982) 249.
17. See, for example F. Figueierido et al, Phys. Rev. **B41** (1989) 4619.
18. S.Chakaravarty, B.I.Halperin, and D.Nelson, *Phys. Rev.* **B39** (1988) 2344.
19. D. L. Cox, unpublished.
20. P. W. Anderson, *Science*, **256** (1992) 1526.
21. A. M. Tsvelick, *J. Phys. Condens. Matt.* **2** (1990) 2833.
22. V. J. Emery and S. Kivelson, preprint.
23. H. Monien, lecture at this workshop.
24. W. Pickett, *Rev. Mod. Phys.* **61** (1989) 433.
25. A. W. W. Ludwig and I. Affleck, *Phys. Rev. Lett.* **67** (1991) 3160.
26. P. C. Hammel; A. Heeger; and M. Mehring; lectures at this workshop.
27. B. Keimer et al, *Phys. Rev. Lett.* **67** (1991) 1930 and M.I.T. preprint (1992).
28. S. Chakravarty, private communication, has shown that an equally good fit can be obtained from an analysis of the correlation length of the non-linear sigma model[18] under the assumption that the holes simply produce a renormalized spin-stiffness parameter.
29. G. Thomas in *High Temperature Superconductivity*, eds. D. P. Tunstall and W.

Barford (Adam Hilger, Bristol, England, 1991).
30. J. W. Allen et al, *Phys. Rev. Lett.* **64** (1990) 595.
31. E. H. Lieb and F. Y. Wu, *Phys. Rev. Lett.* **20** (1968) 1445.
32. See also A. Bianconi and H. Luetgemeier, lectures at this workshop.
33. Y. Tokura, H. Takagi, and S. Uchida, *Nature* **337** (1989) 345.
34. M. Azuma et al, *Nature* **356** (1992) 775.
35. R. J. Cava et al, *Physica* **C165** (1990) 419); J. D. Jorgensen et al, *Phys. Rev.* **B41** (1990) 1863.
36. T. Egami et al, *Physica* **C185-189** (1991) 867.
37.Y. Bertier; P. C. Hammel; H. Luetgemeier; I. Mangelschots; M. Mehring; lectures at this workshop.
38. G. Shirane et al, *Phys. Rev.* **B41** (1990) 6547; J. M. Tranquada et al, *Phys. Rev. Lett.* **64** (1990) 800; J. Rossat-Mignot et al, *Physica* **B163** (1990) 4.
39. P. Burlet, lecture at this workshop.
40. M. Takigawa et al, *Phys. Rev.* **B43** (1991) 247.
41. I. Mangelschots, lectures at this workshop.
42. J. M. Tranquada et al, *Phys. Rev.* **B46** (1992): B. J. Sternlieb, M. Sato, S. Shamoto, G. Shirane, and J. M. Tranquada, unpublished.
43. P. Littlewood, J. Zaanen, G. Aeppli and H. Monien, unpublished; Y. Zha, Q. Si, and K. Levin, unpublished.
44. T. E. Mason et al, *Phys. Rev. Lett.* **68** (1992) 1414; T. R. Thurston et al, preprint.
45. S. M. Hayden, lecture at this workshop.
46. Z. Schlesinger et al, *Phys. Rev. Lett.* **65** (1990) 801; L. D. Rotter et al, *Phys. Rev. Lett.* **67** (1991) 2741.
47. D. B. Tanner and T. Timusk, University of Florida preprint (1992).
48. Y. Watanabe, Z. Z. Wang, S. A. Lyon, N. P. Ong, D. C. Tsui, J. M. Tarascon, and E. Wang, Princeton University preprint, (1992).
49. C. M. Foster, K. F. Voss, T. W. Hagler, D. Mihailovic, A. J. Heeger, M. M. Eddy, W. L. Olson, and E. J. Smith, UCSB preprint, (1992).
50. See also lectures by B. Bucher at this workshop.
51. V. J. Emery in *Highly Conducting One-Dimensional Solids*, eds J. T. De Vreese, R. P. Evrard, and V. E. Van Doren (Plenum, New York, 1979): J. Solyom, *Adv. Phys.* **28** (1979) 201.
52. V. L. Berezinskii, *Pisma Zh. Eksp. Teor. Fiz.* **20** (1974) 628 [*JETP Lett.* **20** (1974) 287]; A. Balatsky and E. Abrahams, *Phys. Rev.* **B45** (1992) 13125; E. Abrahams, A. Balatsky, and J. R. Schrieffer, preprint.
53. J.R.Schrieffer, X.G.Wen, and S.C.Zhang, *Phys. Rev. Lett.* **60** (1988) 944; *Phys. Rev.* **B39** (1989) 11663; A.Kampf and J.R.Schrieffer, *Phys. Rev.* **B42** (1990) 7967.
54. A. E. Ruckenstein and C. M. Varma, *Physica* **C185-189** (1991) 134.

Phase Separation and Photo-induced High Tc Superconductivity in the Cuprates

G. Yu, C.H. Lee and A.J. Heeger
Institute for Polymers and Organic Solids
University of California, Santa Barbara
Santa Barbara, CA 93106, USA

Introduction

Following the discovery of high T_c superconductivity in the cuprates by Bednorz and Muller[1], considerable effort has been focused on the ability to control the carrier density, n_c, by chemical doping; either by mixed valence substitution, as in $La_{2-x}Sr_xCuO_4$ and in $Bi_2Sr_2Ca_{1-x}Y_xCu_2O_8$ or by variation of the oxygen content (δ), as in $YBa_2Cu_3O_{7-\delta}$.[2] Photo-excitation offers an attractive alternative method to vary the concentration of charge carriers with a number of advantages over chemical doping: n_c can be increased without the added complication of a change in chemical composition and crystal structure, and n_c can be increased in a transient manner with subnanosecond resolution by using pulsed lasers. The former is important for it allows one to study the system as a function of n_c without simultaneously changing the degree of disorder; the latter is important for it provides one with the opportunity to probe the dynamics of the electronic system at short times and well away from equilibrium.

Subnanosecond transient photo-excitation has been used to photodope the semiconducting cuprates. The resulting photo-induced transient conductivity in crystals of $YBa_2Cu_3O_{7-\delta}$, in ultrathin epitaxial films of $DyBa_2Cu_3O_{7-\delta}$, and in crystals of $La_2CuO_{4+\delta}$ was used to probe the state of the system following photo-injection of carriers onto the

CuO_2 planes.[3-6] Typical data for $YBa_2Cu_3O_{6.3}$ are presented in Figure 1 where we plot $\rho_{ph}(T)$ vs 1/T at several different levels of incident photon flux. Note that in Figure 1, the photo-induced resistivity, $\rho_{ph}(T)=1/\sigma_{ph}(T)$ where $\sigma_{ph}(T)$ is defined as the ***peak in the transient conductivity*** following transient photo-excitation. For details on the experimental methods and on the transient decay etc., the reader is referred to reference 5.

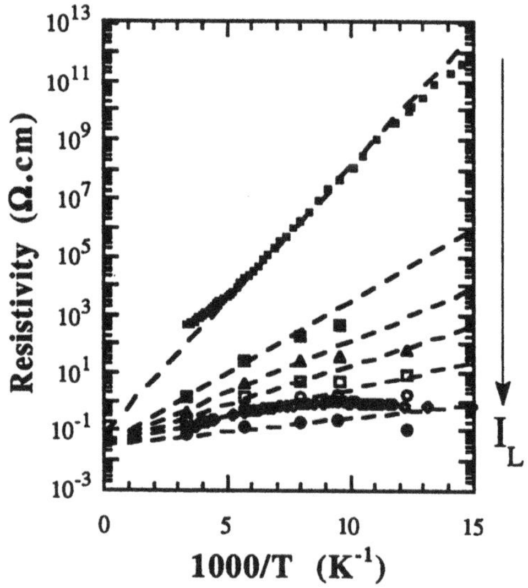

Fig. 1: $\rho_{ph}(T)$ in $YBa_2Cu_3O_{6.3}$ at several photon densities I_L. $I_L=1x10^{13}$ photons/cm^2 (⊞), $I_L=1x10^{14}$ photons/cm^2 (Δ), $I_L=1x10^{15}$ photons/cm^2 (□), $I_L=7x10^{15}$ photons/cm^2 (O), $I_L=1.1x10^{16}$ photons/cm^2 (◇), $I_L=2.3x10^{16}$ photons/cm^2 (●). The dark resistivity $\rho(T)$ in the same sample is also included (■), with a thermal activation energy $E_a \approx 0.18$ eV. The dashed lines are obtained by fitting the high temperature data to $\rho(T)$, $\rho_{ph}(T) = \rho_0 \exp(E_a/k_B T)$.

As shown in Figure 1, the transient photo-induced resistivities at different light levels (and their temperature dependences) in single crystals of $YBa_2Cu_3O_{7-\delta}$ are qualitatively and quantitatively similar to the doping-induced conductivities at different oxygen content, indicative of "photo-doping" over a wide range of resistivities (details about doping-induced conductivities in $DyBa_2Cu_3O_{7-\delta}$ over broad oxygen concentrations can be found in reference 5). Signatures of transient photo-induced transition to metallic behavior are observed at light intensities greater than about $5x10^{15}$ photons/cm^2 (characterized by $E_a \rightarrow 0$, and $\rho(T) \rightarrow \rho_o \approx 3x10^{-2}$ Ωcm).

"Photodoping" Effect in $MBa_2Cu_3O_{7-\delta}$ (M=Y, Dy etc.)

In the case of chemical doping, adding oxygen to the crystal leads to the occupation of oxygen ions on the O1 (and the O5) sites and creates carriers (holes) in the CuO_2 planes. In the case of photodoping, the absorption of photons with energy larger than the charge transfer gap creates electron-hole (e-h) pairs in CuO_2 planes (see Figure 2). We suggest that the electrons are quickly trapped (presumably captured near the O1 and O5 oxygen vacancies), leaving the holes as carriers in the planes. It is supported by the experimental

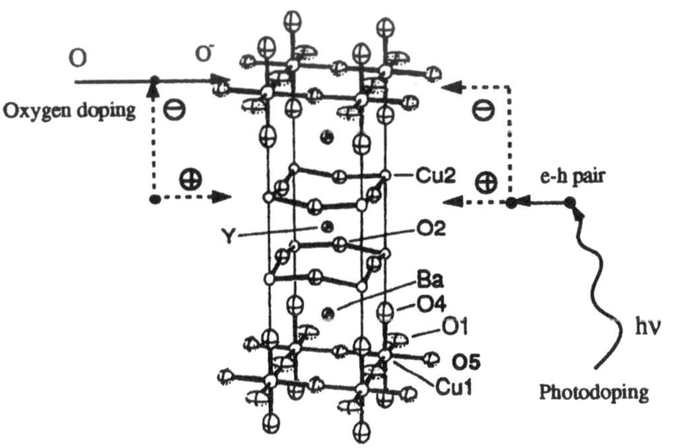

Fig. 2: Comparison between the "photodoping" and the oxygen doping in $YBa_2Cu_3O_{7-\delta}$.

observations of photoinduced absorptions, dynamic decay of the transient photoconductivity and various "dark" transport experiments.[7]

The similarity between chemical doping and photodoping implies that the magnitude of the carrier density plays the dominant role in both the mechanism of charge transport and the M-I transition in the $MBa_2Cu_3O_{7-\delta}$ materials. Moreover, the comparison between the transport results obtained by oxygen doping and by photodoping has provided insight into the mechanism of the M-I transition in these layered cuprates, it has been studied in detail in our early publications.[4, 5]

Minimum in $\rho_{ph}(T)$: phase separation and photoinduced high-T_c superconductivity

By transient "photo-doping", one can produce a sufficient density of carriers in $YBa_2Cu_3O_{6.3}$ to shift the Fermi level to the mobility edge and take the material to the borderline of metallic behavior.[4, 5] The obvious question is whether or not such a transient borderline-metal can condense to the superconducting phase. To have a better understanding, we plot $\rho_{ph}(T)$ on semi-log scales in Figure 3. In addition to $E_a \rightarrow 0$ and $\rho(T) \rightarrow \rho_o$ at $I_L > 5 \times 10^{15}$ photons/cm^2 (where ρ_o equals to the minimum metallic resistivity ρ_{min} in this material; see Figure 1), a minimum appears in $\rho_{ph}(T)$ below 100K. For I_L= 2.3×10^{16} photons/cm^2, the minimum value is a factor of ten below the value extrapolated from $\rho_{ph}(T)$ above 100K. This borderline metallic/super-conducting feature appears only at photon fluxes of 5×10^{15} cm^{-2} and above.

The observation of the minimum in $\rho_{ph}(T)$ at the highest excitation level (Figure 3) is suggestive of the onset of photoinduced transient superconductivity. Curves similar to those in Figure 3 were seen frequently during the period following the discovery of high-T_c superconductivity in resistivity measurements on samples with multi-phase composition.

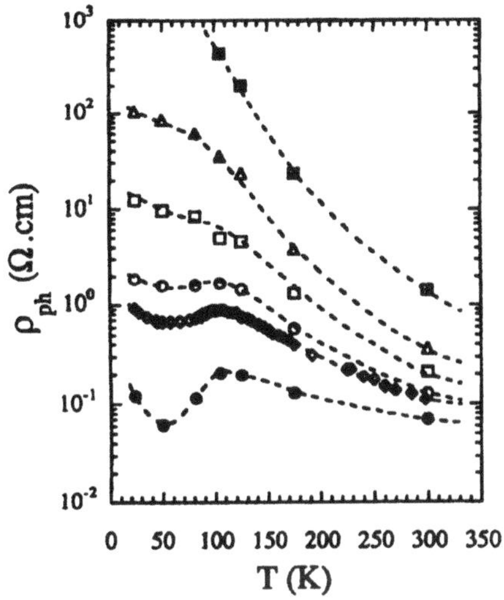

Fig. 3: Temperature dependences of the $\rho_{ph}(T)$ in $YBa_2Cu_3O_{6.3}$ ($E_a \approx 0.18$ eV), at several photon densities. The symbols and the corresponding light intensities are the same as in Fig. 1.

An example of $\rho(T)$ in ceramic $La_2CuO_{4+\delta}$ is given in Figure 4, which was measured by Kastner's group at MIT in 1987.[8] Plotted in Figure 4 are the resistivities of (i) a sample annealed in 130 atm of O_2 at 1225 K and cooled slowly, (ii) an as-prepared sample (cooled in air after synthesis), (iii) a sample annealed in a low flow of Ar at 675 K, and (iv) a sample annealed in a rapid flow of Ar to reduce the oxygen pressure as much as possible. The highly oxygenated ceramic shows a sharp superconducting transition with onset at 38 K, comparable to the transitions in single crystals of $La_2CuO_{4.05}$ and $La_{2-x}Sr_xCuO_4$. The as-prepared sample shows evidence of superconductivity, but with a very broad transition and non-zero resistivity even at very low temperatures. The partially reduced sample has a thermal activation energy, E_a, of approximately 30 meV at high temperatures and displays variable-range-hopping conduction at low temperatures. For the sample reduced under fast

Ar flow, E_a is only about 70 meV, suggesting a considerable density of localized states in the material. In addition to the E_a reduction due to increasing oxygen concentration, a minimum in $\rho(T)$ appears below T_c when the sample approaches the M-I transition (as indicated by the $E_a \to 0$ and the $\rho \to \rho_{min}$).

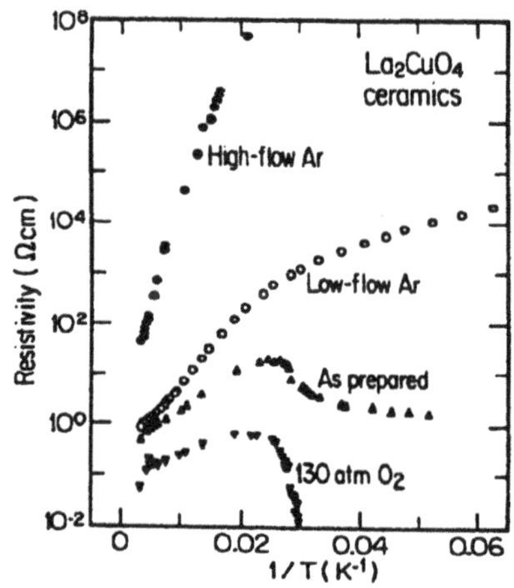

Fig. 4: Resistivity vs 1/T in $La_2CuO_{4+\delta}$ ceramics at several oxygen concentrations; from Kastner and co-workers.[8]

Similar phenomena also exist in another class of well-studied materials — granular superconducting films which exhibit "local superconductivity" below T_c with tunneling between grains.[9-13] In Figure 5, we compare the $\rho_{ph}(T)$ data from photodoped $YBa_2Cu_3O_{6.3}$ (Fig. 5a) and $\rho(T)$ in a granular superconducting film (Fig. 5b). The data in Fig. 5b are for different film thicknesses measured *in-situ* between successive depositions. The amount of material deposited was increased in very fine steps, typically 0.05-0.2 Å were used in the evaporations once initial electrical conductivity across the sample was established. It is well known that, in granular superconducting films, the transport behavior depends strongly on the cluster size. As the thickness (and thus the cluster size) increases, ρ(T) evolves through three major stages: non-superconducting, locally superconducting, and globally superconducting (zero resistivity below T_C). The term "local superconductivity" has been designated to the phenomenon characterized by a minimum in ρ(T) below T_C. More explicitly, the sample resistivity decreases rapidly below T_C, but does not go to zero, increasing again at low temperatures under circumstances where the clusters are

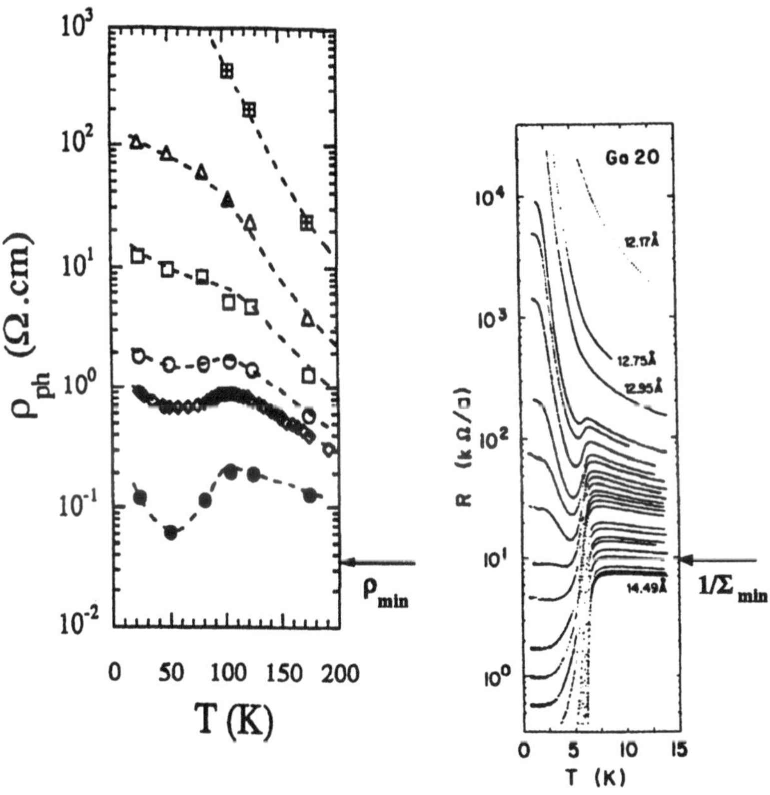

Fig. 5: Comparison between the $\rho_{ph}(T)$ in $YBa_2Cu_3O_{6.3}$ at different optical excitation levels (a), and ρ(T) in granular superconducting films of Ga with different deposition thicknesses (b). The values for $\rho_{min} = 1/\sigma_{min}$ are indicated. The data in Fig. 5b are from Jaeger and co-workers.[9]

not macroscopically connected.[9-11] Global superconducting behavior is achieved when the cluster size reaches the percolation threshold beyond which continuous superconducting paths are formed. At the percolation threshold, the normal-state sheet resistivity of the film falls below the maximum metallic resistivity, $\rho_{min} = 1/\sigma_{min}$.[9]

An important, general feature of $\rho(T)$ in inhomogeneous materials (Fig. 4), $\rho(T)$ in granular superconducting films (Fig. 5b), and $\rho_{ph}(T)$ in $YBa_2Cu_3O_{6.3}$ (Fig. 5a), is that T_c does not vary significantly with average carrier concentration, implying that the carrier concentration in the superconducting regions remains constant at the level of that in samples with optimized T_c. In contrast, T_c is suppressed as the dopant concentration is reduced in homogeneous bulk samples, or when the film thickness is reduced in homogeneous ultrathin films.[11, 14]

The overall similarities in both the magnitude and the shape of the $\rho_{ph}(T)$ curves to the $\rho(T)$ curves obtained from inhomogeneous media imply that the photoinduced charge carriers in $YBa_2Cu_3O_{6.3}$ spontaneously become inhomogeneously distributed after photogeneration; i.e. that phase separation occurs. The behavior of $\rho_{ph}(T)$ at high excitation levels is indicative of phase separation with disconnected metallic/superconducting regions separated by insulating regions (metallic/superconducting droplets): the decrease in $\rho_{ph}(T)$ near T_c is due to the onset of superconductivity, and the increase in $\rho_{ph}(T)$ at lower temperatures is due to the "series resistance" of the insulating regions.

Because of the phase separation, the resistivity is dominated by the insulating regions until the metallic droplets become connected above the percolation threshold. Thus, in the $YBa_2Cu_3O_{6.3}$ experiments, the insulating regions between metallic droplets would be expected to have conductivity intermediate between that of δ=0.7 and the δ corresponding to σ_{min} (near 0.55-0.6). The onset of superconductivity (as indicated by the onset of the resistivity decrease at T_c in Fig. 5a) is insensitive to the excitation level; i.e. the carrier concentration in hole-rich droplets is well into the metallic regime.

Metallic particles encapsulated in an insulating matrix have been investigated for many years.[15-17] If the particles are small and not in direct contact, an activation energy (associated with the capacitance energy of adding one extra charge to a small particle) is necessary for the transfer of charge from one particle to another. This results in local superconductivity below T_c and thermally-activated behavior above T_c. This effect probably contributes to the residual activation energy in the photoinduced conductivity experiments at light intensities above 10^{15} photons/cm^2.

An important distinction can be made among the three systems discussed above: in the photodoping experiment, the local superconductivity results from electronic phase separation—charge density fluctuations and/or inhomogeneous carrier distribution, while in the other cases the phase separation is chemical and structural—an inhomogeneous distribution of the dopants or of the metallic grains.

Other supporting evidence of intrinsic electronic phase separation in HTSC cuprates

There has been growing interest in the possibility of intrinsic electronic phase separation in HTSC cuprates. A review of recent developments in this field has been given by Emery and Kivelson.[18] Theoretical studies of the two-dimensional antiferromagnet have shown that dilute holes are unstable against phase separation into a hole-rich phase and a pure antiferromagnetic insulating phase; no region in the zero-temperature phase diagram was found in which a dilute gas of holes is stable.[19-21] In the presence of the long-range Coulomb interaction (for example when the counterions are uniformly distributed), there is still frustrated phase separation, leading to large-amplitude, low energy fluctuations in the hole-density at intermediate length scales, provided that the dielectric constant is sufficiently large.[18] These theoretical results suggest that phase separation might be an intrinsic feature of the antiferromagnetic cuprates, whether the carriers are injected through changes in oxygen concentration or through photo-doping.

It is also well known that in materials similar to the HTSC oxides (such as in perovskite compounds[22]), chemical phase separation is a common phenomenon; it is typically necessary to thermally anneal specimens at high temperatures to obtain a homogeneous material. At the relatively high temperatures at which the materials are prepared, the increased entropy associated with the dilute holes may partially overcome the tendency for phase separation, leading to a somewhat more uniform oxygen content throughout the $YBa_2Cu_3O_{7-\delta}$ series. Since, however, the photo-excitation is carried out at lower temperatures, entropic contributions to the free energy are less important, and more complete phase separation would be expected.

In HTSC cuprates, structural phase separation has been widely observed. A typical example is $La_2CuO_{4+\delta}$. This material undergoes phase separation when adding oxygen to the lattice, resulting in a hole-poor insulating phase and a hole-rich metallic/super-conducting phase. Bulk superconductivity can be obtained when doping the sample to a level of $\delta \geq 4\%$.[23] A recent study of the magnetization in $La_2CuO_{4+\delta}$ doped below the percolation threshold has provided more insight into this phenomenon.[24] Related observations in the $YBa_2Cu_3O_{7-\delta}$ system have been reported by Veal and co-workers.[25,26]

In the case of photodoping, there is supporting evidence of electronic phase separation of the photoinduced carriers.

(i) Although the holes are initially photogenerated uniformly within the optical absorption depth, hole mobility is inhibited by the antiferromagnetic spin correlations. Diffusion to form the phase separated droplets (as is shown in the calculations with t-J-V model[19-21]) results in locally hole-rich regions with relatively high mobility causing a diffusion limited increase in mobility $\mu(t)$ and, thereby, in photoconductance. The competition between this effect and charge carrier recombination results in the delay of the *peak* photocurrent.[3, 5] As the pump light intensity increases, the hole density in the resulting hole-rich regions is sufficient to shift the Fermi level through the mobility edge, to locally convert these regions into the metallic and superconducting phase, and to decrease the total resistivity of the (partially) phase-separated system.

(ii) The frequencies of the infrared active vibrational (IRAV) modes observed in photoinduced absorption do not correlate with the Raman modes of the insulating phase, but they do correlate with the Raman modes of the metallic phase.[27-31] This is true even in $Tl_2Ba_2Ca_{1-x}Gd_xCu_2O_8$ where the semiconductor (x=0.02) and the metal (x=0) have the same structure.[27, 28, 30] These results imply that photo-doping leads to metallic screening and associated shifts in the phonon frequencies characteristic of the metallic phase, consistent with droplet formation. Isolated holes would not give IRAV modes correlated with the metallic phase, especially when the crystal structure is the same in both cases. Moreover, in spite of the relatively large shifts ($\Delta\omega$~50-100 cm^{-1}), the IRAV are narrow and well-defined indicating a large size for the metallic phase and therefore consistent with phase separation; otherwise a broad featureless band covering the entire $\Delta\omega$ would be observed.

Although there is no direct experimental information on the length scale for the phase separation, the size of the droplets must be at least comparable to or larger than the pairing coherence length. This condition is more easily reached in HTSC cuprates (where $\zeta \approx$ 15Å[32]) than in conventional low T_c materials (where ζ is typically in excess of one thousand Ångstroms). Since in HTSC cuprates, the pairing coherence length is comparable to the Ioffe-Regel localization length, superconductivity is expected once the systems are induced into the metallic phase either by means of chemical doping or by photodoping.

Experiments on $YBa_2Cu_3O_{6.4}$: Evidence of Metastability at High Levels of Photo-excitation

Figure 6 shows a series of transient decay curves following different levels of photo-excitation for a sample with oxygen content even closer to the M-I transition, $YBa_2Cu_3O_{6.4}$; the data were taken at 35K. Since for this sample, $E_a \approx 0.03$ eV, a full set of experiments (for example at temperatures T>50K) were not possible because of the high thermal activation rate at high temperatures. At $I_L = 4.5 \times 10^{14}$ photons/cm^2, the magnitude of the transient photoconductivity is more than an order of magnitude below the value corresponding to the universal sheet conducance ($\sigma_{min} \approx 20$ S/cm; see Figure 1), and the time decay of the conductivity follows a power law ($\approx t^{-\beta}$). With increasing light intensity, the magnitude of σ_{ph} approaches σ_{min}, and the exponent of the power law decreases significantly. Finally, when the pump light intensity is strong enough to photogenerate the metallic state ($\sigma_{ph} \approx \sigma_{min}$), the photoconductivity persists at $\sigma_{ph} \approx \sigma_{min}$ to times in excess

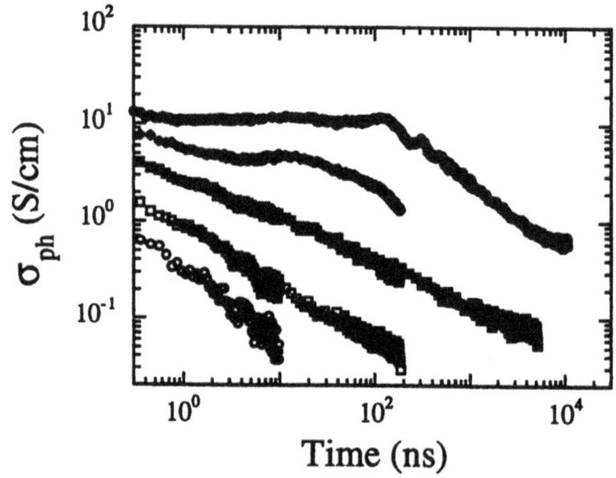

Fig. 6: Transient σ_{ph} in $YBa_2Cu_3O_{6.4}$ at 35K. (O) $I_L = 4.5 \times 10^{14}$ photons/cm^2, (□) $I_L = 8.6 \times 10^{14}$ photons/cm^2, (⊞) $I_L = 2.1 \times 10^{15}$ photons/cm^2, (◇) $I_L = 4.4 \times 10^{15}$ photons/cm^2, (●) $I_L = 1 \times 10^{16}$ photons/cm^2.

of 100 ns, corresponding to enhancement of the lifetime by approximately three orders of magnitude (see Figure 6). Similar phenomena have also been observed in La_2CuO_{4+d}.[6]

The time decay of the photo-induced conductivity is determined by,

$$\sigma(t)=n_c(t)e\mu(t), \tag{1}$$

where both $n_c(t)$ and $\mu(t)$ are time dependent; $n_c(t)\to 0$ as the carriers recombine and the system approaches the ground state; $\mu(t)\to 0$ as a result of the trapping of carriers in localized states.

The power law decay of the transient photoinduced conductivity at low excitation levels is consistent with dispersive transport of the carriers, as expected in disordered systems[33] where the Fermi level lies below E_c in the range of energies for which the electronic states are localized. Since the localized states act as traps, charge transport is realized via a sequence of multiple trapping and release steps among localized states in the presence of an applied electric field.[34, 35] Due to the disorder, the transfer time is a random variable. Thus, the accumulated sequence of these transfer events in the motion of a charge carrier can be viewed as a continuous-time random walk,[36-39] and the resulting photocurrent follows a power law time dependence, $t^{-(1-\beta)}$, where the exponent β is related with the decay power, α, by

$$\alpha = 1-\beta, \tag{2}$$

and β can be expressed as

$$\beta = T/T_o. \tag{3}$$

In the simplified model of an exponential distribution of the trap (localized state) density, T_o specifies the energy range over which the localized trap states can be filled.[33] For example, in the insulating phase of a HTSC oxides, k_BT_o corresponds to the energy between the Fermi level and the mobility edge, $|E_F-E_c|$ (the activation energy in resistivity measurements) so that

$$\beta \approx k_BT/E_a. \tag{4}$$

The data at low excitation levels in Figure 6 are consistent with dispersive transport. Since the thermal activation energy in the dark resistivity is about 33 meV in this sample (the Fermi level of the holes lies above the mobility edge by about 33 meV), at temperatures near 300K (k_BT=26 meV), the density of thermally activated charge carriers at the mobility edge is so large that photoinduced conductivity measurements become essentially impossible (the density of photo-excitations is less than the density of thermal excitations, and the resistivity is too low to support bias voltage without damaging the sample). At 35K (corresponding to 3 meV, so that $\beta \approx k_BT/E_a \approx 0.1$), the density of thermally excited carriers is negligible, and photo-induced conductivity can be observed as shown in Figure 6. The exponent in the time decay of σ_{ph} at the lowest light intensity is $\alpha \approx 0.9$ which is consistent with Eqs. 2-4. Within this model, at a fixed temperature, a decrease in β on increasing the light intensity implies a shift in the Fermi level toward the mobility edge indicating an approach of M-I transition by photoexcitation, consistent with the fact that the peak σ_{ph} approaches σ_{min}.

At higher excitation levels, the photoinduced conductivity persists at $\sigma \approx \sigma_{min}$ to times in excess of 100ns. This remarkable persistence of the photoinduced conductivity at $\sigma \approx \sigma_{min}$ suggests metastability in $n_c(t)$, consistent with the phase separation and the formation of metallic droplets.

Conclusion

Photo-doping of $YBa_2Cu_3O_{6.3}$, $DyBa_2Cu_3O_{6.3}$ and $La_2CuO_{4+\delta}$ to borderline metallic/superconducting levels has been demonstrated. Three specific signatures of the transient photo-induced transition to metallic behavior at light intensities greater than about 5×10^{15} photons/cm^2 were observed:

(1) The activation energy of $\sigma_{ph}(T)$ decreases as a function of the light intensity and approaches zero for $I_L \approx 5 \times 10^{15}$ photons/cm^2;

(2) A minimum appears below 100K in the temperature dependence of the photoresistivity, $\rho_{ph}(T) = 1/\sigma_{ph}(T)$.

(3) Metastability of the photo-excited metallic phase is observed for $YBa_2Cu_3O_{6.4}$ where the lifetime of the photo-excited state is enhanced by three orders of magnitude following photo-excitation to high levels.

The minimum in $\rho_{ph}(T)$ is reminiscent of the onset of superconductivity in inhomogeneous mixed phase HTSC samples, and similar to the "local superconductivity" found in granular metallic films. Such behavior is indicative of phase separation and metallic droplet formation subsequent to photo-excitation with disconnected superconducting regions separated by insulating regions: the decrease in $\rho_{ph}(T)$ near 100K is due to the onset of superconductivity, and the increase in $\rho_{ph}(T)$ at lower temperatures is due to the "series resistance" of the insulating regions.

The data are consistent with disorder induced localization as the origin of the metal-insulator transition. Analysis of the temperature dependence of the dark conductivity of $DyBa_2Cu_3O_{7-\delta}$ for different δ and the temperature dependences of $YBa_2Cu_3O_{6.3}$ and $DyBa_2Cu_3O_{6.3}$ for different I_L, yield minimum surface conductances which are consistent with $\Sigma_{min} \approx e^2/h$, the value predicted for two-dimensions where the minimum metallic conductivity is universal. The data separate into two categories: when $\Sigma > \Sigma_{min}$, the material is metallic, and when $\Sigma < \Sigma_{min}$, the material shows thermally activated behavior.

In conclusion, the experimental results have the following implications:

(1) The photo-excitations phase separate into metallic droplets.

(2) The metallic regions go superconducting below T_c of the heavily doped material.

(3) The photo-generated metallic state is metastable; the lifetime is enhanced by several orders of magnitude in samples with δ near the M-I transition.

(4) The M-I transition is caused principally by disorder-induced localization arising from the random distribution of oxygen at the O1 sites in the lattice.

Perhaps of equal importance is the sheer magnitude of the observed phenomena. The photoinduced conductivity changes span 10 to 15 orders of magnitude, or even greater, depending on the temperature, and the lifetime of the photo-induced conductivity is enhanced by more than three orders of magnitude at high levels of excitation.

Acknowledgement: The photoinduced conductivity and related characterization measurements were carried out at UCSB and supported by an NSF SGER grant (DMR91-02634) and by a UC INCOR Grant (LLNI-Incore 908A95). The ultrathin DBCO films were prepared at Minnesota. The YBCO single crystal samples were prepared at E.I. du Pont de Nemours and Co., Inc. (Central Research and Development Department).

References:

1. J. G. Bednorz and K. A. Muller, Z. Phys. B **64**, 198 (1986).

2. For recent reviews, see, for examples: *Physical Properties of High Temperature Superconductors, Vols. I and II*, ed. by D. M. Ginsberg (World Scientific, Singapore, 1990).

3. G. Yu, A. J. Heeger, G. Stuky, N. Herron, and E. M. McCarron, Sol. St. Comm. **72**, 345 (1989).

4. G. Yu, C. H. Lee, A. J. Heeger, N. Herron, and E. M. McCarron, Phys. Rev. Lett. **67**, 2581 (1991).

5. G. Yu, C. H. Lee, A. J. Heeger, N. Herron, E. M. McCarron, L. Cong, G. C. Spalding, C. A. Nordman, and A. M. Goldman, Phys. Rev. B **45**, 4964 (1992).

6. G. Yu, C. H. Lee, A. J. Heeger, S.-W. Cheong, and Z. Fisk, Physica C **190**, 563 (1992).

7. G. Yu, Ph.D dissertation (University of California at Santa Barbara, 1992).

8. M. A. Kastner, R. J. Birgeneau, C. Y. Chen, Y. M. Chiang, D. R. Gabbe, H. P. Jenssen, T. Junk, C. J. Peters, P. J. Picone, T. Thio, T. R. Thurton and H. L. Tuller, Phys. Rev. B **37**, 111 (1988).

9. H. M. Jaeger, D. B. Haviland, B. G. Orr, and A. M. Goldman, Phys. Rev. B **40**, 182 (1989).

10. D. B. Haviland, H. M. Jaeger, B. G. Orr, and A. M. Goldman, Phys. Rev. B **40**, 719 (1989).

11. D. B. Haviland, Y. Liu, T. Wang, and A. M. Goldman, Physica B **169**, 238 (1991).

12. R. C. Dynes, J. P. Garno, and J. M. Rowell, Phys. Rev. Lett. **40**, 479 (1978).

13. A. E. White, R. C. Dynes, and J. P. Garno, Phys. Rev. B **33**, 3550 (1986).

14. T. Wang, K. M. Beauchamp, D. D. Berkley, B. R. Johnson, J.-X. Liu, J. Znang, and A. M. Goldman, Phys. Rev. B **43**, 8623 (1991).

15. C. A. Neugebauer and M. B. Webb, Appl. Phys. **33**, 74 (1962).

16. R. M. Hill, Proc. R. Soc. London, Ser. A **309**, 377 (1969).

17. C. A. Neugebauer, Thin Solid Films **6**, 443 (1970).

18. V. J. Emery and S. A. Kivelson, preprint. (1992).

19. V. J. Emery, S. A. Kivelson, and H. Q. Lin, Phys. Rev. Lett. **64**, 475 (1990).

20. S. A. Kivelson, V. J. Emery, and H. Q. Lin, Phys. Rev. B **42**, 6523 (1990).

21. M. Grilli, R. Raimondi, C. Castellani, C. DiCastro, and G. Kotliar, Phys. Rev. Lett. **67**, 259 (1991).

22. N. F. Mott and E. A. Davis, in *Electronic Processes in Non-crystalline Materials* (Claredon, Oxford, 1979), Chapter 4.

23. C. Chaillout, J. Chenavas, S. W. Chiang, Z. Fisk, M. Marezio, B. Morosin, and J. E. Schirber, Physica C **170**, 87 (1990).

24. R. K. Kremer, E. Sigmund, V. Hizhnyakov, F. Hentsch, A. Simon, K. A. Muller, and M. Mehring, Z. Phys. B **86**, 319 (1992).

25. B. W. Veel, H. You, A. P. Paulikas, H. Shi, Y. Fang, and J. W. Downey, Phys. Rev. B **42**, 447 (1990).

26. B. W. Veel, A. P. Paulikas, H. You, H. Shi, Y. Fang, and J. W. Downey, Phys. Rev. B **42**, 6305 (1990).

27. C. M. Foster, A. J. Heeger, Y. H. Kim, G. Stucky, and N. Herron, Synth. Met. **33**, 171 (1989).

28. C. M. Foster, Ph.D dissertation (University of California, Santa Barbara, 1990).

29. K. F. Voss, Ph.D dissertation (University of California, Santa Barbara, 1992).

30. D. Mihailovic, C. M. Foster, K. F. Voss, T. Mertelj, I. Poberaj, and N. Herron, Phys. Rev. B **44**, 237 (1991).

31. A recent review is given by: D. B. Tanner and T. Timusk, preprint.

32. B. Batlogg, Physica B **169**, 7 (1991).

33. H. Sher, M. F. Shlesinger, and J. T. Bendler, Phys. Today **44**, 26 (1991).

34. N. F. Mott and E. A. Davis, *Electronic Processes in Non-Crystalline Materials* (Clarendon Press, Oxford, 1979).

35. N. F. Mott, Adv. Phys. **34**, 329 (1985).

36. E. W. Montroll and H. Sher, J. Stat. Phys. **9**, 101 (1973).

37. M. F. Shlesinger, J. Stat. Phys. **10**, 421 (1974).

38. H. Sher and E. W. Montroll, Phys. Rev. B **12**, 2455 (1975).

39. G. Pfister and H. Sher, Adv. Phys. **27**, 747 (1978).

Neutron Scattering Studies of the Spin Dynamics in $La_{2-x}Sr_xCuO_4$

S. M. Hayden[(1)], T. E. Mason[(2,3)], G. Aeppli[(3)], H. A. Mook[(4)], S–W. Cheong[(3)], Z. Fisk[(5)]

[(1)]*H. H. Wills Physics Laboratory, University of Bristol, Tyndall Avenue, Bristol BS8 1TL, UK*

[(2)]*Risø National Laboratory, Roskilde, DK-4000, Denmark*

[(3)]*AT&T Bell Laboratories, Murray Hill, New Jersey 07974, USA*

[(4)]*Oak Ridge National Laboratory, Oak Ridge, Tennessee 37831, USA*

[(5)]*Los Alamos National Laboratory, Los Alamos, New Mexico 87545, USA*

ABSTRACT

Inelastic neutron scattering has determined the modulation vectors and correlation lengths for the low-energy incommensurate magnetic fluctuations observed in $La_{2-x}Sr_xCuO_4$ for $x \gtrsim 0.075$. On entering the superconducting state a strong suppression of the low-energy fluctuations is observed. These results are discussed with reference to a two-dimensional band structure model.

1. Introduction

Neutron scattering can provide valuable information about the electronic correlations in solids through the spin correlation function. It has proved to be highly informative in the high–T_c problem where spin fluctuations have now been detected across the $La_{2-x}(Ba,Sr)_xCuO_4$[1,2] and YBa_2CuO_{6+x}[3] series. We report here experiments on highly doped ($x \gtrsim 0.075$) $La_{2-x}Sr_xCuO_4$, where the incommensurate magnetic scattering has been characterised in the normal[4] and superconducting[5] states. The relationship to an itinerant picture of the carriers is discussed.

2. Preliminaries

Except for small x, the $La_{2-x}(Sr,Ba)_xCuO_4$ system does not show well defined magnetic excitations. 'Diffuse' magnetic correlations can be described in terms of the power spectrum $|m(\boldsymbol{q},\omega)|^2$ of the local spin magnetisation $m(\mathbf{r},t)$. The power spectrum is the Fourier transform in space and time of the autocorrelation function $\langle m(\boldsymbol{r}', 0)m(\boldsymbol{r}'+\boldsymbol{r}, t)\rangle$. Neutron scattering measures this power spectrum directly, where m is the component of the magnetisation perpendicular to the momentum transfer $\mathbf{Q} = \boldsymbol{k}_i - \boldsymbol{k}_f$, $\boldsymbol{k}_i$ and $\boldsymbol{k}_f$ are the incoming and outgoing neutron wavevectors. More formally the neutron scattering cross section is,

$$\frac{d^2\sigma}{d\Omega dE} = \sigma_0 \frac{k_f}{k_i} \sum_{\mu\nu} (\delta_{\mu\nu} - \hat{Q}_\mu \hat{Q}_\nu) \frac{1}{2\pi} \int dt \exp(i\omega t) \langle m_\nu(\boldsymbol{Q},t) m_\mu(-\boldsymbol{Q},0) \rangle. \quad (1)$$

Another quantity of interest is the closely related imaginary part of the generalised susceptibility $\mathrm{Im}\chi(\boldsymbol{q},\omega) = \chi''(\boldsymbol{q},\omega)$. This is a measure of the absorption when the system is probed by a spatially and temporally varying field $\boldsymbol{B}(\boldsymbol{r},t)$ (eg due to the magnetic moment of the neutron). The power spectrum (derived from the unsymmetrised correlation function) and the response function $\chi''(\boldsymbol{q},\omega)$ are related through the fluctuation–dissipation theorem,

$$|m(q,\omega)|^2 \propto [n(\omega) + 1]\chi''(\boldsymbol{q}, \omega), \qquad (2)$$

where the Bose-Einstein factor $n(\omega)+1=\{1-\exp(-\hbar\omega/k_B T)\}^{-1}$.

3. Experiments

Throughout the paper, we label reciprocal space using a (nearly) square-lattice notation appropriate to the CuO_2 planes, the real space separation of the Cu ions is one distance unit, $a_0 = 3.8$ Å, as in Fig. 1(a). Fig. 1(b) shows the position of the reciprocal lattice points due to the ionic lattice, these have coordinates ($m2\pi$, $n2\pi$), where $\boldsymbol{Q} = (Q_x, Q_y)$. The unit–cell doubling associated with the antiferromagnetism in La_2CuO_4 introduces new points at positions ($(2m+1)\pi$, $(2n+1)\pi$). We will not be concerned with the component of $\boldsymbol{Q}$ perpendicular to the CuO_2 planes, since the strong anisotropy in the magnetic interactions leads to no observable variation of the scattering described here with this component. Neutron scattering experiments have shown that for small doping in $La_{2-x}(Sr,Ba)_xCuO_4$ and $YBa_2Cu_3O_{6+x}$ the antiferromagnetic long–range order is replaced by commensurate fluctuations[2,3] with a correlation length of the order of the separation of the dopant atoms. At higher doping levels, incommensurate fluctuations have been reported[2] in $La_{2-x}Sr_xCuO_4$ but not for $YBa_2Cu_3O_{6+x}$.

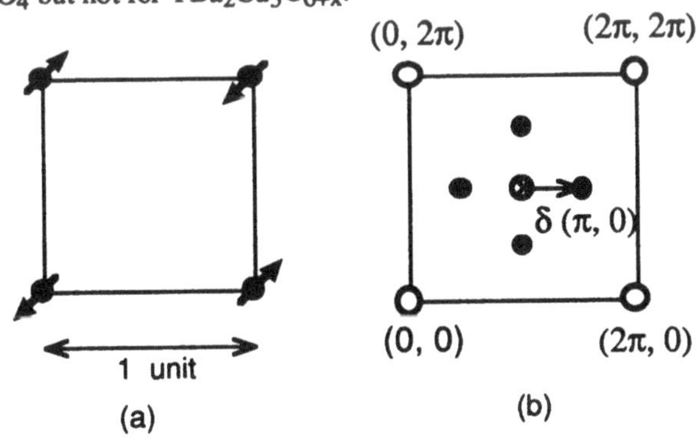

Fig. 1. (a) The lattice of Cu atoms in a nearly square CuO_2 plane. Throughout this paper we measure real space distance in units of a_0 =3.8 Å and reciprocal space in units of ($1/a_0$). (b) Reciprocal space with open circles representing the positions of lattice points due to the ionic lattice; the crossed circle represents the extra point introduced due to the doubling of the unit cell by antiferromagnetic order in La_2CuO_4. The closed circles represent (not to scale) the positions of the incommensurate peaks described in the text.

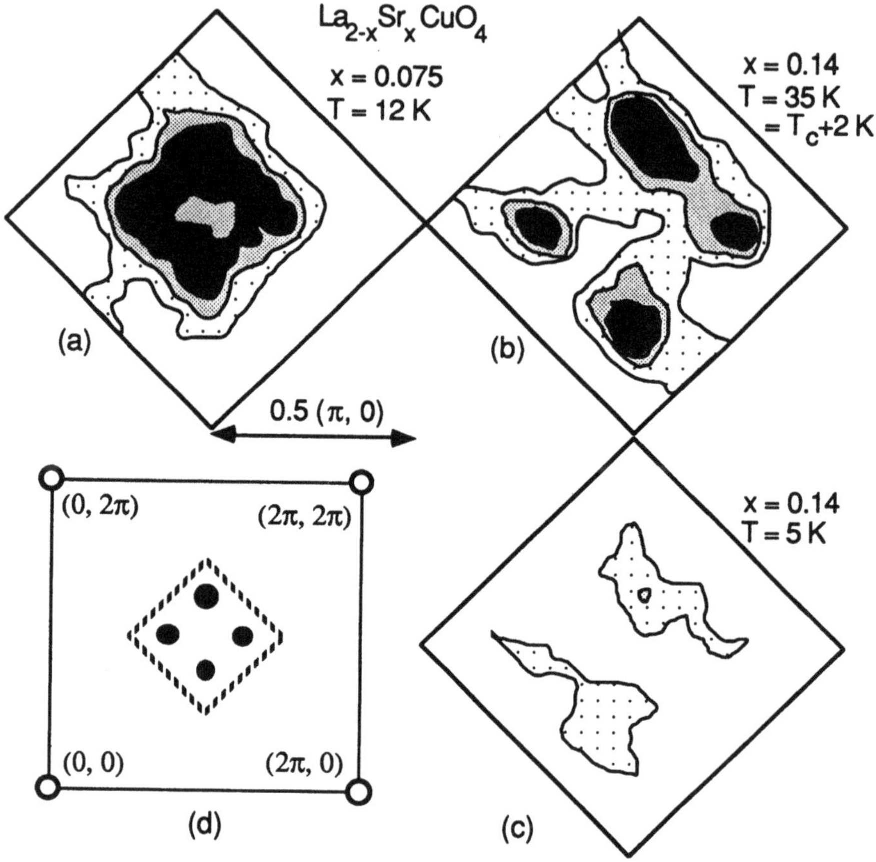

Fig. 2. The magnetic scattering observed around Q = (π, π) at low energies in $La_{2-x}Sr_xCuO_4$. (a) Contour plots of scattering collected at $\hbar\omega$ = 1 meV for x = 0.075 from Cheong et al.[4] The dark regions represent the areas with the highest scattered intensity. Some spurious scattering is present near (0.9π, 1.1π). This could be eliminated under different experimental conditions. (b) and (c) show scattering collected[6] for $\hbar\omega$ = 2 meV and x =0.14 for (b) T = 35 K = T_c + 2K and (c) T = 5 K. For $T < T_c$, $\chi''(q,\omega)$ is strongly suppressed (see text). (d) The region within the dashed square represents the area covered by the maps in (a)–(c).

3.1. Normal State

Fig. 2(a) shows the form of the incommensurate scattering at low energy, $\hbar\omega$ = 1 meV, near the (π, π) position in $La_{1.925}Sr_{0.075}CuO_4$[4]. From Fig. 2(a), it can be seen that the scattering shows square-like symmetry, with the strongest scattering at the corners of the square. The scattering in Fig. 2(a) can be well described by four resolution-corrected

Gaussian peaks centred on the corners of the square, ie the positions $(\pi, \pi) \pm \delta(\pi, 0)$ and $(\pi, \pi) \pm \delta(0, \pi)$. The best fit to the data gives $\delta = 0.14 \pm 0.02$ and the half-width-at-half-maximum (HWHM) of the Gaussians $\xi^{-1} = 0.055 \pm 0.003$ Å^{-1}. Fig. 2(b) shows a contour map[6] for the $x = 0.14$ composition. The peaks move out from (π, π) with increased doping.

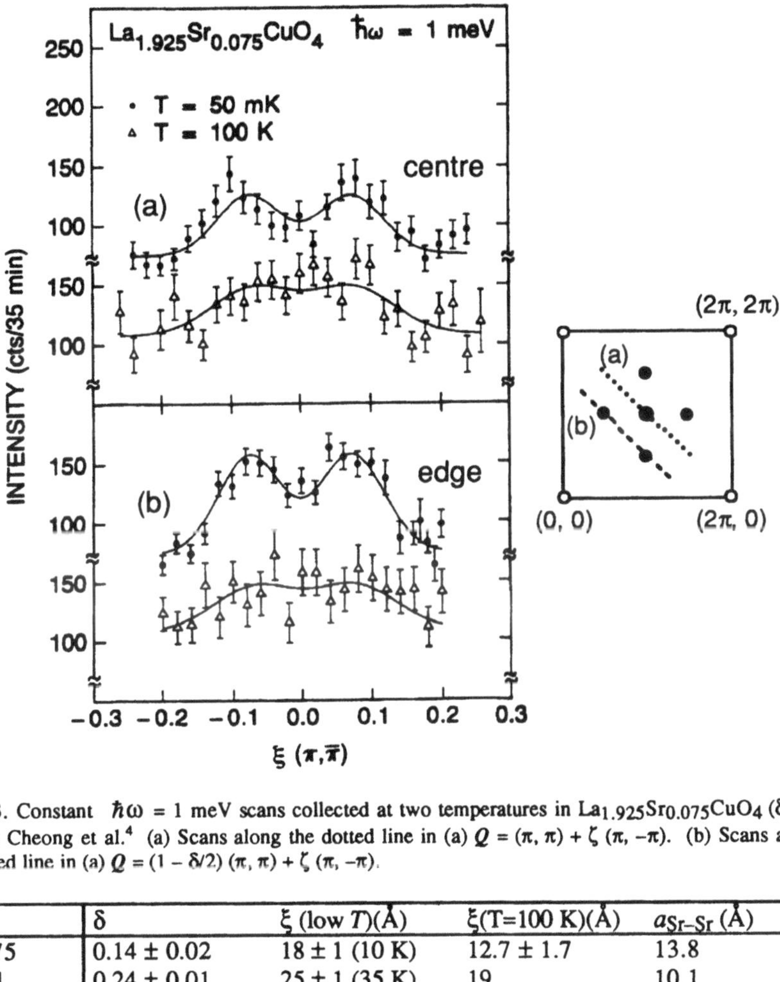

Fig.3. Constant $\hbar\omega$ = 1 meV scans collected at two temperatures in $La_{1.925}Sr_{0.075}CuO_4$ (δ = 0.14) from Cheong et al.[4] (a) Scans along the dotted line in (a) $Q = (\pi, \pi) + \zeta\,(\pi, -\pi)$. (b) Scans along the dashed line in (a) $Q = (1 - \delta/2)\,(\pi, \pi) + \zeta\,(\pi, -\pi)$.

x	δ	ξ (low T)(Å)	ξ(T=100 K)(Å)	a_{Sr-Sr} (Å)
0.075	0.14 ± 0.02	18 ± 1 (10 K)	12.7 ± 1.7	13.8
0.14	0.24 ± 0.01	25 ± 1 (35 K)	19	10.1

Table 1. Wavevectors (δ) and inverse half-width-at-half-maximum widths (ξ) of the low energy ($\hbar\omega$ = 1 meV) incommensurate peaks in $La_{2-x}Sr_xCuO_4$. Values determined from refs. 4 and 5.

Of great interest is the effect of temperature on the scattering, Fig. 3 shows $\hbar\omega = 1$ meV scans parallel to the $(\pi, -\pi)$ direction[4]: (i) through the centre of the box and (ii) along the edge for $T = 0.05$ K and $T = 100$ K. The effect of warming to 100 K is to cause a 'filling in' of the middle of both scans. Thus, qualitatively, the four-peak structure becomes a broad maximum centred on (π, π), with the intensity at (π, π) increasing on warming as previously observed by Shirane et al.[7] A quantitative analysis of the peak positions and widths is given in table 1.

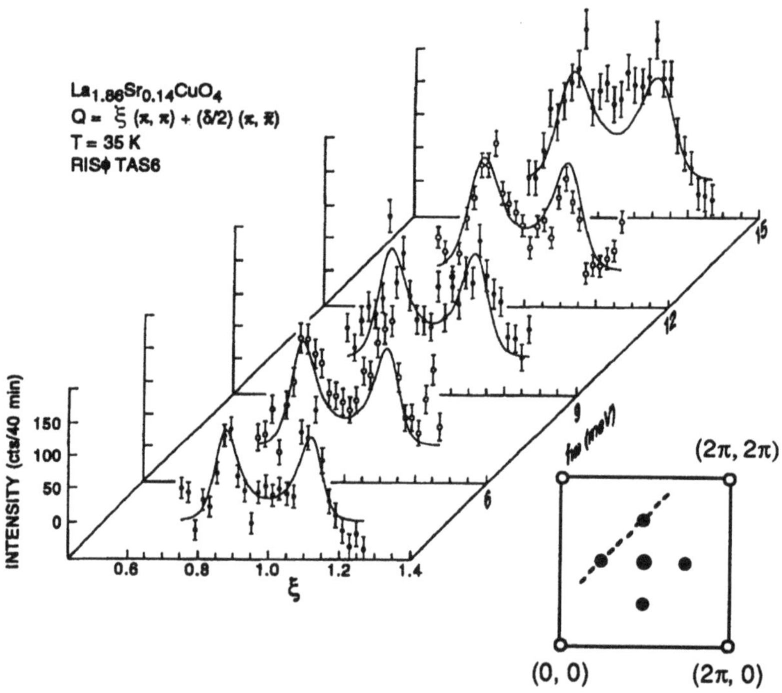

Fig. 4. A series of constant-$\hbar\omega$ scans along the line $Q = \xi\,(\pi, \pi) + (\delta/2)(\pi, -\pi)$ for different energy transfers in $La_{1.86}Sr_{0.14}CuO_4$ ($\delta = 0.24$) from Mason et al.[5]

Fig. 4 shows a series of constant-$\hbar\omega$ scans at $T = 35\text{ K} = T_c + 2$ K made parallel to the (π, π) direction through the peak positions $(\pi-\delta\pi, \pi)$ and $(\pi, \pi+\delta\pi)$ for $La_{1.86}Sr_{0.14}CuO_4$ from ref. 5. At the lowest energies two sharp peaks are observed, as above. As $\hbar\omega$ is increased the peaks become broader as the region between the peaks 'fills in'. Thus, the higher-energy fluctuations show shorter correlation lengths.

3.2. Superconducting State

One of the key issues in this field is how the correlations between carriers are affected by the superconductivity. While nuclear magnetic resonance techniques provide momentum averaged information at very low frequencies, neutron scattering is unique in that it gives the full response function in the frequency ($\hbar\omega \lesssim \Delta$) and momentum range relevant to the superconductivity. Fig. 5 shows that, as is appropriate for magnetic fluctuations derived from carriers near a Fermi surface, there is a suppression in the low-$\hbar\omega$ scattering on cooling through T_C. For $\hbar\omega$ = 3 meV, it is found that the drop in $|m(q,\omega)|^2$ is *not* due to a reduction in the Bose-Einstein factor (which drops by 33% from

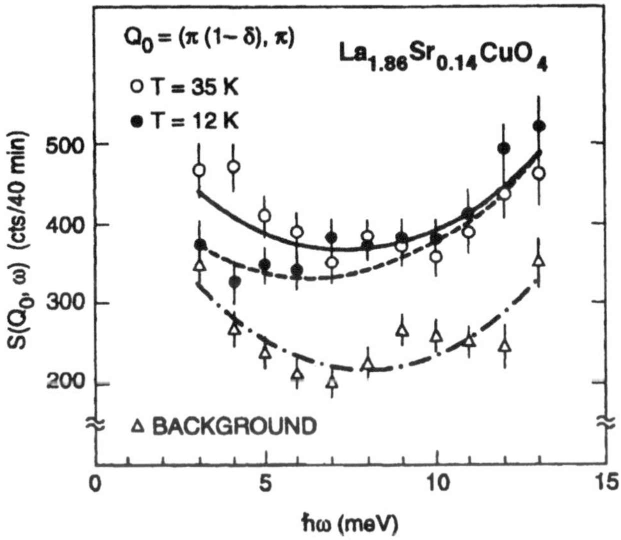

Fig. 5. Constant $Q = (\pi, \pi) - \delta(\pi, 0)$ scans in $La_{1.86}Sr_{0.14}CuO_4$ for T = 35 K $\gtrsim T_c$ = 33 K and T = 12 K from Mason et al.[5]

T = 35 K to 12 K) but to a decrease in $\chi''(q, \omega)$. While $\chi''(q, \omega)$ at elevated $\hbar\omega$ is reduced on going through T_C, it is probably not completely eliminated. The q dependence of what remains is given by Fig. 2(c), which is indistinguishable from a reduced-intensity version of the scattering above T_C. Thus, any scattering remaining below T_c, at the frequencies probed so far, does not have a radically different shape to that seen at the same $\hbar\omega$ for $T \gtrsim T_c$

4. Discussion

In order to understand the observed magnetic scattering of $La_{2-x}Sr_xCuO_4$ for $x \gtrsim$ 0.075, where the incommensurate peaks are present, it is useful to make a phenomenological model for $\chi''(q, \omega)$. The paramagnetic scattering[8] of itinerant antiferromagnet Cr and Cr-V alloys bears much similarity to $La_{2-x}Sr_xCuO_4$. In the Cr case, 3D Fermi-surface nesting gives rise to a *incommensurate* magnetic instability. Noakes et al.[8] have successfully analysed the paramagnetic scattering of Cr and Cr-V alloys using a dynamic susceptibility derived for a two-band incommensurate

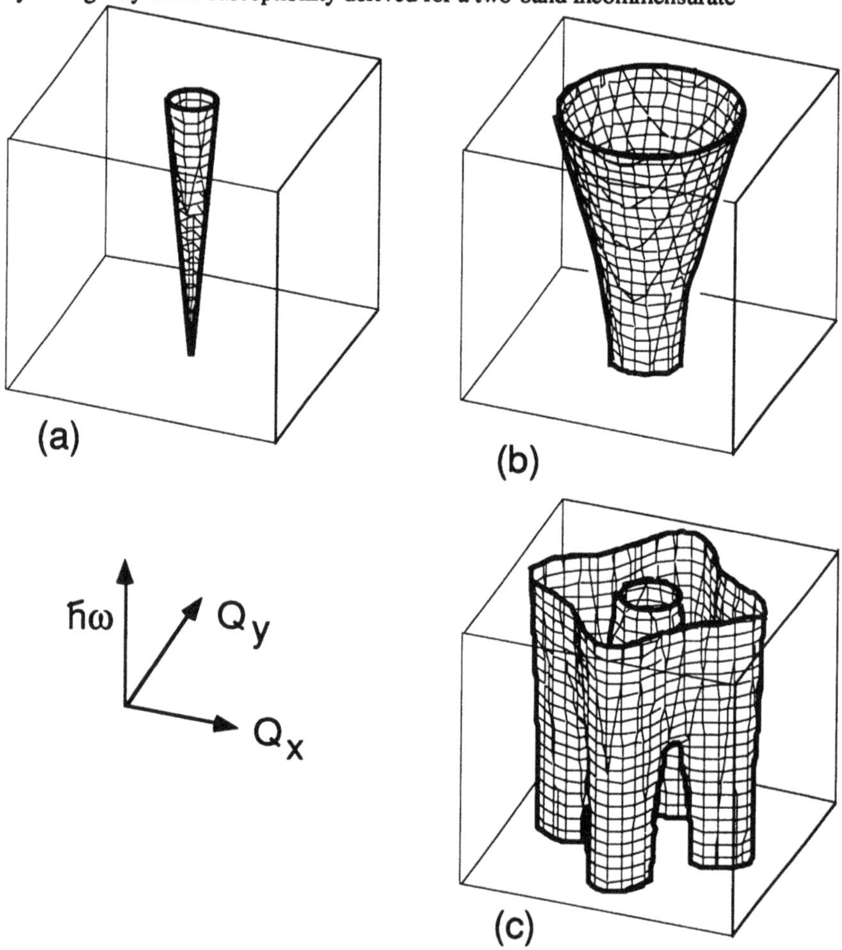

Fig. 6. Visualisation of the evolution of the experimental determined $\chi''(q,\omega)$ across the $La_{2-x}(Ba,Sr)_xCuO_4$ series. The figures show the 0.25 of maximum contour plotted for each energy ω. (a) Spin wave poles in La_2CuO_4. (b) Magnetic excitations in $La_{1.95}Ba_{0.05}CuO_4$ from Hayden et al.[15] (c) Incommensurate spin fluctuations in $La_{1.86}Sr_{0.14}CuO_4$ for $T = 35$ K from Mason et al.[5] (Eqs. 3,4 of this paper). The volumes enclosed by the boxes are $0.6\pi < Q_x, Q_y < 1.4\pi$ and $0 < \hbar\omega < 50$ meV for (a), (b) and $0 < \hbar\omega < 20$ meV for (c).

antiferromagnet. Mason et al.[5] used a phenomenological form[9] closely related to Noakes et al,

$$\chi(\mathbf{Q},\omega)=\frac{\chi^0}{i\omega+A^2\left(\kappa^2+R(\mathbf{Q})\right)};\quad \chi''(\mathbf{Q},\omega)=\frac{\chi^0\omega A}{\omega^2+A^4\left(\kappa^2+R(\mathbf{Q})\right)^2},\tag{3}$$

where

$$R(Q)=\frac{[(q_x-q_y)^2-(\pi\delta)^2]^2+[(q_x+q_y)^2-(\pi\delta)^2]^2}{2a_0^2\,4\pi^2\delta^2},\tag{4}$$

and $\boldsymbol{q}=(q_x,q_y)=\boldsymbol{Q}-(\pi,\pi)$. Fitting the scans in Fig. 4 yielded values for the parameters $A^2 = 3858 \pm 210$ meVÅ and $\kappa = 0.063 \pm 0.002$ Å^{-1} for $T = 35$ K. A representation of this functional form with these fitted parameters is shown in Fig. 6(c). Fig. 6(c) summarizes the observed scattering in $La_{1.86}Sr_{0.14}CuO_4$ for T ≈ 35 K. At low energies, $\hbar\omega \sim$ 1 meV, there are four well defined peaks. With increasing $\hbar\omega$, the fluctuations broaden in q with a characteristic energy scale $A^2\kappa^2 = 15$ meV.

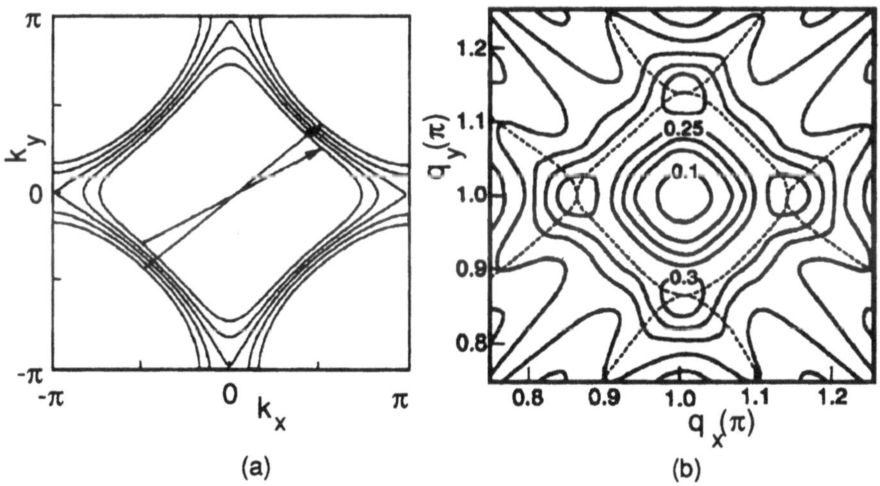

Fig. 7. (a) Fermi surface contours in a rigid 3-band model for $La_{2-x}Sr_xCuO_4$. The arrows illustrates how mutually tangential pieces of Fermi surface can give rise to singularities in $\chi''(\boldsymbol{q},\omega)$. (b) Contour plot of $\chi''(\boldsymbol{q},\omega = 1$ meV) for $\varepsilon_d - \varepsilon_p = 2.4$ eV, $t = -1.5$ eV, $t' = 0.3$ eV and $n = 0.43$. The dashed lines mark lines of singularities for FS "touching". Both figures kindly provided by P. B. Littlewood (see ref. 10)

In 2D space, appropriate to $La_{2-x}Sr_xCuO_4$, even a normal Fermi liquid with a circular Fermi surface has singularities[10] in $\chi''(\boldsymbol{q},\omega)$ as $\omega \to 0$. A more complicated 2D Fermi surface (FS) has singularities for $\boldsymbol{q} = \boldsymbol{q}_s$, where $\boldsymbol{q}_s$ connects two parts of the FS whose tangents are parallel. Several groups[10-12] have made calculations of $\chi''(\boldsymbol{q},\omega)$ using band structure models with parameters derived from realistic band structure calculations.

Fig. 7(a) shows some typical Fermi surfaces for different Fermi energies[10]. Fig. 7(b) shows $\chi''(q,\omega)$ calculated from the Lindhard function by Littlewood and coworkers[10] for $x = 0.14$. It shows incommensurate peaks which have approximately the correct positions (and variation with x), however, a modest q–independent enhancement of the susceptibility must be included to obtain the measured value for $\chi(q \to 0, \omega \to 0)$. Further support for an itinerant picture of the carriers comes from the widths at small ω of the observed scattering. Since $\xi^{-1} < a_{\mathrm{Sr\text{-}Sr}}^{-1}$ (see table 1), $\chi''(q,\omega)$ displays features on an inverse length scale considerably shorter than that determined by the Sr–Sr separation. This suggests that the local effects of the Sr atoms have been screened out.

The calculations of $\chi''(q,\omega)$ discussed above show that the presence of strong magnetic interactions is *not* required to explain the existence and positions of the incommensurate peaks. However, the energy and temperature dependence is a different matter. The Lindhard "non-interacting" susceptibility $\chi''(q,\omega,T)$ determined using realistic energy bands yields essentially no variation of χ'' over the temperature and energy ranges investigated experimentally, $0 \lesssim T \lesssim 100$ K and $1 \lesssim \hbar\omega \lesssim 15$ meV. This is at variance with the data in Fig. 4. However, models incorporating magnetic interactions[11,12] or (self energy) corrections to the electronic states[10] can yield the appropriate variations.

The formation of a BCS-type superconducting state leads to a gap in the quasiparticle dispersion, Δ. Within the Lindhard picture described above, magnetic excitations correspond to the formation of electron-hole pairs. In the superconducting state, these can only be created if $\hbar\omega > 2\Delta$. Thus for $T < T_c$, one would expect a strong suppression of $\chi''(q,\omega)$ for $\hbar\omega < 2\Delta$. Unconventional pairing or the existence of a finite lifetime to the quasiparticle states ("gap-less superconductivity") can lead to a reduction in this suppression. Fig. 5 shows a strong suppression of $\chi''(q,\omega)$ on entering the superconducting state[5] in $La_{1.86}Sr_{0.14}CuO_4$ ($T_c = 33$ K) for $\hbar\omega \lesssim 6$ meV, however, $\chi''(q,\omega)$ does not disappear completely for $T < T_c$ and $\hbar\omega > 3$ meV.

In summary, for low doping, $x \lesssim 0.02$, the magnetic excitation spectrum $|m(q,\omega)|^2$ in $La_{2-x}(Ba,Sr)_xCuO_4$ is characteristic of an insulating antiferromagnet with strong coupling[1,13,14] ($J = 0.13$ eV) within the CuO_2 planes. Doping causes a disruption of the long range antiferromagnetic order. At high doping ($x \gtrsim 0.075$) the observed spin fluctuation spectrum can be interpreted naturally within an itinerant model for the carriers. The existence of incommensurate peaks in $\chi''(q,\omega)$ follows naturally from Fermi surface "touching".

5. Acknowledgements

We are grateful P. B. Littlewood, H. Monien and J. Zaanen for helpful discussions and K. N. Clausen and J. L. Martinez for fruitful collaborations. This work has been supported in part by the SERC, the Danish Research Academy, US DOE (Contract No. DE-ACOS-8140R21400) and NATO.

6. References

[1]G Shirane, Y Endoh, R J Birgeneau, M A Kastner, Y Hidaka, M Oda, M Suzuki and T Murakami, Phys. Rev. Lett. **59**, 1613 (1987).
[2]T. R. Thurston, R. J. Birgeneau, M. A. Kastner, N. W. Preyer, G. Shirane, Y. Fujii, K. Yamada, Y. Endoh, K. Kakurai, M. Matsuda, Y. Hidaka and T. Murakami, Phys. Rev. B **40**, 4585 (1989).
[3]J. Rossat-Mignod, L. P. Regnault, C. Vettier, P. Burlet, J. Y. Henry and G. Lapertot, Physica B **169**, 58 (1991).
[4]S.-W. Cheong, G. Aeppli, T. E. Mason, H. Mook, S. M. Hayden, P. C. Canfield, Z. Fisk, K. N. Clausen and J. L. Martinez, Phys. Rev. Lett. **67**, 1791 (1991).
[5]T. E. Mason, G. Aeppli and H. A. Mook, Phys. Rev. Lett. **68**, 1414 (1992).
[6]G. Aeppli, S. M. Hayden, T. E. Mason and H. A. Mook, (unpublished).
[7]G. Shirane, R. J. Birgeneau, Y. Endoh, P. Gehring, M. A. Kastner, K. Kitazawa, H. Kojima, I. Tanaka, T. R. Thurston and K. Yamada, Phys. Rev. Lett. **63**, 330 (1989).
[8]D. R. Noakes, T. M. Holden, E. Fawcett and P. C. de Camargo, Phys. Rev. Lett. **65**, 369 (1990).
[9]A. J. Millis, H. Monien and D. Pines, Phys. Rev. B **42**, 167 (1990).
[10]P. B. Littlewood, J. Zaanen, G. Aeppli and H. Monien (unpublished); P. B. Littlewood, in *Strongly Interacting Fermions and High T_C Superconductivity*, proceedings of Les Houches 1991 Summer School, edited by B. Douçot.
[11]N. Bulut, D. W. Hone, D. J. Scalapino and N. E. Bickers, Phys. Rev. B **41**, 1797 (1990).
[12]J. P. Lu, Q. Si, J. H. Kim and K. Levin, Physica C **179**, 191 (1991).
[13]G. Aeppli, S. M. Hayden, H. A. Mook, Z. Fisk, S-W. Cheong, D. Rytz, J. P. Remeika, G. P. Espinosa and A. S. Cooper, Phys. Rev. Lett. **62**, 2052 (1989).
[14]S. M. Hayden, G. Aeppli, R. Osborn, A. D. Taylor, P. G. Perring, S.-W. Cheong and Z. Fisk, Phys. Rev. Lett. **67**, 3622 (1991).
[15]S. M. Hayden, G. Aeppli, H. Mook, D. Rytz, M. F. Hundley and Z. Fisk, Phys. Rev. Lett. **66**, 821 (1991).

Percolative phase separation and high T_c superconductivity

E.Sigmund, V.Hizhnyakov[†], G.Seibold
Institut für Theoretische Physik der Universität Stuttgart,
Pfaffenwaldring 57, 7000 Stuttgart 80, FRG
[†]Institute of Physics, Estonian Academy of Sciences,
Riia 142, Tartu, Estonia

Abstract

When doping the antiferromagnetically (AF) ordered high-T_c cuprates with holes magnetic polarons (spinpolarized clusters) are formed. At higher hole concentrations and due to cluster diffusion a phase separation is formed by establishing large fractal or percolative clusters. This percolation picture allows to understand experimentally obtained results for phase separation as well as magnetic and conductive phase diagrams.

1 Introduction

Doping or oxydation of the antiferromagnetically (AF) ordered perovskite compounds like La_2CuO_4 and $YBa_2Cu_3O_6$, which are the parent materials for high-T_c superconductors, leads to the creation of holes in the CuO_2 planes. The dynamics of the charge carriers is governed by strong correlation effects and the Pauli exclusion principle. Therefore the double occupancy of a state with two particles with antiparallel and parallel spin alignement is forbidden in the considered systems. Due to these effects a hole in the CuO_2 plane rearranges the antiferromagnetic order in its nearest vicinity forming a spin polarized cluster (magnetic polaron or ferron) [1]. The cluster has only low mobility whereas the hole inside the cluster can move freely. As a result, when increasing the hole concentration and due to cluster diffusion the clusters start to overlap and a (fractal) percolation network is built up. This leads to the destruction of the AF order and to the appearance of metallic-like conductivity, or below T_c to superconductivity, within the percolation network [2]. Such a mechanism allows to explain [3] the phase diagrams of $La_{2-x}Sr_xCuO_4$ and $YBa_2Cu_3O_{6+\delta}$ in quantitative agreement with experiment giving support to spin cluster formation in these materials. The spin cluster model also allows to explain recent experimental data on percolative phase separation in weakly doped $La_2CuO_{4+\delta}$ and $La_{2-x}Sr_xCuO_4$ samples [4]. Also the magnetic susceptibility measurements strongly show the existence of small ferromagnetic (superparamagnetic) particles associated with holes [3].

2 Hole dynamics in CuO_2 planes

Our considerations are based on the three band Hubbard model (see e.g. Emery [5]) which takes the $d_{x^2-y^2}$ band of the Cu ions and the p_σ bands of the oxygen ions into account. The model hamiltonian reads

$$H = \sum_\sigma H_{0\sigma} + H_{int} \tag{1}$$

where

$$H_{0\sigma} = \epsilon_d \sum_n n^d_{n\sigma} + \epsilon_p \sum_m n^p_{m\sigma} + T^\pm \sum_{nm} (d^\dagger_{n\sigma} p_{m\sigma} + h.c.) \tag{2}$$

$$H_{int} = U \sum_n n^d_{n\sigma} n^d_{n-\sigma} \tag{3}$$

$n^d_{n\sigma}$ and $n^p_{m\sigma}$ are the electronic occupation numbers of the $d_{x^2-y^2}$ and p_σ orbitals, respectively, $d^\dagger(d)$ and $p^\dagger(p)$ are electronic creation (annihilation) operators in the corresponding Cu and O orbitals obeying the usual commutation rules for fermion operators. The spin index σ indicates the spin directions ($\sigma = \uparrow\downarrow$). Only the transfer between Cu and the four nearest oxygen ions is taken into account. The transfer integral $T^\pm$ changes sign due to the two possibilities of overlap. H_{int} describes the Coulomb repulsion for two electrons on the same Cu site with opposite spins. Band structure calculations indicate the following parameter values: $U \approx 8eV$, $\epsilon = \epsilon_p - \epsilon_d \approx 3eV$, $T \approx 1eV$. The effect of the strong particle-particle (Hubbard) repulsion is treated by two different approaches.

2.1 First approach

In our first approach the calculation is based on a Hartree-Fock (mean-field) approximation of the original three-band Hubbard hamiltonian. Using the identity

$$\begin{aligned} n_{m\sigma} n_{m-\sigma} = \; & (n_{m\sigma} - \langle n_{m\sigma}\rangle)(n_{m-\sigma} - \langle n_{m-\sigma}\rangle) + \\ & + \langle n_{m-\sigma}\rangle n_{m\sigma} + \langle n_{m\sigma}\rangle n_{m-\sigma} - \langle n_{m\sigma}\rangle\langle n_{m-\sigma}\rangle \end{aligned} \tag{4}$$

and neglecting fluctuations around the expectation values one obtains the mean-field approximation of the three-band Hubbard model

$$H^{MF} = \sum_{\sigma} H_{0\sigma}^{MF} + H_{int}^{MF} \tag{5}$$

where

$$H_{0\sigma}^{MF} = \sum_{n} \left(\epsilon_d + U\langle n_{n-\sigma}^d \rangle\right) n_{n\sigma}^d + \epsilon_p \sum_{m} n_{m\sigma}^p + T^{\pm} \sum_{nm} \left(d_{n\sigma}^{\dagger} p_{m\sigma} + h.c.\right) \tag{6}$$

$$H_{int}^{MF} = -U \sum_{n} \langle n_{n\sigma}^d \rangle \langle n_{n-\sigma}^d \rangle \tag{7}$$

Due to the symmetry of the hamiltonian with respect to the spin direction, one possible solution which always exists is that with $\langle n_{n\sigma}^d \rangle = \langle n_{n-\sigma}^d \rangle$, i.e. the nonmagnetic solution. For the half-filled case it corresponds to a metal with completely nested Fermi surface. Such state is expected to be unstable against charge fluctuations. However, if one allows the spin densities within a fixed spin state to be unequal at neighbouring Cu sites one finds a new solution corresponding to the antiferromagnetic (AF) ordered situation [6].

In order to describe the AF ordered state we introduce two Cu sublattices (the magnetic unit cell contains two CuO_2 units). The components of the state vector corresponding to the two different Cu ions in a unit cell are denoted by d_1 and d_2. The corresponding diagonal elements of the hamiltonian (5) read

$$\epsilon_{1\sigma} = \epsilon_d + U\langle n_{1-\sigma} \rangle \tag{8}$$

$$\epsilon_{2\sigma} = \epsilon_d + U\langle n_{2-\sigma} \rangle \tag{9}$$

From the symmetry of the AF state one obtains immediately the following relations

$$\langle n_{1\uparrow} \rangle = \langle n_{2\downarrow} \rangle \tag{10}$$

$$\langle n_{1\downarrow} \rangle = \langle n_{2\uparrow} \rangle \tag{11}$$

Therefore we can use the notations $\langle n_{\uparrow} \rangle$ and $\langle n_{\downarrow} \rangle$ and consider these quantities as modulated with twice the lattice period.

2.2 Diagonalization

We find it convenient to perform the following calculations in the hole picture which is given by replacing electron creation (annihilation) operators by hole annihilation (creation) operators and observing the Fermi commutation rules. This yields for the mean-field hamiltonian (5)

$$H^{MF} = \sum_{\sigma} H_{0\sigma}^{MF} + H_{int}^{MF} + const. \tag{12}$$

with

$$H^{MF}_{0\sigma} = \sum_n \left(-\epsilon_d + U\left(\langle n^d_{n-\sigma}\rangle - 1\right)\right) n^d_{n\sigma} + \epsilon_p \sum_m n^p_{m\sigma} \tag{13}$$
$$+T^{\mp} \sum_{nm} \left(d^\dagger_{n\sigma} p_{m\sigma} + h.c.\right)$$
$$H^{MF}_{int} = -U \sum_n \langle n^d_{n\sigma}\rangle\langle n^d_{n-\sigma}\rangle \tag{14}$$

where all operators act on hole states. Consider the fixed spin direction of a hole, e.g. $\sigma = \uparrow$. Then

$$\epsilon_1 = -\epsilon_d + U(\langle n_\downarrow\rangle - 1) \tag{15}$$
$$\epsilon_2 = -\epsilon_d + U(\langle n_\uparrow\rangle - 1) \tag{16}$$

We write the state vector as

$$\Phi_n(\mathbf{k}) = \begin{pmatrix} d_1(\mathbf{k}) \\ d_2(\mathbf{k}) \\ p_{x1}(\mathbf{k}) \\ p_{y1}(\mathbf{k}) \\ p_{x2}(\mathbf{k}) \\ p_{y2}(\mathbf{k}) \end{pmatrix} e^{i\mathbf{k}\mathbf{R_n}} \tag{17}$$

which yields for the hamiltonian (13 in **k**-space

$$H_{\mathbf{kk}} = \begin{pmatrix} \epsilon_1 & 0 & -2Ta_x & 2Ta_y & 2Ta^*_x & -2Ta^*_y \\ 0 & \epsilon_2 & 2Ta^*_x & -2Ta^*_y & -2Ta_x & 2Ta_y \\ -2Ta^*_x & 2Ta_x & \epsilon_p & 0 & 0 & 0 \\ 2Ta^*_y & -2Ta_y & 0 & \epsilon_p & 0 & 0 \\ 2Ta_x & -2Ta^*_x & 0 & 0 & \epsilon_p & 0 \\ -2Ta_y & 2Ta^*_y & 0 & 0 & 0 & \epsilon_p \end{pmatrix} \tag{18}$$

where we have used the abbreviation $a_{x/y} = \frac{1}{2}e^{ik_{x/y}\frac{a}{2}}$. Introducing new p-states

$$\begin{pmatrix} p_1 \\ p_2 \\ q_1 \\ q_2 \end{pmatrix} = \begin{pmatrix} a_x & -a_y & -a^*_x & a^*_y \\ b^*_x & -b^*_y & -b_x & b_y \\ a^*_y & a^*_x & -a_y & -a_x \\ b_y & b_x & -b^*_y & -b^*_x \end{pmatrix} \begin{pmatrix} p_{1x} \\ p_{1y} \\ p_{2x} \\ p_{2y} \end{pmatrix} \tag{19}$$

with

$$b_{x/y} = \frac{a_{x/y} - \frac{1}{2}V_k a^*_{x/y}}{\sqrt{1-\frac{1}{4}V_k^2}} \tag{20}$$

$$V_k = \cos k_x a + \cos k_y a \tag{21}$$

one finds that states q_1 and q_2 are decoupled from all other states while in the basis of the four remaining states the matrix (18) transforms to

$$H_{\mathbf{kk}} = \begin{pmatrix} \epsilon_1 & 0 & -TV_k & -2T\sqrt{1-\frac{1}{4}V_k^2} \\ 0 & \epsilon_2 & 2T & 0 \\ -TV_k & 2T & \epsilon_p & 0 \\ -2T\sqrt{1-\frac{1}{4}V_k^2} & 0 & 0 & \epsilon_p \end{pmatrix} \tag{22}$$

The secular equation reads

$$Det[E - H_{\mathbf{kk}}] = D_1(E)D_2(E) - 4T^4V_k^2 \tag{23}$$

where

$$D_i(E) = (E - \epsilon_p)(E - \epsilon_i) - 4T^2 \tag{24}$$

Eq. (23) describes four bands where in the case of CuO_2 only the lowest band is filled with holes. A pecular property of the model is the appearance of nonanalytical critical points along the line $k_x + k_y = \pi$ which is also the maximum of the lowest band. These critical points lead to a divergent behaviour in the density of states of the type $\frac{1}{\sqrt{E}} \ln \frac{1}{E}$, while normal logarithmic singularities disappear.

2 .3 Green's Functions

The diagonal elements of the one-particle Green's functions for the two Cu sites in the doubled unit cell are given by

$$G_{ii} = \sum_{\mathbf{k}} (E - H_{\mathbf{kk}})^{-1}_{ii} = \sum_{\mathbf{k}} \frac{|\{E - H_{\mathbf{kk}}\}|_{ii}}{det|E - H_{\mathbf{kk}}|} \tag{25}$$

where $|\{A\}|_{ii}$ denotes the adjoint subdeterminant of elememt (ii) of the matrix A. Making use of Eqs. (10) and (11) we write the Green's function on Cu site 1 as $G_{11} = G_\uparrow$ and on Cu site 2 as $G_{22} = G_\downarrow$ respectively. After a straightforward calculation one can express G_σ in terms of the Green's function G^{2-D} of a simple quadratic lattice which yields

$$G_\uparrow = (E-\epsilon_p)\sqrt{\frac{D_2}{D_1}}G^{2-D}(\sqrt{D(E)}) \tag{26}$$

$$G_\downarrow = (E-\epsilon_p)\sqrt{\frac{D_1}{D_2}}G^{2-D}(\sqrt{D(E)}) \tag{27}$$

where $D(E) = D_1(E)D_2(E)\ G^{2-D}$ is derived in [12] and is given by a simple elliptic integral of the first kind **IK**. For the Green's functions we finally get

$$G_\uparrow = 2\frac{E-\epsilon_p}{\pi D_1}\mathbf{IK}\left(\frac{4T^2}{\sqrt{D(E)}}\right) \tag{28}$$

$$G_\downarrow = 2\frac{E-\epsilon_p}{\pi D_2}\mathbf{IK}\left(\frac{4T^2}{\sqrt{D(E)}}\right) \tag{29}$$

where $\frac{D(E)}{16T^4} > 1$. To obtain the spectral functions in the band region and the gap one has to use the analytical continuations of the elliptic integral (cf. [13]). The resulting expressions read as

$$G_\uparrow = \frac{E-\epsilon_p}{2\pi T^2}\sqrt{\frac{D_2}{D_1}}\left[\mathbf{IK}\left(\frac{\sqrt{D(E)}}{4T^2}\right) + i\mathbf{IK}\left(\sqrt{1-\frac{D(E)}{16T^4}}\right)\right] \tag{30}$$

$$G_\downarrow = \frac{E-\epsilon_p}{2\pi T^2}\sqrt{\frac{D_1}{D_2}}\left[\mathbf{IK}\left(\frac{\sqrt{D(E)}}{4T^2}\right) + i\mathbf{IK}\left(\sqrt{1-\frac{D(E)}{16T^4}}\right)\right] \tag{31}$$

where $0 < \frac{D(E)}{16T^4} < 1$

$$G_\uparrow = 2\frac{E-\epsilon_p}{\pi D_1}\frac{1}{\sqrt{1-\frac{16T^4}{D(E)}}}\mathbf{IK}\left(\frac{1}{\sqrt{1-\frac{D(E)}{16T^4}}}\right) \tag{32}$$

$$G_\downarrow = 2\frac{E-\epsilon_p}{\pi D_2}\frac{1}{\sqrt{1-\frac{16T^4}{D(E)}}}\mathbf{IK}\left(\frac{1}{\sqrt{1-\frac{D(E)}{16T^4}}}\right) \tag{33}$$

where $\frac{D(E)}{16T^4} < 0$

In fig.(1) we have plotted the real and imaginary part of $G_\uparrow$ and $G_\downarrow$ up to the second hole band. Without going into detail we note that the nondiagonal elements G_{ij}^σ can be obtained by simple recurrence relations provided that the diagonal elements and one nondiagonal element are known (cf. [13]). The expectation values $\langle n_\uparrow \rangle$ and $\langle n_\downarrow \rangle$ are immediately found from the selfconsistent equations a)

b)

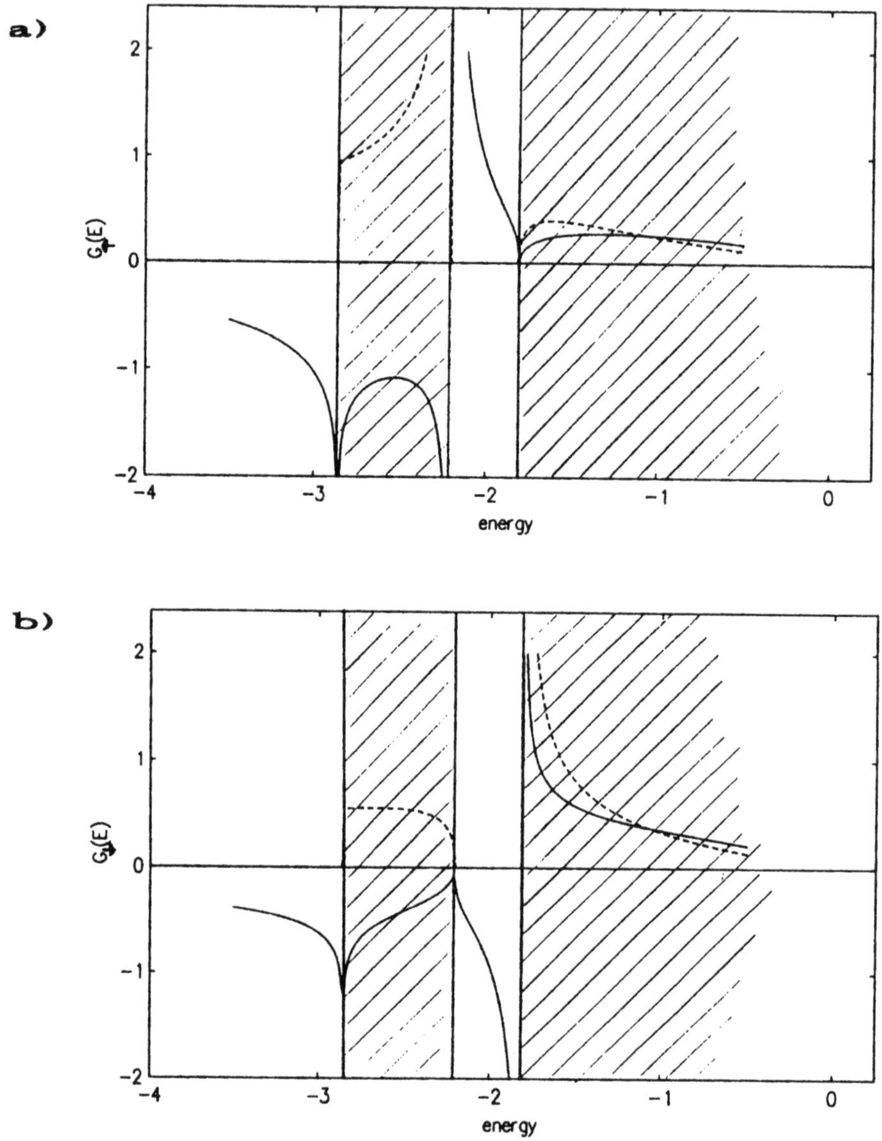

Figure 1: Local Green functions for spin-up (a) and spin-down (b) projections of the wave-function on Cu sites. The shaded areas indicate the lower Hubbard band and the lower part of the bonding oxygen band. Solid line:Real part; Dashed line:Imaginary part of the Green functions

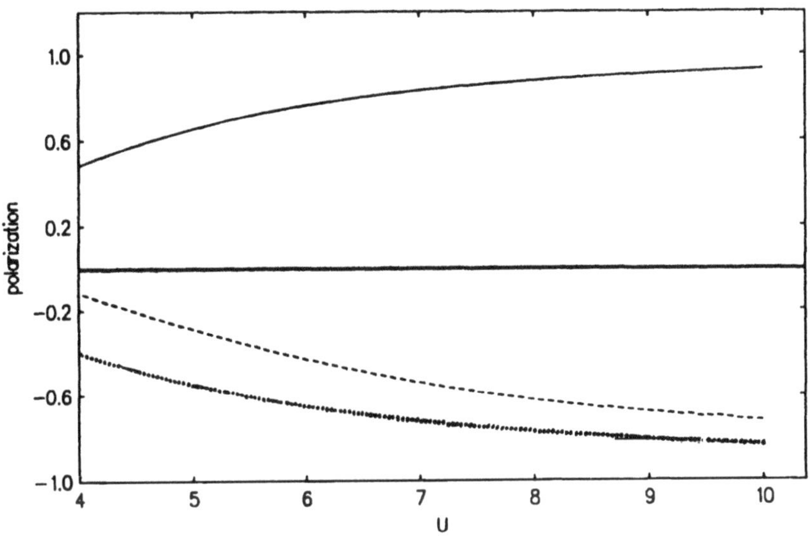

Figure 2: Spin-polarization $\langle n_\uparrow \rangle - \langle n_\downarrow \rangle$ on Cu sites versus the Hubbard repulsion parameter U ($\epsilon = 3, T = 1$). Solid line: Undoped case; Dashed line: Polarization on the perturbed Cu-site (one turned spin, one additional hole); Dotted line: Polarization on the perturbed Cu-sites (two turned spins, one additional hole)

$$\langle n_\uparrow \rangle = \frac{1}{\pi} \int dE \, Im(G_\uparrow(E)) \tag{34}$$

$$\langle n_\downarrow \rangle = \frac{1}{\pi} \int dE \, Im(G_\downarrow(E)) \tag{35}$$

where the integration is restricted to the lowest hole band. As can be seen from fig.(2), for real values of parameters the ground state is almost completely polarized although the total Cu spin is considerably smaller due to the p-d hybridization. The ground state properties of the totally AF lattice have been intensively discussed in [6].

2.4 Doped system

With doping two main scenarios have to be considered

1. The ideal AF order remains unchanged and the additional holes fill up the second band. However, since the Coulomb repulsion depends on the filling, doping leads to the decrease of the total spin at every site which results in a charge redistribution and particulary in the decreasing of the charge transfer gap. This scenario ends up with a metal-insulator transition (MIT) at $\langle n_\uparrow \rangle = \langle n_\downarrow \rangle$. As found by Oleś and Zaanen [6], such a MIT would occur at a hole concentration of $\delta \approx 0.24$.

2. Due to the nonlinear nature of the mean-field equations one can expect (in analogy with e.g. electron-lattice polaron theory) a spontanous symmetry breaking, leading to the creation of localized excitations. The new states arising from such symmetry breaking should be determined selfconsistently.

We have found that the second scenario leads to a lower ground state energy which was already proved by us in the context of percolative phase seperation by use of unitary transformations. First we assume that the Cu spin at site m = 0, which has originally spin direction, say $\uparrow$, can spontaneously fluctuate, changing the spin values to $\langle n_{\uparrow}^{l}\rangle$ and $\langle n_{\downarrow}^{l}\rangle$. These values are treated selfconsistently. This problem is mathematically similar to the problem of magnetic impurities (e.g. Wolff model) and can be treated within the standard localized perturbation theory. The perturbation of the hamiltonian (13) is given by

$$V = V^{\uparrow} + V^{\downarrow} \tag{36}$$

where

$$V_{mm'}^{\downarrow} = U\delta_{m0}\delta_{m'0}(\langle n_{\uparrow}^{l}\rangle - \langle n_{\uparrow}\rangle d_{0\downarrow}^{*}d_{0\downarrow} \tag{37}$$

$$V_{mm'}^{\uparrow} = U\delta_{m0}\delta_{m'0} + (\langle n_{\downarrow}^{l}\rangle - \langle n_{\downarrow}\rangle d_{0\uparrow}^{*}d_{0\uparrow} \tag{38}$$

The new Green's functions for each spin direction can be found from Dyson's equation

$$G^{\sigma}(mm') = G_0^{\sigma}(mm') + G_0^{\sigma}(m0)V^{\sigma}G^{\sigma}(0m') \tag{39}$$

where G_0 denotes the Green's function of the undisturbed lattice. The solution of (39) is

$$G^{\sigma}(mm') = G_0^{\sigma}(mm') + V^{\sigma}\frac{G_0^{\sigma}(m0)G_0^{\sigma}(0m')}{1 - V^{\sigma}G_0^{\sigma}(00)} \tag{40}$$

The equations determining the spin distribution at the disturbed lattice site then read as

$$\langle n_{\sigma}\rangle = \frac{1}{\pi}\int_{-\infty}^{E_F} dE\, Im\, G^{\sigma}(00, E) \tag{41}$$

where E_F is the chemical potential, depending on the number of additional holes. The localized states are the solution of Eqs.

$$1 = V^{\sigma}G_0^{\sigma}(E), \quad \sigma = \uparrow, \downarrow \tag{42}$$

As can be concluded from the real part of $G(E)$ in fig.(1), Eq.(42) has only solutions with $V^{\uparrow} < 0$ and $V^{\downarrow} > 0$ (beside the trivial solution $V^{\sigma} = 0$). It is seen that there exist for any values of parameters three localized states, one originating from the change of the $\langle n_\uparrow \rangle$ value and two due to the perturbation of $\langle n_\downarrow \rangle$. For one additional hole the Fermi level coincides with the upper localized level in the gap. Fig.(2) shows the dependence of the polarization at site $\mathbf{m} = 0$ on the repulsion parameter U in comparison with the polarization at the undisturbed sites. Obviously, there exists a reversed spin polarization the value of which is increasing with U. Thus we showed that the mean-field hamiltonian (12) for a doped system has solutions with local fluctuations (mainly reversion) of Cu-spins. The total change of energy consists of two parts:

1. Due to the perturbation (36) of the hamiltonian (13) the energy change is given by

$$\Delta E^{loc} = \sum_{\sigma} \int_{-\infty}^{E_F} dE\, E\, \Delta N_\sigma(E) + V^{ht} \tag{43}$$

$\Delta N_\sigma(E)$ is the change in the density of states whereas V^{ht} is already given in (36). We have to consider two contributions to $\Delta N_\sigma(E)$. The first is the change of density in the bands mainly from the band edges, the second is the appearance of singularities in $\Delta N_\sigma(E)$ outside the bands due to the formation of localized levels. The expression for the change in the density of states is derived in [14] and reads

$$\Delta N_\sigma(E) = \frac{1}{\pi}\frac{d}{dE} Im\, ln\left[1 - V^{-\sigma} G_0^\sigma(E)\right] \tag{44}$$

Evaluating (44) one finds that every local state gives a contribution equal to one, besides the change of the local density in the bands. Inserting eq.(44) into (43) immediately leads to

$$\Delta E^{loc} = \sum_{l} E_l + \frac{1}{\pi}\sum_{\sigma} \int_{E_0}^{E_1} \arctan \frac{V^{-\sigma} Im\, G^\sigma(00)}{1 - V^{-\sigma} Re\, G^\sigma(00)} \tag{45}$$

where E_l denote the energies of the occupied localized states measured from the band edges.

2. The energy change resulting from the perturbation of (14) reads as

$$\Delta E^{int} = U(\langle n_\uparrow \rangle \langle n_\downarrow \rangle - (\langle n_\uparrow^l \rangle \langle n_\downarrow^l \rangle) \tag{46}$$

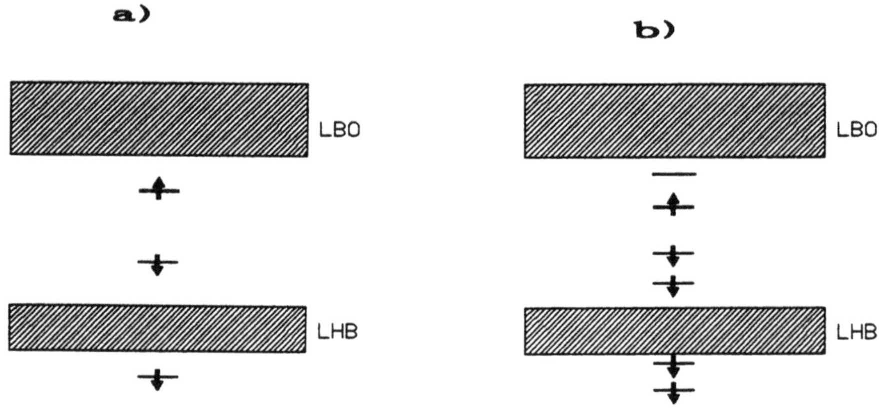

Figure 3: Sketch of the band scheme for one additional hole in the case of one (a) and two (b) turned spins which originally had mainly spin-up mean-values. The additional hole occupies a localized level of the spin-up subsystem with mainly oxygen character.LHB: Lower Hubbard band; LBO: Lower part of the bonding oxygen band

2 .5 Results

In fig.(4) we have plotted the binding energy of one additional hole in a CuO_2 lattice with one flipped spin $-(\Delta E^{int} + \Delta E^{loc})$ versus the repulsion parameter U for two values of $\epsilon = \epsilon_p - \epsilon_d$. The behaviour in the large U-limit coincides with the results obtained in an earlier work by use of unitary transformations /citeHIZ. In fact, for large U the mean-field ground state is described by a well polarized lattice so that fluctuations around the expectation values can be neglected. The decrease in the binding energy for smaller U is due to the different behaviour of E^{loc} and E^{int}. For weak perturbations, the localization energy is small whereas in this case E^{int} gives a very large contribution due to the strong deviation of $\langle n^l_\sigma \rangle$ from the ground state values $\langle n_{-\sigma} \rangle$. Of course in this parameter region the results from a mean-field calculation are questionable since a large interaction term means large deviations from the approximation (4). For increasing U E^{int} becomes smaller and finally is dominated by E^{loc}. In fig.(3) we have sketched the electronic obtained for one turned spin.

Up to now we have only considered the symmetry breaking of the ground state by one flipped spin. Nevertheless, for a critical value of U the situation with two turned spins (diagonally neighboured) gives a larger binding energy as can be seen in fig.(4). This behaviour is in agreement with calculations of Auerbach and Larson [7] for the t-J-model. In the mean-field picture two turned spins maximally produce four localized states in the CT-gap and two below the first hole band (fig.(3)). These levels can be filled up with one or two holes respectively. The results are summarized in table (1). Obviously the polarization of the perturbed spins is much larger in the case of two

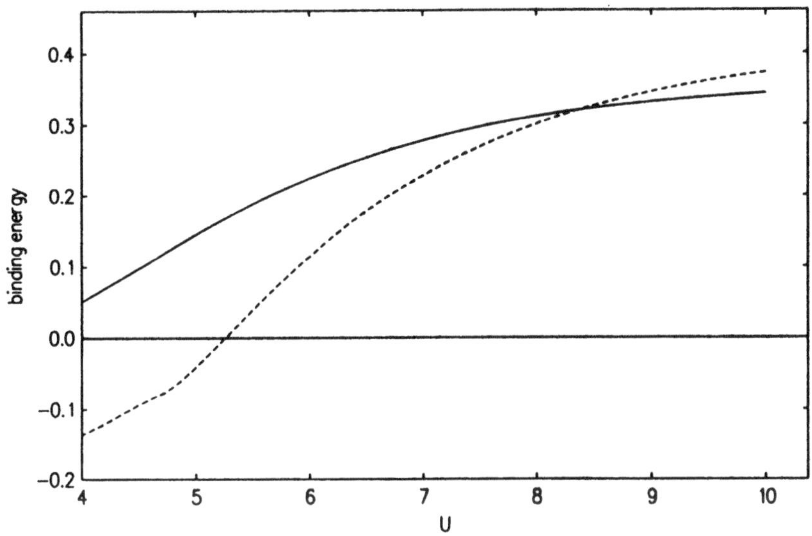

Figure 4: Binding energy for one additional hole in dependence of the Hubbard repulsion parameter U ($\epsilon = 3, T = 1$). Solid line:One turned spin; Dashed line: Two turned spins

turned spins and one additional hole than in the case of only one turned spin. This means that the interaction energy for two turned spins is always smaller. Doping a system with two holes and additionally turning two Cu spins leads to a polarization of the perturbed spins comparable to the situation of one flipped spin and one hole. The energy of this two-turned-spin cluster with two holes is at least for larger U the same as for two separated clusters with one turned spin and one hole. In the limit $U \rightarrow \infty$ and one additional hole the ground state should be totally ferromagnetically ordered according to Nagaoka's theorem [15].

2.6 Second approach

This approach makes use of the small value of the difference ϵ between the oxygen and the Cu electronic on-site energies compared with the on-site Hubbard repulsion energy U on Cu sites. A modified Schrieffer-Wolff transformation [2] is applied to the hamiltonian (1) allowing to eliminate those oxygen-copper hybridization terms which change the double occupancy of the Cu ions. The transformed hamiltonian is treated in Hartree-Fock approximation. For the undoped case the structure of the bands and the ground state energy are calculated once more a) for the paramagnetically (non-spin-ordered) and b) for the antiferromagnetically ordered structure. The case (b) yields a lower energy. Considering a doubled elementary cell in the non-doped

Table 1: The binding energy and the Cu-spin polarization for the undoped $\langle n_\uparrow\rangle, \langle n_\downarrow\rangle$ and doped $\langle n^l_\uparrow\rangle, \langle n^l_\downarrow\rangle$ cases with one and to turned spins in dependence of U and ϵ.

				1 turned spin, 1 hole			2 turned spins,1 hole			2 turned spins,2 holes		
$\frac{U}{T}$	$\frac{\epsilon}{T}$	$\langle n^0_\uparrow\rangle$	$\langle n^0_\downarrow\rangle$	$\langle n^l_\uparrow\rangle$	$\langle n^l_\downarrow\rangle$	$-\Delta E$	$\langle n^l_\uparrow\rangle$	$\langle n^l_\downarrow\rangle$	$-\Delta E$	$\langle n^l_\uparrow\rangle$	$\langle n^l_\downarrow\rangle$	$-\Delta E$
4	2	0.66	3E-2	0.31	0.57	0.12	0.12	0.65	0.04	0.32	0.55	0.38
6	2	0.84	5E-3	0.29	0.76	0.30	0.11	0.82	0.24	0.29	0.75	0.63
8	2	0.91	2E-3	0.25	0.86	0.37	0.11	0.89	0.37	0.26	0.86	0.74
6	3	0.77	8E-3	0.25	0.68	0.22	0.10	0.75	0.11	0.25	0.67	0.51
8	3	0.88	2E-3	0.21	0.83	0.31	0.09	0.86	0.30	0.21	0.83	0.64
10	3	0.93	8E-4	0.18	0.90	0.34	0.08	0.92	0.37	0.18	0.90	0.68

AF-ordered state we again find two oxygen bands being decoupled. In addition the lowest Cu band, filled in hole representation and described by the wave function $|d_2\rangle$, remains also decoupled and dispersless; its energy is ϵ (in the first approach this band has finite although small band width $\approx \frac{T^4}{U^3}$). Thus we have three bands with finite dispersion of the order of $\approx \frac{T^2}{U}$, which we describe by the matrix-hamiltonian in the basis of $|d_1\rangle$, $|p_1\rangle$ and $|p_2\rangle$:

$$H_{\mathbf{kk}} = \begin{pmatrix} \epsilon_1 & 2T & 0 \\ 2T & 2t(1-\frac{V_k^2}{4}) & -tV_k \\ 0 & -tV_k & 2t(1-\frac{V_k^2}{4} \end{pmatrix} \tag{47}$$

where $t = \frac{T^2}{U-\epsilon}$. The secular equation, determining the dispersion of these bands reads

$$(E-\epsilon)[E^2 - 4t^2(1+\frac{V_k^2}{4})] + 4T^2[E - 2t(1-\frac{V_k^2}{4})] = 0 \tag{48}$$

This dispersion and, consequently, the corresponding hole transport comes from a Kondo-type interaction $\approx S_{pm}S_{dl} - \frac{1}{4}n_{pm}n_{dl}$ of the localized Cu holes with coherent oxygen holes in the CuO_4 plaquettes (S_{dn} are the corresponding spin operators). When an extra hole is added to the system two different situations have been considered

1. The hole will occupy the top of the highest filled band and AF order persists.
2. The hole forms a localized spin-polarized state (magnetic polaron); in the vicinity of the hole the spins are allowed to differ from the spin values of the AF ordered structure.

For the considered parameter range 2.) leads to the lower energy. In this situation due to the local spin deviation from the AF order the hole can move rather freely inside the cluster, gaining hybridization energy $\approx T$. This energy can be calculated by Green's functions from the secular equation

$$det|(E-\epsilon)1-\sum_{m,n}V(M,m)G(m-n,E)V(n,N)|=0 \tag{49}$$

where G(E) is the Green function of hole states in the AF ordered lattice. V(M,m)≈ T specifies the Cu-O hybridization interaction in dependence of the Cu spin polarization. M and N refer to the positions of the disturbed spins, while m and n refer to the positions and the orbitals of the surrounding oxygen atoms. The value of the local (disturbed) spin polarization $\langle n_\alpha \rangle$ is also estimated selfconsistently from Eq.(41). In the limit $\frac{U-\epsilon}{T} \gg 1$ the clusters are totally spin- polarized (ferron-type hole clusters). For $\frac{U}{T} \approx 8$ and $\epsilon \approx 2$ the most stable clusters contain 2 to 4 turned Cu spins. We stress that unlikely with the first approach in which the local perturbation is determined by the Hubbard repulsion U, in the second approach the local perturbation is given by the hybridization interaction T. Nevertheless, despite of this remarkable difference of the two used approaches, the obtained results are in good agreement with each other; for large U they practically coincide.

3 The barrier between ferron and free-particle states

Besides the chemical doping or oxydation in pure La_2CuO_4 and YBa_2CuO_6 the holes can be created via optical excitation across the insulating gap. Such transitions are known to correspond to the charge transfer from O^{2-} p-orbitals to $Cu^{2+}d_{x^2-y^2}$- orbitals. Since the mobility of the extra electron in the Cu^{2+} sublattice with AF ordered spins is expected to be much lower than that of a hole in the nonmagnetic O^{2-} sublattice, the optically created electron and hole should separate and at least for some time exist independently. Within this time the hole tends to become self-trapped forming a ferron state. Indeed, the experimental investigations [4, 10] of the photoinduced absorption in the materials under consideration display the induced broad band absorption centered around 0.5eV with lifetime ≈ 1ms which is attributed to the optical creation of localized in-gap excitations. The similar absorption band is also observed for small chemical doping. Since the photoinduced changes of the absorption coefficient in undoped materials resemble the absorption spectrum of doped (high T_c) systems it is naturally to assume that the localized species responsible for the photoinduced absorption are just the ferron hole states. We show here that the ferron state is separated from the free-hole states by a barrier which should strongly influence the hole kinetic processes. We consider the energy of a hole wave-packet in two cases:

1. in an ideal AF ordered CuO_2 lattice;

2. in the same lattice but with one turned Cu spin.

Assuming that the width of the hole bands ($\approx \tilde{t} = t + t^l$, t^l is the O-O transfer) is much smaller than the distance between them ($\approx T$) one may consider the lowest hole band only. We choose the wave-packet for the free and ferron-type hole states respectively as follows

$$|\Psi_L\rangle = \frac{1}{2L} \sum_{m_x, m_y = 0} \exp^{-\frac{|m_x| + |m_y|}{2L}} c^{\dagger}_{m_x, m_y, \sigma} |0\rangle \tag{50}$$

$$|\Psi^f_L\rangle = a_L |\Psi(0)\rangle + \sqrt{1 - a_L^2} |\Psi_L\rangle \tag{51}$$

Here $m_{x,y}$ refer to the positions of the uO_4 plaquettes in the x-y plane,

$$c_{m\sigma} = \sin\alpha d_{m\sigma} + \frac{1}{2}\cos\alpha \sum_i p_{i\sigma} \tag{52}$$

$|\Psi(0)\rangle$ is the wave function of the ferron kernel, $\sin\alpha \approx \frac{\pi}{4}, a_L$ is a variational parameter. The energies of these packets for $L \gg 1$ equal

$$E^{(0)}_L \approx -\tilde{t} + \frac{\tilde{t}}{8L^2} \tag{53}$$

$$E^f_L = E_S + E^{(0)}_L - \frac{T^2}{4L^2(1.1T + 3t)} \tag{54}$$

where $E_S \approx 0.1eV$ is the energy required to reverse a spin in the AF ordered lattice.

Comparing these two energies one obeserves that the free hole state is always energetically favourable for large-size wave packets (L $\rightarrow \infty$). However, if the condition $\tilde{t} < \frac{T^2}{1.1T+3t)}$, i.e. $\tilde{t} < 0.42T$ is fulfilled, the wave-packet energy increases with the rise of its size. It means that the localized wave-packet, i.e. the ferron state, possesses the lowest energy. The height of the barrier separating free and ferron hole state is

$$\delta_{ex} \approx \frac{1}{2} E_S [1.1(\frac{\tilde{t}}{T}) + 3(\frac{\tilde{t}}{T})^2] \tag{55}$$

This expression clearly displays the physics of the barrier:

Suppose that the hole is created in the free (extended) state (what is usually the case for optical interband excitation). The free hole state will not transform to the localized ferron state by first turning a Cu spin since the latter process costs a large amount of magnetic energy. Instead, the hole state will be squeezed until the localization itself will provide the energy for turning a Cu spin and, hence, the final hole relaxation to the ferron state. The barrier is absent for $\tilde{t} = 0$ since in this case the particle is localized from the very beginning and the formation of a ferron state leads to its partial delocalization [1]. We want to emphasize that only the existence of a barrier leads to a metastable behaviour of the free hole states the lifetimes of which can be rather large especially at low temperatures. Accepting the values of parameters mentioned above (i.e.$E_S \approx 0.1eV, T \approx 1.5eV$) and taking $\frac{\bar{t}}{T} \approx 0.3$ one obtains the estimate $\delta_{ex} \approx 15 - 30meV$. The barrier of such height should remarkably influence the relaxation of the optically created free hole into the ferron-type state at low temperatures and should lead to a strong temperature dependence of this process. Experimentally, the existence and the properties of the barrier can probably be deduced from the short-time kinetics of the photoinduced mid-IR absorption feature in pure $LaCuO_4$ and $YBa_2Cu_3O_6$.

4 Percolation Thresholds

The hole-clusters are practically immobile due to the small probability of spin-flip processes which have to accopmpany the motion of the clusters. On increasing the concentration of the holes, and with it that of the immobile clusters, at a critical concentration $c_{crit.}$ the clusters build up a percolation network along the lines of which the holes can move freely, and the system goes from the insulating to the conducting state [16, 17]. Within the percolation model the magnetic properties of the perovskite-type materials can be understood by taking the three-dimensional magnetic ordering into account which is characterized by strong in-plane and weak inter-plane interactions. Three different percolation thresholds are obtained [17]:

1. Percolation lines in the plane are built up from clusters of both spin directions. This limit characterizes the destruction of long-range AF order caused by the strong in-plane magnetic interactions and corresponds to a hole concentration

$$c_M^{(1)} \approx \frac{2.2}{8N} \tag{56}$$

2. Percolation lines in the planes are built up from clusters with the same spin orientation. This limit is important for the motion of the charge carriers and characterizes the transition to electrical conductivity corresponding to a hole concentration

$$c_c \approx \frac{4.4}{7N} \tag{57}$$

3. At higher concentrations the three-dimensional inter-plane percolation of AF-ordered areas will be destroyed. At this second magnetic threshold the system finally loses also the weak long-range AF order

$$c_M^{(2)} \approx \frac{15.8}{18N} \tag{58}$$

These considerations lead to the following relation between the three critical concentrations:

$$\mathbf{c_M^{(1)} \geq 1.4c_c \approx 3.2c_M^{(2)}} \tag{59}$$

For $La_{2-x}(Sr,Ba)_xCuO_4$, where the hole concentration c equals 2x, this leads to $x_M^{(1)} \approx 0.02, x_c \approx 0.05, x_M^{(2)} > 0.08$ in agreement with experiment [11]. In the case of $YBa_2Cu_3O_{6+x}$ a phase transition from tetragonal to orthorombic structure appears at $x \approx 0.4$ connected with the ordering of oxygens in the chains. This leads to the fact that for $x < 0.4$ most of the holes are kept in the chains, whereas at $x \approx 0.4$ a large number of holes are introduced into the planes, bringing the system immediately above the threshold for destruction of strong AF order and the beginning of electrical conductivity.

5 Experimental evidence for percolative phase separation

Field dependent magnetization measurements in slightly doped (oxygen enriched as well as Sr doped) La_2CuO_4 samples show the formation of conducting phases by the existence of a percolative phase separation [3].The essential results are summarized in fig.(5). Below 37 K, the sample of $La_2CuO_{4+\delta}$ exhibits a sharp transition from the paramagnetic to a diamagnetic magnetization indicating the presence of superconductivity. The diamagnetic fraction in the magnetization strongly depends on the temperature T_S to which the sample was rapidly quenched. Below T_S magnetization data were acquired during slow cooling temperature scans. Note that a sudden decrease of the diamagnetic fraction occurs at values of T_S in the range of 200-250 K. Surprisingly, the same sample quenched to 5 K (Fig.(5), uppermost curve) shows a very small diamagnetic fraction only. This clearly proves that superconductivity can almost completely be suppressed by rapid temperature quenching of the sample. If, however, cooling was performed slowly the diamagnetic shielding signal became maximal. As shown in [3] the formation of superconducting phases can also be initiated by application of sizeable magnetic fields. The results obtained for the Sr doped samples can be seen in Fig.(6) showing an increasing diamagnetic fraction with higher Sr contents. The most obvious feature is the distinct separation between the field cooled (applied measuring field of 90G) and the zero-field cooled magnetization both for the $La_2CuO_{4+\delta}$ and the Sr doped samples in the same temperature range. It is important to note that this temperature range is well above any of the Neel temperatures in the Sr doped samples. We find that the $La_2CuO_{4+\delta}$ sample shows quantitatively similar behaviour as the $La_{1.98}Sr_{0.02}CuO_4$ sample. All these measurements clearly prove that in the investigated $La_2CuO_{4+\delta}$ samples diffusion processes of magnetic quasi particles take place which can be influenced either by thermal treatment (fast and slow cooling) or by magnetic fields. Not all the quasiparticles, however, contribute to the conducting phase which is of fractal nature but rather coexist with single magnetic quasiparticles which give rise to the observed paramagnetic behaviour. The

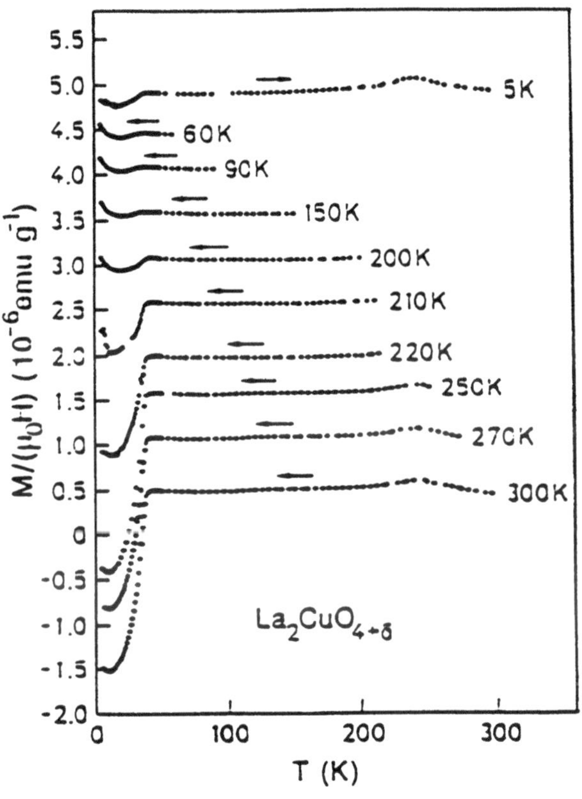

Figure 5: Magnetization of a $La_2CuO_{4+\delta}$ sample as a function of temperature. The starting temperatures T_S to which the sample was quenched from room temperature are indicated. The uppermost curve was obtained after quenching the sample to 5K and slowly heating to room temperature. Beginning from the lowest data set each curve was shifted upwards by a value of $5 \times 10 - 7 emu\, g^{-1}$ compared to the preceeding one. The measuring field was 90 G, arrows indicate the direction of temperature change during the measurements.

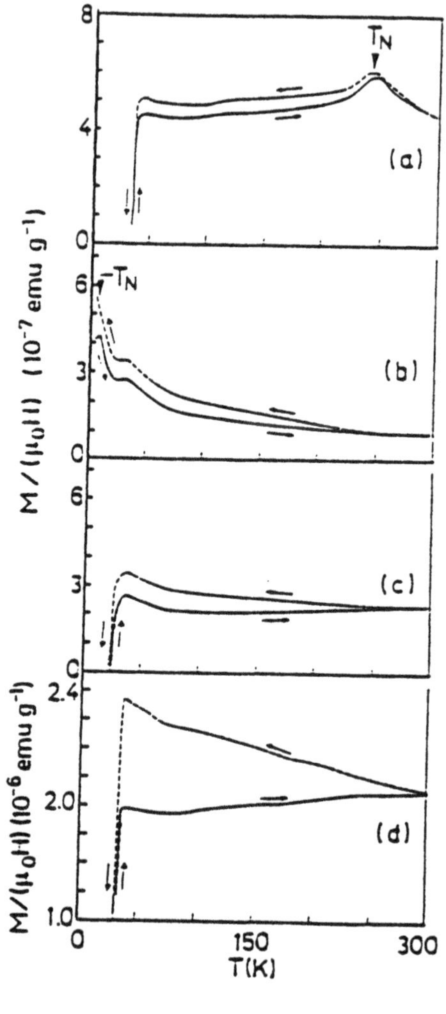

Figure 6: fc and zfc magnetization of the $LaCuO_{4+\delta}$ sample and moderately Sr doped samples $La_{2-x}Sr_xCuO_4$ as a function of temperature. Arrows indicate the direction of the temperature variation. The zfc magnetization curve was measured after the sample had been cooled slowly from room temperature to 5K by using the same temperature interval step sequence, including equilibriation as in the subsequent heating (or cooling) cycles and with applied magnetic field. The Neel temperature T_N for the $La_2CuO_{4+\delta}$ samples and for $La_{1.94}Sr_{0.06}CuO_4$ are indicated. Note: the zero of ordinate for $La_{1.94}Sr_{0.06}CuO_4$ has been suppressed.

equilibrium between these two subsystems can be shifted by thermal treatment of the samples. This coexistence and the by thermal treatment activated redistribution effects can clearly be observed in $La(Sr)CuO_4$ samples too. With increasing temperature thermal fluctuations gradually destroy the conducting (superconducting) phase breaking it up into magnetic quasiparticles as can be seen by the monotonic increase in the paramagnetic signal (Fig.6c and d, lower curves).

6 Conclusion

We showed that in the MF approximation of the three band Hubbard model doping with holes leads to the creation of ferromagnetically ordered clusters. This process is more pronounced when taking the influence of dopant ions (e.g. Sr, Ba, O etc.) into account. The occuring attractive interaction between these charge compensators and the holes in addition supports the hole-localization. The size of the clusters depends on the Coulomb repulsion on the Cu sites. Doping of the CuO_2 planes leads to a separation of the system into hole rich spin clusters (ferrons) and hole free AF ordered regions. The metal insulator transition (MIT) is provided by the percolation of clusters with the same spin direction. However, in the vicinity of the MIT there still exists long range AF order in regions between the percolation lines in accordance with experiment.

References

[1] V.Hizhnyakov,E.Sigmund, *Physica* C**156**,(1988) 655

[2] V.Hizhnyakov,E.Sigmund,M.Schneider,*Phys.Rev.* B**44**, (1991) 795

[3] R.K.Kremer,E.Sigmund,V.Hizhnyakov,F.Hentsch,A.Simon, K.A.Müller,M.Mehring,*Z.Phys.B Cond.Matter* **86**, (1992) 319

[4] C.Taliani,A.Pal,G.Ruani,R.Zamboni,X.Wei and Z.V.Vardeny in *Electronic Properties of High* $\mathbf{T_c}$ *Superconductors and Related Compounds* ed. by H.Kuzmany,M.Mehring and J.Fink, Springer Series on Solid State Science, Berlin 1990

[5] V.J.Emery,*Phys.Rev.Lett.* **58**, 2794 (1987)

[6] A.M.Oleś,J.Zaanen,*Phys.Rev.* B**39**,(1989) 9175

[7] A.Auerbach,B.E.Larson,*Phys.Rev.Lett.* **66** (1991) 2262

[8] A.P.Kampf,*Phys.Rev.* B**44**, (1991) 2637

[9] J.D.Jorgensen et al.,*Phys.Rev.* B**38**, (1988) 11337

[10] D.Mihailovic,C.M.Foster,K.Voss,A.Heeger,preprint

[11] A.Weidinger,C.Niederring,A.Golnik,R.Simon, E.Recknagel,*Phys.Rev.Lett.*62,(1989) 102

[12] E.N.Economou,*Green's Functions in Quantum Physics*, Springer Verlag (1990)

[13] T.Morita,*J.Math.Phys.* **12**,(1971) 1744

[14] T.L.Einstein,J.R.Schrieffer,*Phys.Rev.* B**7**, (1973) 3629

[15] Y.Nagaoka,*Phys.Rev.* B**147**,(1966) 392

[16] V.Hizhnyakov,E.Sigmund,G.Zavt*Phys.Rev.* B**44**, (1991) 12639

[17] V.Hizhnyakov,E.Sigmund, *Materials Letters* **9**, (1990) 425

PHASE SEPARATION IN CUPRATE SUPERCONDUCTORS FROM NMR AND MICROWAVE ABSORPTION MEASUREMENTS

M. Mehring, M. Baehr, P. Gergen, J. Groß, C. Kessler and N. Winzek
2. Physikalisches Institut, Universität Stuttgart,
D-7000 Stuttgart 80, Germany

Abstract

Examples for phase separation in cuprate superconductors are presented. We distinguish "chemical" and "electronic" phase separation. Furthermore a distinction between "static" and "dynamic" phase separation is made. Experimental evidence for these different types of phase separation includes NMR/NQR as well as microwave absorption measurements.

1. Introduction

One of the most fascinating aspects of high-T_c superconductors based on cuprates is besides their large T_c values the fact that the starting material is an insulating antiferromagnet which upon doping becomes conducting and superconducting[1]. It is observed that the ordering temperature T_N (Néel temperature) of the antiferromagnetic starting material decreases rapidly with increased doping, progressing to an electronic state which has no longer any static antiferromagnetic order. The resulting new electronic state is a mixed state of antiferromagnetically correlated spins and hole charges. This complex state is believed to be built from Copper d-orbitals and Oxygen p-orbitals of the CuO_2 planes in the cuprate superconductors. Three different types of cuprate superconductors will be considered here, namely $La_{2-x}Sr_xCuO_{4+\delta}$ (LSCO), $YBa_2Cu_3O_{6+\delta}$ (YBCO) and $Tl_2Ba_2CaCu_2O_{8-\delta}$ (TBCO). Doping in these materials proceeds either by oxygenation (parameter δ) or by oxydation with other electron withdrawing ions (Sr, Tl $\rightarrow$ Ca).

In the standard model "hole-doping" proceeds via the oxygen orbitals in the CuO_2 plane and is considered to lead to a homogeneous electronic state. In the antiferromagnetic state doping is visualized as a reduction in correlation length, maintaining, however, a uniform state with reduced Néel temperature T_N. Upon further doping the Néel state vanishes and a spin glass state and with increased doping a (homogeneous) conducting state appears which becomes superconducting below T_c.

This standard model has its difficulties with some recent experimental observations in particular in the low doping regime. It has been realized already some time ago that this homogeneous standard model is not appropriate even from a simple theoretical point of view[2,3,4]. The electronic state of a single hole in an antiferromagnetic surrounding is pretty localized for the parameters believed to be relevant for the CuO_2

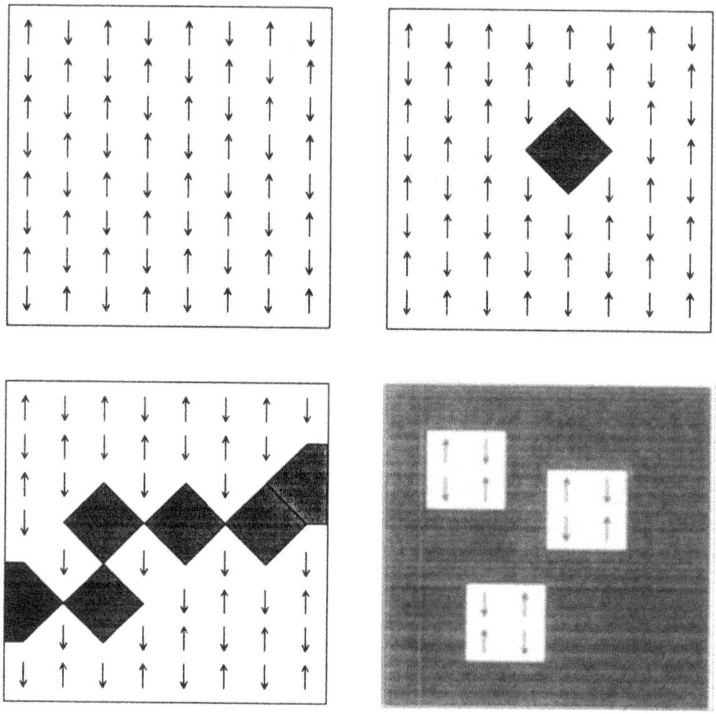

Fig. 1: *Hole doping into an antiferromagnetic background. The metallic phase is sketched as gray regions.*

plane[5]. Patches of localized holes are therefore expected to appear upon further doping as is sketched in Fig. 1. Each hole is surrounded by a "magnetic polarization disk" whose spin structure is ferromagnetic in the most simplest picture[5], but may take a complicated spiral structure under more rigorous theoretical treatments. Only when the "magnetic polarization disks" start to percolate, conductivity is observed. Beyond the percolation threshold there are still two separate phases, namely the antiferromagnetic phase and the conducting phase. When doping proceeds the antiferromagnetic phase is reduced in weight more and more. The question arises if and when the system becomes homogeneous at all.

This certainly is a question of the timescale. We will distinguish in the following stationary versus dynamic phase separation. The distinction will first be based on experimental observations. Later in section 4 we will set up more rigorous definitions:

(i) *stationary phase separation*
The experimental observations are consistent with a spatially inhomogeneous system, i. e. different parts of the sample can be attributed to different phases. Any exchange between these phases does not occur on the characteristic timescale of the measurement.

(ii) *dynamic phase separation*
On the characteristic timescale of the experiment the system appears to be spatially homogeneous, i. e. different phases cannot be distinguished by the particular experimental technique. However, distinct phases do exist in the sample with the lifetime of each phase exceeding the inverse of the characteristic energy spread and with a spatial extension larger than the characteristic length scale (typically some correlation lengths) of the particular phase.

Examples of these different scenarios will be discussed in the following sections. Before we do so, we want to make a distinction with respect to localization phenomena sometimes observed in disordered systems. In Fig. 2 we sketch the different scenarios in a one-dimensional graph. Consider a disordered metallic system with delocalized electrons at a temperature well above any localization temperature T_l. The electronic states are extended and the system appears to be homogeneous (Fig. 2 a). When the temperature is lowered down to T_l, localized states appreciably modify the electronic properties of the electrons which are still delocalized (Fig. 2 b). Experimental parameters show already signs of localization like broadening phenomena. Upon further cooling ($T \ll T_l$, Fig. 2 c) the electrons eventually become completely localized leading to an inhomogeneous system with randomly localized electronic states of narrow spatial extension. Phase separation, however, is a completely different phenomenon which appears at a temperature below T_{ph} and separates the sample into spatially extended different areas with different phases as is sketched in Fig. 2 d.

Another distinction has to be made with regard to chemically or electronically driven phase separations.

(a) *chemically driven phase separation*
Clustering of ions and atoms into large aggregates due to lattice forces can introduce inhomogeneities in the system which attract or repel the electrons, eventually leading to electron (hole) rich clusters embedded in electron (hole) poor clusters. Percolation of the electron rich clusters may lead to the phase separation.

(b) *electronically driven phase separation*
If the driving force for the phase separation is the attraction of the electrons (holes) leading to electron rich clusters, we will call this an electronically driven phase separation. However, due to charge neutrality, the electronic clustering may be accompanied by ion diffusion in order to minimize the energy.

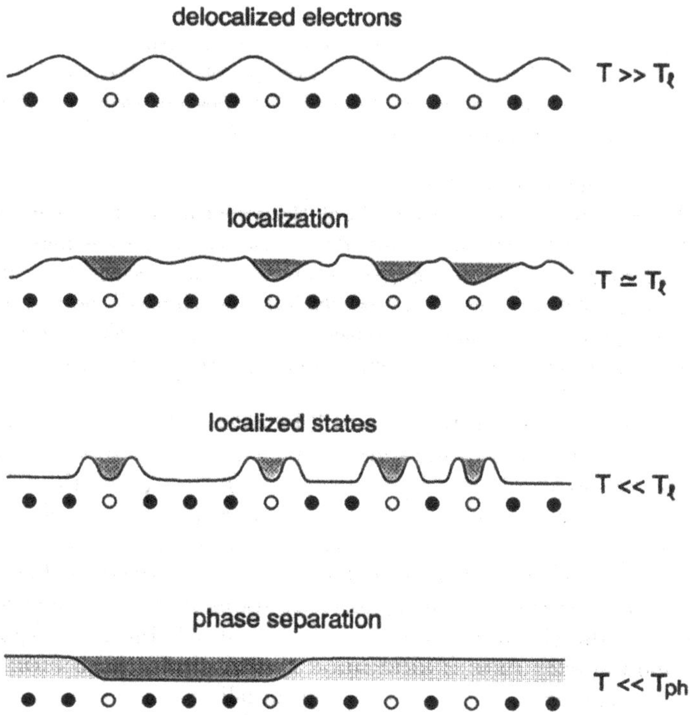

Fig. 2: *(a) - (c) progressive localization in disordered systems leading to randomly distributed states of small extension. (d) phase separation in contrast leads to spatially extended regions of 'homogenous' phases.*

In general it will be difficult to distinguish between these two cases.

2. Proton Induced Phase Separation in YBCO

Let us start from highly doped YBCO, namely $YBa_2Cu_3O_7$ which is superconducting and is considered to be homogeneous. So far no inhomogeneity of the electronic state has been observed. Exposing the sample to H_2 gas at 500 K, it is expected that the system is reduced (electron holes become filled) resulting in a reduced hole concentration of the CuO_2 plane. If the doping proceeded in a random fashion one would expect that T_c varies in a similar fashion as is observed for oxygen doping. On the contrary it is observed by ^{1}H-NMR that a phase separation occurs into a proton rich antiferromagnetic phase and a proton poor superconducting phase as is evident from Fig. 3[6]. Two spectral features can clearly be distinguished, namely a broad, almost rectangular spectral line due to the local magnetic field at the proton site caused by the antiferomagnetically ordered Cu spins and a narrow central line due to protons in the superconducting phase[6].

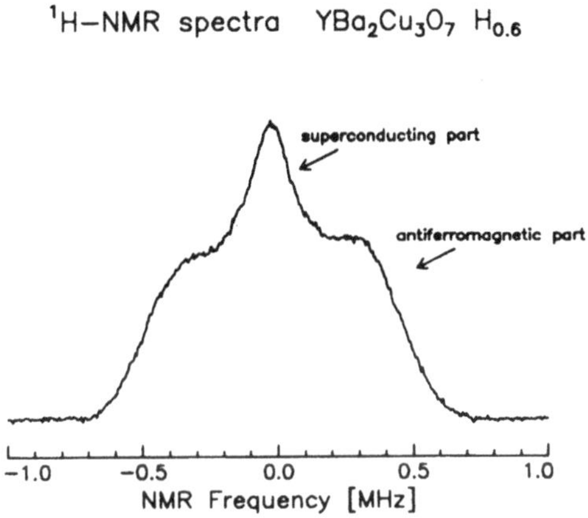

Fig. 3: *^{1}H-NMR spectra of $YBa_2Cu_3O_7H_{0.6}$ at 330 MHz and 20 K showing a broad rectangular spectrum (antiferromagnetic phase) and a narrow asymmetric spectrum (superconducting phase).*

The linewidth of the proton spectrum resulting from the antiferromagnetic phase varies with temperature like the order parameter and vanishes at the Néel temperature $T_N \approx 420$ K as is seen in Fig. 4a. The protons in the superconducting phase also show a temperature dependent linewidth (see Fig. 4b) which is, however, much smaller and decreases with increasing temperature and vanishes at the superconducting transition temperature T_c. This proves that phase separation occurs under proton doping. The inhomogeneity of the sample is clearly seen from the spectral features.

We are left with the question if this phase separation is electronically or chemically driven. In order to clarify if this point temperature cycling experiments were performed[6]. It turned out that the phase separation occurred already at the doping temperature and is accompanied by oxygen loss of the sample. It was therefore concluded that the phase separation is chemically induced in this case.

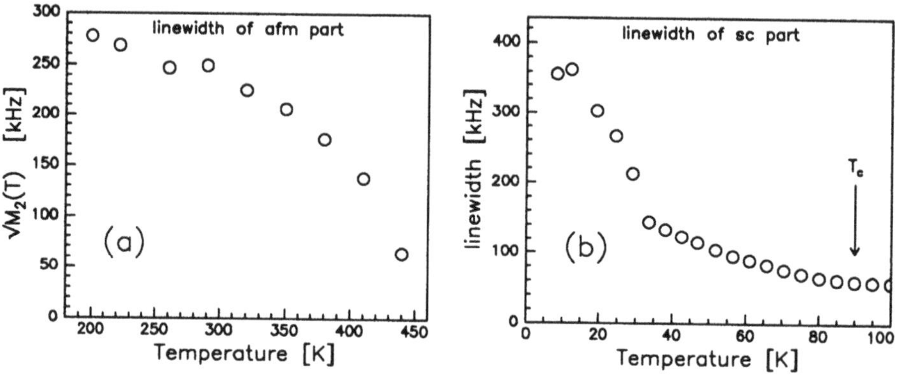

Fig. 4: *(a)* ^{1}*H-NMR linewidth of AFM spectrum approaching the Néel temperature (b) linewidth of vortex-lattice broadened superconducting line below* T_c.

3. Phase Separation in $La_{2-x}Sr_xCuO_{4-\delta}$

Doping of La_2CuO_4 with Sr, Ba initiated the era of high-T_c superconductivity[1]. Soon it was found that also oxygen doping leads to superconducting La_2CuO_4 samples. It came, however, as a surprise that even as-prepared samples of $La_2CuO_{4+\delta}$ are partially superconducting with δ as small as 0.017 [7]. Obviously these as-grown samples contain a small amount of superconducting phase. It was even more surprising that the amount of superconducting phase in these samples could be varied drastically for constant oxygen content just by slow and fast cooling and the application of static magnetic fields[7]. It was demonstrated that in order for the phase separation to occur, particular slow cooling was necessary in the temperature range above 200 K. When the sample was quenched from room-temperature only a very small fraction of the superconducting phase was observed. These observations are consistent with the scenario where hole rich clusters diffuse through the sample and start to aggregate slightly above 200 K. If the cooling through this temperature regime is too fast, only small hole rich aggregates are formed, whereas under slow cooling conditions a percolative phase separation can occur[7]. Evidently this phase separation is accompanied by oxygen diffusion[8].

In the following we want to show by NQR (nuclear quadrupole resonance) of ^{139}La in as-prepared $La_{2-x}Sr_xCuO_{4+\delta}$ that specific spectral and relaxational features may be attributed to the hole rich superconducting phase. Since the oxygen content in our samples is rather small (estimated to be 0.017 from the Néel temperature), we adjust the level of hole doping by the Sr content. In any case we keep the level of doping below x=0.05. At low temperatures the ^{139}La NQR spectrum of the 5/2 - 7/2 transition at 19 MHz is split into two lines due to the local magnetic field caused by the antiferromagnetically ordered Cu spins. For small doping this feature is dominant and it is difficult to distinguish the doped regions from the undoped ones. However, by applying spin echo spectroscopy at small and large pulse spacings we were able to disentangle two different spectral features shown in Fig. 5. For low doping concentrations a broad spectral feature between the two antiferromagnetic peaks can be seen to appear in Fig. 5. Its T_2 relaxation process is rather fast with respect to the dominant antiferromagnetic part. This relaxation process is of magnetic origin and is indicative of low lying magnetic excitations. We attribute the central peak feature to the doped clusters. This is consistent with the observation in Fig. 5 that its contribution increases with the Sr concentration. Note that it becomes dominant already for 2 % Sr doping indicating that each dopant (Sr) affects a large number of sites, i. e. the hole is extended over many copper sites (typically 15).

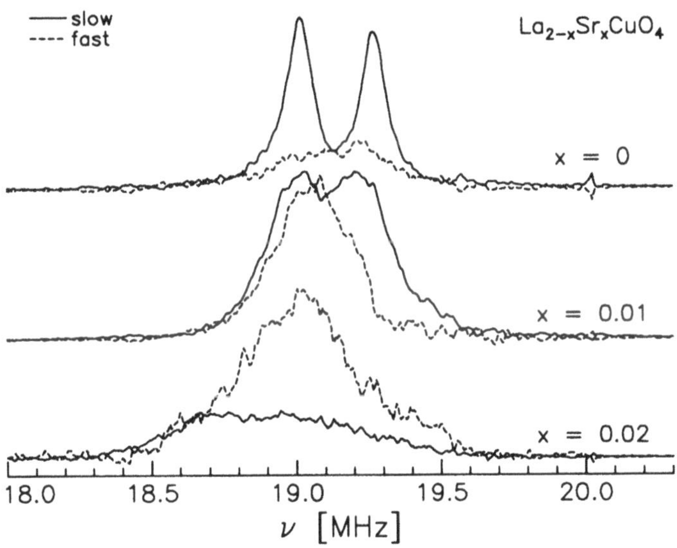

Fig. 5: ^{139}*La-NQR spectra for different Sr concentration at 10 K, fast and slow relaxing parts. The fast relaxing part is attributed to the hole rich phase.*

The spectral features seen in Fig. 5 do not depend on fast or slow cooling, i. e. they consist of small hole rich clusters pinned by the Sr^{++} counterion. Diffusion of Sr below 300 K is not expected to occur. In order to investigate the dynamic nature of the these hole rich clusters we have performed T_2 measurements versus temperature as is shown for 1 % Sr doping in Fig. 6. Note that two peaks in the relaxation rate appear, one at about 60 K and a much larger peak at 6 K. The peak at 60 K is attributed to the reduced Néel temperature in the 1 % Sr doped material and is observed only for those dopings where still an antiferromagnetic ordering exists. The 6 K peak, however, is still visible for Sr dopings above 2 % where no Néel temperature exists anymore. Its height increases with Sr doping concentration up to 3 % whereupon the peak starts to decrease with increased Sr doping as is evident from Fig. 7. The exact cause for this relaxation peak is unclear at the moment. It has been observed before in nominally undoped La_2CuO_4 [16]. Its appearance is very reminiscent of similar peaks observed at magnetic phase transitions. We therefore propose that a 3D magnetic ordering of the hole rich clusters occurs at this temperature. As mentioned before the relaxation process is unusually strong and indicates low-lying magnetic excitations. Further investigation will have to clarify this point. Hole rich regions in excess oxygen treated La_2CuO_4 also lead to extra peaks in the ^{139}La NMR/NQR spectrum. For details see Hammel et al.[18] also in this issue.

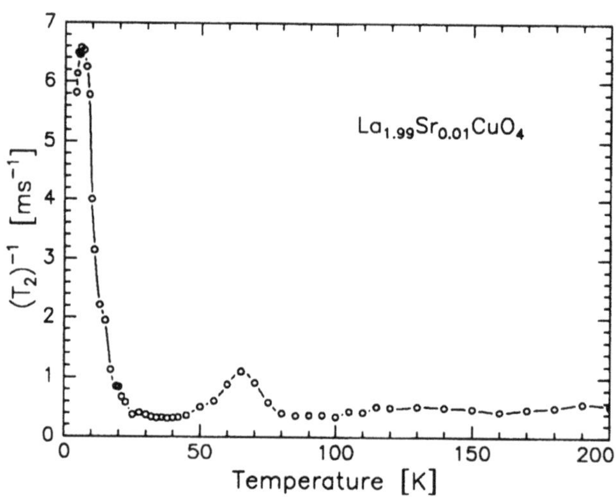

Fig. 6: *^{139}La-NQR $1/T_2$ relaxation rate versus temperature T in 1 % Sr doped LSCO. See text for discussion of the peaks.*

One of the most sensitive techniques for observing small amounts of superconducting phase in an otherwise non-superconducting environment is the microwave (MW) absorption technique in small magnetic fields. The sample is placed in a microwave cavity at about 9 GHz which is connected to a standard ESR (electron spin resonance) spectrometer. A small external magnetic field up to $\pm 10\,\mathrm{mT}$ and a modulation field parallel to the external field are applied to the sample. The change in microwave

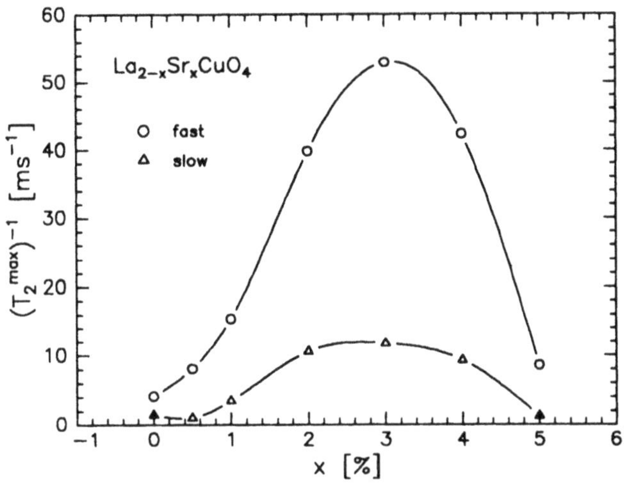

Fig. 7: *Doping (x) dependence of the relaxation rate* $1/T_2$ *at* $T = 6\,K$.

(MW) absorption is detected via lock-in detection when varying the external field[17]. We have applied this technique to different La_2CuO_4 samples either as-prepared with some excess oxygen or slightly Sr doped. Let us begin with the as-prepared samples. Fig. 8 shows the microwave absorption signal for two different cooling rates. The shape of the signal is characteristic for a granular superconductor in the superconducting state. The signal vanishes above T_c. It is caused by weak links or Josephson junctions connecting large superconducting areas. Note that the signal only appears when at least part of the sample is superconducting. Small hole clusters which are too small for superconductivity to occur do not give rise to this kind of MW absorption signal.

The different signal height in Fig. 8 is caused by different cooling processes. It is observed that the signal is larger for slow cooling and becomes smaller for rapid cooling. This is consistent with the earlier observation from susceptibility measurements[7] that a phase separation occurs by slowly cooling as-prepared La_2CuO_4.

In fact the type of cooling process is very important for the size and shape of the superconducting regions. Different cooling runs can result in a different topology of the superconducting phase. An example of this is shown in Fig. 9. The oscillatory feature of the MW absorption is indicative of an ordered superconducting phase which was created unintentionally. The signal character is similar to the sharp structures which appear in single crystals under certain conditions. Obviously fairly uniform superconducting regions were formed during the cooling process.

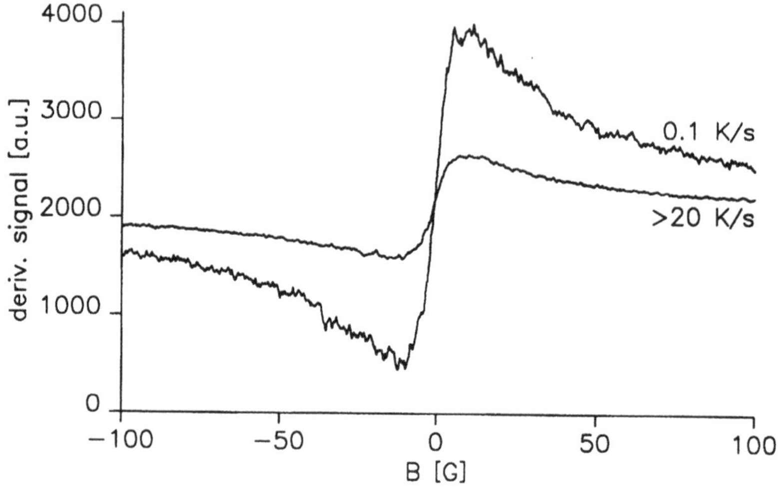

Fig. 8: *MW absorption signal for two different cooling rates, 0.1 K/s and >20 K/s, respectively.*

In contrast 1–3 % Sr doped samples do not show any MW absorption signal, although they contain the same amount of excess oxygen as the as-prepared samples. Only for larger doping a superconducting MW absorption signal is observed. The shape of this signal is quite different from the one for oxygen doping. Fig. 10 shows a typical example. The broad feature extending to relatively large magnetic fields indicates very small superconducting clusters. This leads to the conclusion that a phase separation does not occur with Sr doping. Obviously can superconducting clusters only be formed if the concentration of hole doped clusters is large enough for overlap to occur in contrast to the oxygen doping where even for small concentrations of excess oxygen a phase separation in a hole rich and a hole poor phase takes place.

In order to investigate at which temperature phase separation for $La_2CuO_{4+\delta}$ becomes effective we show in Fig. 11 the recovery of the MW absorption signal at different annealing temperatures. The experiment was performed in the following way. The sample was rapidly quenched to low temperatures (10 K) and a MW absorption signal was recorded. This signal served as a reference for the non-annealed state marked 'R' in Fig 11. Then the temperature was raised to some value (the annealing temperature) where the sample was kept for some time t (10 min.). After this the sample was cooled again to the measuring temperature (10 K, slow cooling) and a MW absorption signal was recorded. Rapidly lowering the temperature has an effect on the signal intensity but not on the main features of the recovery behaviour. The height

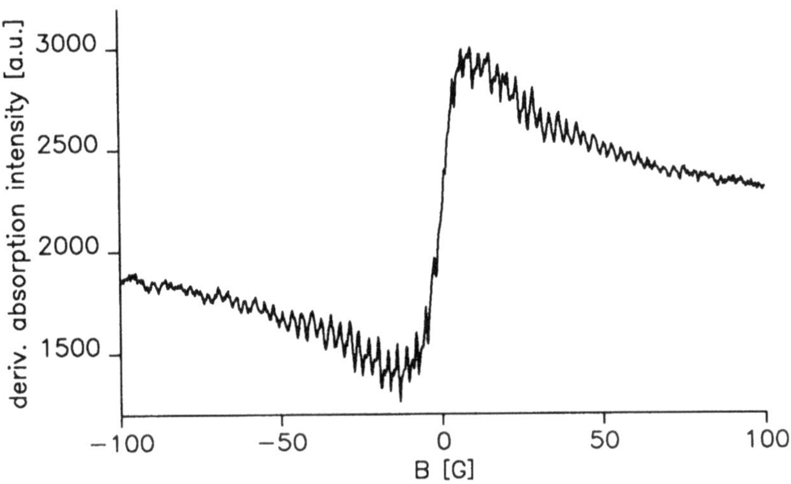

Fig. 9: *MW absorption signal of $La_2CuO_{4+\delta}$ showing narrow lines superimposed on the broader signal.*

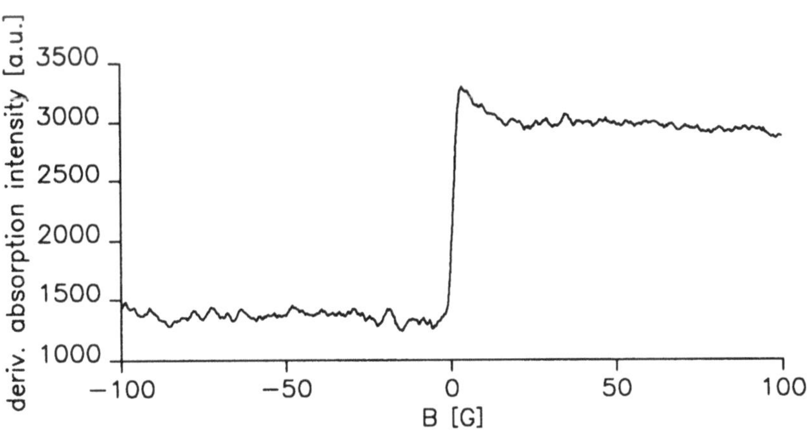

Fig. 10: *MW absorption signal of a 5% Sr doped sample. The absolute intensity is about 200 times smaller than that of $La_2CuO_{4+\delta}$.*

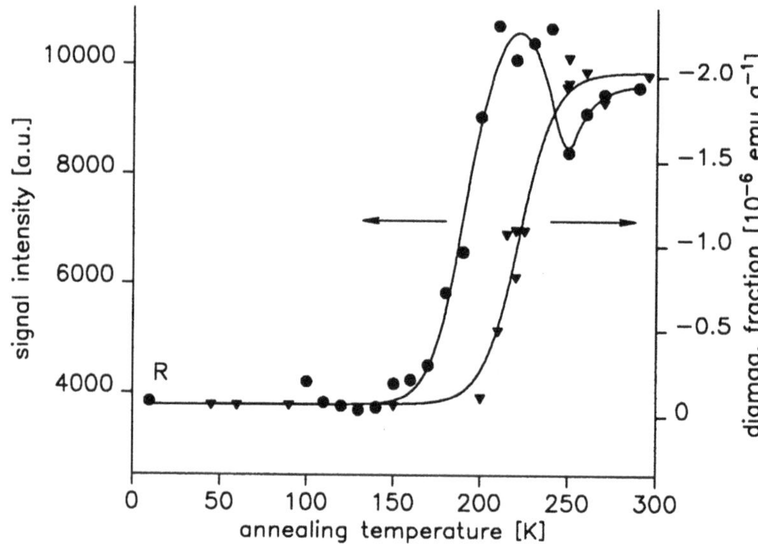

Fig. 11: *Isochronal annealing curve for as-prepared $La_2CuO_{4+\delta}$. Full circles are MW absorption data, triangles are taken from susceptibility measurements[7]. 'R' indicates the reference (non-annealed) intensity. The solid lines are guides to the eye.*

of this signal (Peak-peak-intensity) and the reference signal are plotted versus T in Fig. 11. Note that annealing, i. e. phase separation sets in around 170 K and reaches its maximum at 220 K. For comparison Fig. 11 also contains the annealing data taken from susceptibility measurements obtained earlier for a different sample[7]. The similarity of the results gained from two different experimental techniques and despite the different samples strongly indicates that the critical temperature range from 200–240 K in La_2CuO_4 is unique.

4. Dynamic Phase Separation

The dynamic phase separation as defined in the introduction can be visualized by referring to a simple example. Let us consider fog. It consists of very fine water clusters which move in a type of diffusive or Brownian motion through air. Definitely two different phases, namely water droplets and air (N_2, O_2) are present. By looking with the bare eye the system "fog" looks homogeneous. Coarse graining performed by the eye leads to the impression that only a single phase exists. The "measurement" performed by the eye occurs on a much slower timescale (25 Hz) than the diffusive motion of the water droplets. The "measurement" performed by the eye therefore leads to a spatial and temporal coarse graining. Note that the characteristic measuring time is not the observation time. It is rather the inherent time resolution of the eye during which two different events can be distinguished.

This principle has to be obeyed in every measurement. It therefore seems to depend on the type of measurement performed if a phase separation can be classified as static or dynamic. This situation is rather unsatisfactory and a definition is requested which is independent of the measurement. We are therefore led to the following definitions. Let us suppose that two different phases exist, namely phase A and B with correlation length ξ_A, ξ_B and correlation time τ_A, τ_B. These parameters characterize the spatial and temporal correlation of the particles within each phase. The different phases A and B form clusters (see Fig. 1 and Fig. 12) which are mixed into each other with spatial extensions l_A, l_B and lifetimes t_A, t_B. We say that a phase exists if the lifetime of the cluster is larger than the correlation time and the spatial extension is larger than the correlation length.

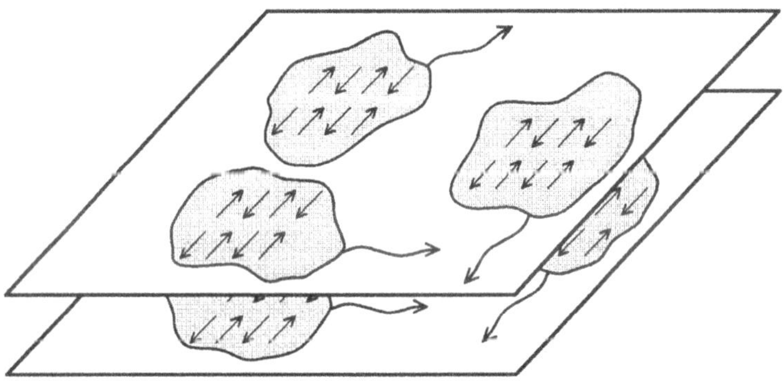

Fig. 12: *Snapshot picture of dynamic phase separation. Clusters of the highly mobile minority phase are sketched.*

Another lengthscale refers to the average separation d_A, d_B of the clusters A and B, respectively. Since the thermal excitations cause the clusters to diffuse, another timescale τ_D is set by the time required for the clusters to diffuse over a distance $\mathrm{Max}(d_A, d_B)$.

(i) *static phase separation*

$$\tau_A \ll \tau_D \ll t_A; \quad \xi_A \ll l_A$$
$$\tau_B \ll \tau_D \ll t_B; \quad \xi_B \ll l_B$$

(ii) *dynamic phase separation*

$$\tau_A < t_A < \tau_D; \quad \xi_A < l_A < d_A$$
$$\tau_B < t_B < \tau_D; \quad \xi_B < l_B < d_B$$

With this definition in mind we refer to the $YBa_2CuO_{6+\delta}$ cuprate superconductors in the intermediate doping regime ($0.5 < \delta < 0.9$) as being in the dynamic phase separation regime. Guided by Fig. 1, we visualize two different phases, namely a metallic (superconducting) phase with large hole concentration and a hole poor antiferromagnetic phase which forms antiferromagnetically correlated clusters diffusing through the metallic phase. We assume that the conditions (ii) apply, since most of the experimental techniques show a homogeneous system. This is in particular true for NMR/NQR experiments which we will discuss in the following.

It is assumed that the spin-lattice relaxation processes of the nuclear spins are dominated by fluctuating hyperfine fields[11]. The relaxation rate $1/T_1$ can be expressed by the equation

$$\frac{1}{T_1} = \hbar k_B T \gamma_n^2 A_{iso}^2 I_K \tag{1}$$

with

$$I_K = \sum_q F(q) \frac{\mathrm{Im}\chi^{+-}(q,\omega_0)}{\hbar\omega_0} \tag{2}$$

where the dynamics of the electron spins is contained in the dynamic susceptibility. The isotropic hyperfine interaction has been extracted from the sum. The formfactor which can have some structure in q-space depending on the nuclear site with respect to the electron spins contains in our definition also the anisotropic hyperfine interactions. Details can be found in a recent review[9].

Here we make the assumption that in cuprate superconductors in the intermediate doping regime, two different phases coexist dynamically where the spatial extension l_A of the antiferromagnetic clusters is only slightly larger than the correlation length ξ_A. We further assume that for nuclei whose formfactor is such that antiferromagnetic $Q = (\pi, \pi)$ fluctuations are not filtered out, the relaxation process is dominated by these fluctuations. This has been proven for Cu(2) and Tl(1) (see review[9]). Within a simple model of a two channel scattering process of the electrons, the dynamic susceptibility can be expressed as[3, 19]

$$\mathrm{Im}\chi^{+-}(q,\omega_0) = \chi_0 \left[\tanh\left(\frac{\omega+\Delta}{2T}\right) + \tanh\left(\frac{\omega-\Delta}{2T}\right)\right]$$

which reduces in the low frequency limit, relevant for NMR spin lattice relaxation, to

$$\frac{\mathrm{Im}\chi^{+-}(q,\omega_0)}{\hbar\omega_0} = \frac{\chi_0}{T}[1 - \tanh^2(\frac{\Delta}{2T})]$$

In order to allow for additional relaxation of the nuclear spins by dynamical processes within the cluster it was proposed to use the following form[9]

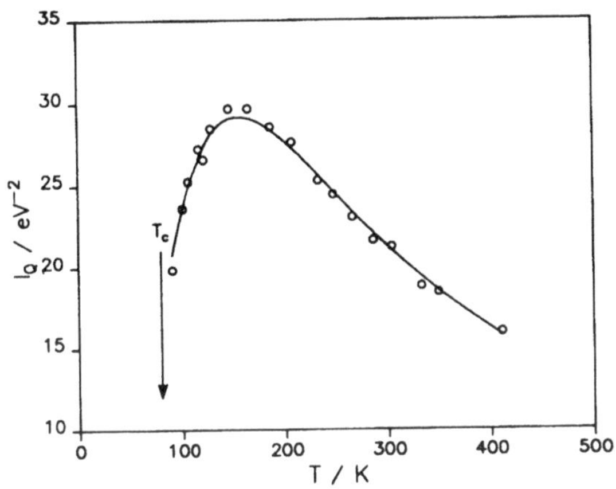

Fig. 13: *Fit of the ^{63}Cu(2) relaxation data of YBCO(124)*[13] *to the spin gap expression Eq. (3).*

$$I_K = A_0\left(\frac{1}{T}\right)^{\alpha}[1 - \tanh^2(\frac{\Delta}{2T})] \tag{3}$$

with $\alpha = 1.5$ and some arbitrary constant A_0. Fig. 13 shows a fit of the relaxation data of YBCO(124) obtained by Brinkmann et al.[13] to Eq. (3). A value of $\Delta = 280\,\mathrm{K}$ was obtained for the spin gap energy. Similar values were obtained for a series of Tl-cuprate superconductors[10] as shown in Fig. 14.

The most important parameter here is Δ which represents the spin gap energy which is required to flip a spin in the antiferromagnetic cluster. Such a spin gap has indeed been observed in neutron scattering experiments recently[19, 20]. It is evident from Figs. 13 and 14 that the relaxation data can be explained quite consistently within this approach. It is important to note that, although the spin gap energy Δ is on the same order of magnitude as a possible BCS gap, it does exist already above T_c. It is therefore quite natural to consider it as a degree of freedom within the antiferromagnetic phase. The contribution from the metallic phase is considered to be rather weak

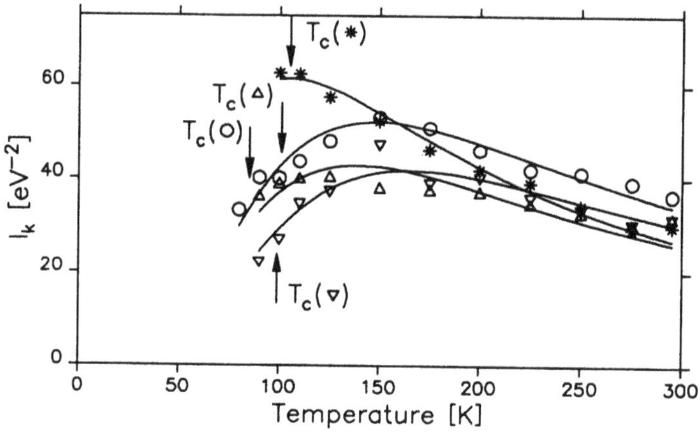

Fig. 14: *Fit of ^{205}Tl relaxation data to the spin gap expression Eq. (3).*

because of its much larger energy scale and correspondingly much smaller spectral density. For those nuclei at symmetric sites where the formfactor filters the antiferromagnetic fluctuations like O(2,3) and ^{89}Y, the relaxation is caused basically by the metallic phase. It is therefore not surprising that the temperature dependence of their relaxation is quite similar to that of ordinary metals[9]. Moreover, the high degree of cancellation of the antiferromagnetic fluctuations at the O(2,3) and Y sites demands a strong antiferromagnetic correlation of eight neighbouring copper spins (in the case of Y), which is very unlikely to occur in a highly "diluted" antiferromagnet, but is expected in antiferromagnetic clusters.

This is in distinct contrast to the "standard spin lattice relaxation model" advanced by Millis et al.[14] and Bulut et al.[15] It is still to be proved that the model of dynamic phase separation advocated here can explain all the observed NMR and neutron scattering phenomena. A model for the q-dependent susceptibility based on phase separation will be published elsewhere[21].

5. Conclusions

We have discussed NQR/NMR and microwave absorption measurements in terms of a static and dynamic phase separation model for cuprate superconductors. The distinction between "chemical" and "electronic" phase separation has been made. Although in the case of ^1H in YBCO and excess oxygen in La_2CuO_4 the counterions H^+ and O^- aggregate in some regions of the sample it is not clear, however, if this phase separation is driven by the chemical gradient or by the electronic forces. In any case a separation into hole rich and hole poor regions occurs. In both cases we consider these phase separations to be static. A dynamic phase separation model has been invoked to explain the unusual spin lattice relaxation phenomena in cuprate superconductors in the intermediate doping regime. It is characterized by a dynamic spin gap, whose energy has been determined from the temperature dependent relaxation data. In order to verify if these concepts really apply further theoretical and experimental investigations are required.

Acknowledgements. We gratefully acknowledge the encouragement by K. A. Müller to follow this line of thought. Discussions on phase separations with E. Sigmund helped to clarify certain points. We thank B. Nebendahl for performing part of the MW experiments. A financial support by the Bundesministerium für Forschung und Technologie (BMFT) and the Ministerium für Wissenschaft und Kunst, Baden-Württemberg is highly appreciated.

References

[1]J. G. Bednorz and K. A. Müller, *Z. Phys. B – Condensed Matter* **64** (1986) 189; J. G. Bednorz, M. Takashige and K. A. Müller, *Europhys. Lett.* **3** (1987) 379.

[2]V. Hizhnyakov and E. Sigmund, *Physica C* **156** (1988) 655.

[3]V. J. Emery, S. A. Kivelson and H. Q. Lin, *Phys. Rev. Lett.* **04** (1990) 475; J. C. Phillips, *Phys. Rev.* **B39** (1989) 7356, *Phys. Rev.* **B40** (1989) 7348, *Phys. Rev. Lett.* **64** (1990) 1605;
M. Marder, N. Papanicolaou and G. C. Psaltakis, *Phys. Rev.* **B41** (1990) 6940;
C. Castellani, M. Grilli and G. Kotliar, *Preprint*.

[4]T. Thio, T. R. Thurston, N. W. Preyer , P. J. Picone, M. A. Kastner, H. P. Jenssen, D. R. Gabbe, C. Y. Chen, R. J. Birgeneau and A. Aharony, *Phys. Rev.* **B38** (1988) 905.

[5]V. Hizhnyakov and E. Sigmund, *Mat. Lett.* **9** (1990) 425;
V. Hizhnyakov, N. N. Kristoffel and E. Sigmund, *Physica C* **161** (1989) 435;
E. Sigmund and V. Hizhnyakov, *Electronic properties of high-T_c superconductors and related compounds.* In: *Springer Series in Solid-State Sciences*, Vol. **99**, p. 366. (Springer, Berlin, Heidelberg, New York, 1991).

[6]J. Groß and M. Mehring (to be published).

[7]R. K. Kremer, E. Sigmund, V. Hizhnyakov, F. Hentsch, K. A. Müller and M. Mehring, *Z. Phys. B – Condensed Matter* **86** (1992) 319.

[8]J.-C. Grenier, A. Wattiaux, N. Lagueyte, J. C. Park, E. Marquestaut, J. Etourneau and M. Pouchard, *Physica C* **173** (1991) 139;

J.-C. Grenier, A. Wattiaux, J.-P. Doumerc, P. Dordor, L. Fournes, J.-P. Chaminade and M. Pouchard, *J. Solid State Chem.* **96** (1992) 20.

[9]M. Mehring, *Appl. Magn. Res.* **3** (1992) in press

[10]N. Winzek and M. Mehring, *Appl. Magn. Res.* **3** (1992) in press

[11]A. Abragam, *The Principles of Nuclear Magnetism* (Clarendon Press, Oxford, 1961);
C. P. Slichter, *Principles of Magnetic Resonance* (Springer, Berlin, 1989), 3rd ed.

[12]J. Emery and S. A. Kivelson (to be published).

[13]H. Zimmerman, M. Mali, J. Roos, D. Brinkman, J. Karpinski, E. Kaldis and S. Rusiecki, *Physica C* **159** (1989) 681;
H. Zimmerman, M. Mali, I. Mangelschots, J. Roos, L. Paul, D. Brinkman, J. Karpinski, S. Rusiecki and E. Kaldis, *J. Less-Common Metals* **164/5** (1990) 138.

[14]A. J. Millis, H. Monien and D. Pines, *Phys. Rev.* **B42** (1990) 167;
H. Monien, D. Pines and M. Takigawa, *Phys. Rev.* **B43** (1991) 258.

[15]N. Bulut, D. Hone, D. J. Scalapino and N. E. Bickers, *Phys. Rev. Lett.* **64** (1990) 2723;
N. Bulut and D. J. Scalapino *Phys. Rev. Lett.* **67** (1991) 2898; *Phys. Rev. Lett.* **68** (1992) 706.

[16]F. Hentsch H. Seidel, M. Mehring, J. G. Bednorz and K. A. Müller, *Physica C* **153-155** (1988) 727.

[17]K. A. Müller, M. Pomerantz C. M. Knoedler and D. Abraham, *Phys. Rev. Lett.* **45** (1980) 832-835;
K. W. Blazey, K. A. Müller, J. G. Bednorz, W. Berlinger, G. Amoretti, E. Buluggiu, A. Vera and F. C. Matacotta, *Phys. Rev. B* **36** (1987) 7241-7243.

[18]P. C. Hammel, A. P. Reyes, Z. Fisk, M. Takigawa, J. D. Thompson, R. H. Heffner, S. W. Cheong and J. E. Schirber *Phys. Rev. B* **42**, (1990) 6781.

[19]J. M. Tranquada in *Miami Workshop on Electronic Structure and Mechanisms for High Temperature Superconductivity, January 1991* eds. J. Ashkenazi and G. Vezzoli (Plenum, New York, 1992)

[20]J. Rossat-Mignod, L. P. Regnault, C. Vettier, P. Burlet, J. Y. Henry and G. Lapertod *Physica B* **169** (1991) 58;
J. Rossat-Mignod, L. P. Regnault, C. Vettier, P. Bourges, P. Burlet, J. Bossy, J. Y. Henry and G. Lapertod *Physica C* **185.9** (1991) 86.

[21]M. Letz, E. Sigmund and M. Mehring (to be published).

ELECTRONIC STRUCTURE AND PHASE SEPARATION IN SUPERCONDUCTING CUPRATES

C. Di Castro, and M. Grilli

Dipartimento di Fisica, Università "La Sapienza"
2, Piazzale Aldo Moro, 00185 Roma, Italy.

Abstract

Different models depending on the type of short range interaction (magnetic, excitonic...) and on the number of local hole orbitals involved, all show phase separation of electronic origin and nearby superconductive instabilities, suggesting a possible relation between the two phenomena.

In addition to the above specific problem, the full understanding of the electronic structure in cuprates is of general relevance for the understanding of high T_c superconductivity. A relatively simple band structure follows from the involvement of two d orbitals of the copper atoms, of the two p_σ orbitals of the planar oxygen atoms and of the p_z orbital of the apical oxygen in the presence of a strong repulsive Hubbard interaction on the d-orbitals.

As a generic feature of the high T_c cuprates, a single Cu-O band formed by an admixture of states with b_1 and a_1 symmetry crosses the Fermi level. The relative occupancy and the reduced bandwidth are in agreement with the experiments.

We finally present the correlation between the considered electronic structure and the maximum critical temperature attained in the different cuprate families.

1 Introduction

High temperature superconducting oxides [1] are almost two dimensional systems. The chrystallografic analysis, the transport experiments [2] and the band structure calculations (using the density functional in Local Density Approximation (LDA)) [3] all evidenciate a strong anisotropy. The materials are formed by copper and oxygen (CuO_2) planes intercalated with rare-earth and/or oxygen planes. There is now general agreement on the fact that the Cu-O planes are the most relevant structures as far as the electronic properties are concerned. In particular the hybridization of the orbitals closest to the Fermi level, the $3d_{x^2-y^2}$ (d_x) and $3d_{3z^2-r^2}$ (d_z) orbitals of the copper and the $2p_x$ and $2p_y$ of the oxygen, in the Cu-O planes give rise to the bands with largest dispersion, therefore playing a major role in transport or generally electronic phenomena.

One hole per unit planar (CuO_2) cell is present in the absence of doping and band structure calculations show the Fermi level to be in the middle of a well dispersed band. Nevertheless high T_c materials display an insulating antiferromagnetic (AFM) behaviour at half-filling. By doping substituting the out of plane components of the material, electrons or, more frequently, holes can be added to the system. In this case the AFM rapidly disappears and the system becomes metallic. Electrons introduced by n-doping mostly reside on copper d_x orbitals, while the extra holes due to p-doping have been shown [4] to go into states with a large amplitude on oxygen. These facts, together with the experimental evidence from X-ray Absorption Spectroscopy (XAS) and Photoemission Spectroscopy (PES) clearly indicate the presence of strong Coulomb repulsion (U=8-10 eV) among holes on copper. It is this repulsion together with the difference in the copper-oxygen energy level, $\varepsilon_p^0 - \varepsilon_d^0$, that confers a charge-transfer insulating character to the system at half-filling [5]. The presence of a strong interaction and of various (insulating-AFM, metallic, superconducting) phases close to each other renders difficult theoretical treatments as well as the determination of a simple theoretical model which can capture the essential physics. In particular three (naturally interwoven) problems are not yet settled: i) the nature of the electronic liquid in the normal state ii) the electronic structure and iii) the relevance of the various interactions in determining the pairing mechanism.

The layered structure of the high temperature superconducting oxides conferring a remarkable two dimensional character to these materials and the very strong electron-

electron interactions present in these systems have raised some questions on the very existence of a Landau Fermi Liquid (FL) normal state in these oxides. Moreover the most relevant interactions leading to pairing in these systems could be of purely electronic origin. Thus a clear distinction between the various energy scales cannot be done. This situation is in contrast with the traditional view of metallic-superconducting materials, which was essentially based on the large difference between the energy scales of the electronic repulsion and of the pairing potential. This big difference allowed a two step procedure in which the strong electronic interaction was first effectively described by the Landau Fermi Liquid (FL) theory, leading to a small residual interaction between quasiparticles. This last one is usually neglected. However it competes against the weak phononic pairing potential responsible for the Cooper instability. In this case the weakness of these terms justified mean field treatments and the (at least) qualitative answers were believed to be reliable.

In high T_c oxides, the repulsive and the attractive forces both being of electronic origin, do not allow the above mentioned distinction between different energy scales. This fact makes difficult the theoretical treatment. The so far unproven hypothesis that high temperature superconducting oxides form a non conventional (non-FL) normal state has some support from the anomalous behaviours of many quantities like the planar resistivity, the optical conductivity, the intensity of the Raman scattering, the tunneling conductance as a function of voltage, the nuclear relaxation rate and the Hall coefficient. The unusual phenomenology stimulated the appearance of various theoretical proposals assuming rather "exotic" normal states for these systems. The still open debate on the FL vs non-FL nature of the normal state is surely of crucial importance in the understanding of high T_c superconductors (HT_cSC) and it has attracted considerable attention in the theoretical scenario. Nevertheless in this paper we will only focus our attention on the two other fundamental above mentioned problems: the one concerning the electronic structure of these materials and the determination of the most relevant interactions for pair formation, that are present therein. Independently from the actual pairing mechanism, which will turn out to be relevant for the real superconducting oxides, the models and the formalism we shall consider have their own validity in the understanding of strongly correlated electron systems.

The aim of the present work is therefore to present a particular (necessarily incomplete) path in the rich and multiform scenario of theoretical models based on electronic degrees of freedom. In particular the starting point will be the three band Hubbard model [6]-[8] including the copper d_x orbital and the oxygen $p-\sigma$ orbitals in the Cu-O planes. Having as ingredients copper-oxygen charge transfer energy, copper-oxygen hopping, and Coulomb-repulsion on copper, the three band Hubbard model is constructed to capture both the strong copper-oxygen hybridization and the strong correlations on the copper ions, and through superexchange it gives rise to antiferromagnetism.

The three band model can be simplified to an effective single band model according to the proposal, first put forward by Zhang and Rice [9], which will briefly be discussed in the next section. This simplification cannot be done in at least two cases: i) when the spins of the mobile holes on oxygen and the spins of the localized holes on copper exist as separate degrees of freedom coupled via a Kondo-like coupling [12]-[14]; ii) when a nearest neighbor Coulomb repulsion between holes on copper and on oxygen is present. In this latter case the model contains charge transfer excitons as a relevant feature[15]-[23].

Superexchange, Kondo coupling and charge transfer excitons have been proposed as possible electronic mechanisms for the pairing.

The three band model can be further generalized by the inclusion of other orbitals like the d_z of the copper [24] and p_z of the out of plane oxygen ions [25, 26]. This extension is motivated firstly by the experimental observation of a sizable occupation of states with this z-oriented symmetry and secondly by the theoretical search for different T_c behaviours in the different classes of compounds despite their similar in-plane electronic structure. This has led to the recent discovery of a correlation [25, 26] between the maximum critical temperature for each class of compounds and the amount of states with z symmetry, indicating a possible relevance of these states in reducing the effective pairing.

The presence of phase separation in various high T_c materials has recently been discovered [27]. The existence of phase separation in the real materials and in various classes of theoretical models with short range interactions [28, 29, 14, 22, 23] is an interesting point deserving careful investigation. Thus, despite the differences between the various models and the variety of interactions considered, a common feature will nevertheless arise: the competition between the kinetic energy (tending to delocalize the holes) and the various short range interactions (favored when the holes are localized) originates phase separation as it has been recently reviewed [30]. Furthermore this phenomenon seems to be related with superconductivity.

In the next section we will present the three band Hubbard model as a starting point and the technical tools which will be used in the following. In section 3 a magnetic Heisenberg coupling between the copper spin will be added to the three band Hubbard model. The resulting extended tJ model will be described giving particular emphasis to the close relation between the kinetic energy-magnetic energy competition resulting in phase separation in this model. In section 4 we will generalize the model introducing the Kondo-like magnetic coupling. The general feature of this model is the interaction between localized spins and a diluted set of mobile holes. Also in this framework phase separation will occur and its connection with superconductivity will be established within this model. In section 5, instead, a Coulomb nearest neighbor repulsion will be introduced, the related physics will be discussed and, like in the previous models, phase separation and superconducting instabilities will be again analyzed.

Finally we will generalize the extended Hubbard model by introducing the d_z orbital of the copper and the p_z orbital of the out of plane oxygen ions (section 6). In this context th above mentioned semiphenomenological correlation between the maximum critical temperature in the various classes of materials and the symmetry of the hole states will be presented.

2 The three band Hubbard model and the slave boson technique

a) The model

The three band Hubbard Model [7],[8] will be our starting point of our journey in the land of electronic models of high temperature superconductors. The Hamiltonian reads

$$
\begin{aligned}
H = & - \sum_{i,\pm\eta,\sigma} \left(t_{ii\pm\frac{\eta}{2}} \tilde{d}^{\dagger}_{i\sigma} p^{\eta}_{i\pm\frac{\eta}{2}\sigma} + h.c. \right) + \sum_{i,j,\sigma} \left(t_{p_{ij}} {p^{x}}^{\dagger}_{i\sigma} p^{y}_{j\sigma} + h.c. \right) + \\
& + \varepsilon^{0}_{d} \sum_{i,\sigma} \tilde{d}^{\dagger}_{i\sigma} \tilde{d}_{i\sigma} + \varepsilon^{0}_{p} \sum_{i,\sigma} {p^{\eta}}^{\dagger}_{i+\frac{\eta}{2}\sigma} {p^{\eta}}_{i+\frac{\eta}{2}\sigma} + U_d \sum_{i} \tilde{d}^{\dagger}_{i\uparrow} \tilde{d}_{i\uparrow} \tilde{d}^{\dagger}_{i\downarrow} \tilde{d}_{i\downarrow}
\end{aligned}
\qquad (1)
$$

with η = x, y. We work in the hole representation in which the vacuum state is the $3d^{10}$ configuration for the copper and the $2p^6$ for the oxygen. The $\tilde{d}_{i,\sigma}$ ($\tilde{d}^{\dagger}_{i,\sigma}$) operators are annihilation (creation) operators for holes in the $3d_{x^2-y^2}$ orbitals of the Cu atoms, while $p^{x,y}_{j,\sigma}(p^{x,y\dagger}_{j,\sigma})$ are the corresponding operators for the $2p_x$, $2p_y$ orbitals on the O sites and σ labels the spin. ε^0_p and ε^0_d, are the bare atomic levels on O and Cu respectively. The first two terms are of kinetic nature and describe the p-d and p-p orbital hybridization respectively (i,j in the second term label nearest neighbor oxygen sites). Together with the third and fourth terms, describing the atomic energies of holes on Cu and O sites, they define a three band tight-binding model in second quantized form. The last term gives a local Hubbard repulsion U_d whenever two holes meet in the same copper site. This interaction is very large in copper oxides and is responsible for the insulating character of these materials at half-filling (one hole per CuO_2 cell) provided the charge transfer excitation energy $\varepsilon^0_p - \varepsilon^0_d$, is large enough.

Being U_d the largest energy scale in the system, it is often assumed $U_d \to \infty$, so that the last term in eq.(1) is replaced by the no double occupancy constraint on copper sites $\sum_\sigma \tilde{d}^{\dagger}_{i\sigma}\tilde{d}_{i\sigma} \leq 1$. This anholonomous constraint is difficult to be treated, but this difficulty can be dealt with by means of the slave boson technique.

b) The slave boson technique [31]

The Fock space has to be suitably enlarged by introducing additional bosonic degrees of freedom describing the absence of a hole on a site. Then, defining a new vacuum state $|0>_{new}$, one can describe the absence of holes on a given site i, by saying that an "emptyness" boson is present on that site

$$|0>_{old} \to b_i^{\dagger}|0>_{new} .$$

In this new language, to create a hole, one must first destroy the emptyness boson and then create a fermion

$$\tilde{d}^{\dagger}_{i\sigma}|0>_{old} \to d^{\dagger}_{i\sigma} b_i b_i^{\dagger}|0>_{new} = d^{\dagger}_{i\sigma}|0>_{new} .$$

The advantage of this procedure is that, since site i is either empty ($\sum_\sigma d^{\dagger}_{i\sigma}d_{i\sigma} = 0,\ b_i^{\dagger}b_i = 1$) or singly occupied ($\sum_\sigma d^{\dagger}_{i\sigma}d_{i\sigma} = 1,\ b_i^{\dagger}b_i = 0$), the constraint assumes the holonomous form of a completeness relation

$$b_i^{\dagger}b_i + \sum_\sigma d^{\dagger}_{i,\sigma}d_{i,\sigma} = 1$$

so that it can be treated by means of a Lagrange multiplier field λ_i. The introduction of fictitious degrees of freedom in of no consequence for the physics since the constraint reduces the accessible Fock space to the regions corresponding to physically meaningful states.

At this point the model is still overwhelmingly difficult to be solved, but a mean field solution can be tried assuming constant values b_0 and λ_0 for the field b_i and λ_i respectively. This defines a mean field free energy of the form

$$F = N_{cells}\left[\lambda\left(b_0^2 - 1\right) - \frac{2T}{N_{cells}}\sum_{k,l}\ln\left(1 + e^{\frac{\mu - E_l(k)}{T}}\right)\right], \tag{2}$$

$E_l(k)$ are the bands of the system obtained by diagonalizing for each point in k-space the Hamiltonian matrix

$$H = \sum_{k\sigma}\begin{pmatrix} d^\dagger_{k\sigma} \\ p^{x\dagger}_{k\sigma} \\ p^{y\dagger}_{k\sigma}\end{pmatrix}\begin{pmatrix} \varepsilon_d & -2b_0 t\sin\frac{k_x}{2} & -2b_0 t\sin\frac{k_y}{2} \\ -2b_0 t\sin\frac{k_x}{2} & \varepsilon_p & 2t_p\beta_k \\ -2b_0 t\sin\frac{k_y}{2} & 2t_p\beta_k & \varepsilon_p \end{pmatrix}\begin{pmatrix} d_{k\sigma} \\ p^x_{k\sigma} \\ p^y_{k\sigma}\end{pmatrix}; \tag{3}$$

with $\varepsilon_d = \varepsilon_d^0 + \lambda_0$ and $\beta_k \equiv 2\sin(k_x/2)\sin(k_y/2)$. The suffix 0 has been dropped in ε_p, which is not renormalized, having assumed that the interaction is only present in d sites.

The free energy has to be minimized with respect to b_0 and λ_0 together with the condition

$$\frac{\partial F}{\partial \mu} = N_{cells}\,(1 + \delta), \tag{4}$$

fixing to $\frac{1}{2}(1+\delta)$ (δ is the doping with respect to half-filling) the number of holes per spin per cell. The solution of the self-consistency equations gives the mean field values b_0 and λ_0 to be inserted into eq.(3), allowing the determination of a renormalized band structure. As it can be easily seen from the expression (3), the hopping constant t is multiplicatively renormalized by b_0, whereas λ_0 shifts the bare atomic level of the copper ε_d^0. The reason for these renormalizations can be understood by tracing back the presence of b_0 and λ_0 to the existence of the b_i and λ_i bosons keeping track of the strong U_d interaction. Due to the constraint the value of b_0 is always less or at most equal to one. Therefore the substitution $t \rightarrow tb_0$ amounts to a reduction of the kinetic energy of the holes. This is a perfectly reasonable effect of the strong interaction. On the other hand, if $\delta < 1$, only the lowest band of predominant copper character[1] is occupied, but the constraint of no double occupancy on copper requires that the additional holes introduced by doping the system, have to be accomodated on oxygen states. Therefore a sufficient amount of copper-oxygen admixture has to be present despite the reduction of the hopping parameter due to b_0. The role of λ_0 is just to approach the copper level to the oxygen one in order to allow enough hybridization between these levels thereby conferring the right amount of p character to the lowest band in order to accomodate the additional holes without violating the constraint.

The effect of the interaction in this slave boson mean field description is therefore to renormalize the parameters of the non interacting tight binding Hamiltonian, giving a renormalized band structure which describes the coherent motion of the quasiparticles. In this context the system is still a Fermi liquid, the spectral weight of the quasiparticle poles being related to b_0. This can be seen in the simple case with $t_p = 0$. Considering first the single particle Green function of the physical holes, which are formed by an admixture of $\tilde{d}$ and p particles due to the hybridization term present in the model Hamiltonian (1) in the absence of interaction, we obtain

$$\begin{aligned}\tilde{G}_{11}(k,\omega_n) &= \text{FT}\left[Tc_1^\dagger(r,t)c_1(0,0)\right] = \tilde{u}_k^2 G_{\tilde{d}\tilde{d}}(k,\omega_n) + \tilde{v}_k^2 G_{pp}(k,\omega_n) + 2\tilde{u}_k\tilde{v}_k G_{p\tilde{d}}(k,\omega_n) \\ &= \frac{1}{i\omega_n + \mu - \tilde{E}_1(k)}, \end{aligned} \tag{5}$$

(FT means Fourier transform in r-space and time and T is the time ordering operator) where the Matsubara finite temperature formalism has been used and $\tilde{E}_1(k)$ is the lowest free particle band arising from the diagonalization via the transformation $\tilde{c}_{1k} = \tilde{u}_k d_k + \tilde{v}_k p_k$ relating the lowest band operator $\tilde{c}_{1k}$ to the orbital operators $\tilde{d}_k$ and p_k. This latter one is the bonding combination of p_k^x, p_k^y oxygen orbitals, $p_k = (\sin(k_x/2)p_k^x + \sin(k_y/2)p_k^y)/\gamma_k$, which, in the $t_p = 0$ limit, is the only combination that mixes with the $d_{x^2-y^2}$ orbitals of the copper. The $\tilde{u}_k$ and $\tilde{v}_k$ coefficient are given by

$$\tilde{u}_k^2(\tilde{v}_k^2) = \frac{1}{2}\left(1 + (-)\frac{\varepsilon_p^0 - \varepsilon_d^0}{R_k^0}\right) \tag{6}$$

where $R_k^{0^2} \equiv (\varepsilon_p^0 - \varepsilon_d^0)^2 + 16t^2\gamma_k^2$, and $\gamma_k^2 \equiv \sin(k_x/2)^2 + \sin(k_y/2)^2$.

---

[1]We always assume $\varepsilon_d^0 < \varepsilon_p^0$ because, being the oxygen more elettronegative than copper, it is more difficult to accomodate holes on it.

To consider the effect of the interaction within the slave boson formalism, the particles become an admixture of $d_i b_i$ and p_i operators. In mean field approximation, ($b_i \approx b_0$) the interaction effects give rise to a wave function renormalization factor for the copper operators $\tilde{d} = b_0 d$ and to a tight-binding parameter renormalization $t_{pd} \to t_{pd} b_0$ and $\varepsilon_d^0 \to \varepsilon_d^0 + \lambda = \varepsilon_d$. Thus the interacting case is recovered by performing this substitutions into (5) obtaining

$$\tilde{G}_{11}(k,\omega_n) = b_0^2 u_k^2 G_{dd}(k,\omega_n) + v_k^2 G_{pp}(k,\omega_n) + 2b_0 u_k v_k G_{pd}(k,\omega_n) = \frac{z_k}{i\omega_n + \mu - E_1(k)}, \quad (7)$$

where the u_k and v_k coefficients and $E_1(k)$ are obtained from the expressions for the non-interacting case with renormalized parameters.

To determine the spectral weight z_k it is necessary to compare $\tilde{G}_{11}$ with the lowest band quasiparticle Green function. This can be obtained by defining the mean field lowest band quasiparticle operators $c_{1k} = u_k d_k + v_k p_k$ so that the Green function for the quasiparticles in the first band is

$$\begin{aligned} G_{11}(k,\omega) &= \frac{1}{i\omega + \mu - E_1(k)} \\ &= u_k^2 G_{dd}(k,\omega) + v_k^2 G_{pp}(k,\omega) + 2u_k v_k G_{pd}(k,\omega). \end{aligned} \quad (8)$$

This relation shows that b_0 relates $\tilde{G}_{11}(k,\omega)$ with $G_{11}(k,\omega)$. In particular, since, as we shall see, at small doping near the insulating regime $b_0 \ll 1$ so that $u_k^2 \approx 1$ and $v_k^2 \approx \frac{4t^2}{\epsilon_p^0 - \epsilon_d} b_0^2$, one has

$$\tilde{G}_{11}(k,\omega) \approx b_0^2 G_{11}(k,\omega) \quad (9)$$

This relation shows that at mean field level b_0^2 plays the role of the quasiparticle spectral weight in the physical particles $\tilde{G}_{11}$ Green function.

An interesting feature of the present model and the related treatment is that it exhibits a metal-insulator transition [7] in some region of the parameters δ, t, t_p and $\varepsilon \equiv \varepsilon_p^0 - \varepsilon_d^0$. This behaviour is particularly relevant in view of a possible modelization of the HT_c oxides where a metal-insulator transition is observed as a function of doping. We now briefly report the behaviour of the model again assuming for simplicity $t_p = 0$ [7]. If very few holes are present in the system ($\delta \approx -1$), the interaction term U_d plays no role because the probability of two holes to meet in the same copper site is very small. In this situation the band structure is not renormalized, $b_0 = 1$ and $\lambda = 0$. When holes are added to the system two different situations can occur. If the hopping t is sufficiently large compared with the level separation ($\varepsilon < \varepsilon_c \equiv 3.34t$ in the simple $t_p = 0$ case) the Cu-O mixing is strong, the lowest band has a sufficient p character so that the interaction on copper sites is not so efficient to give rise to insulating behaviour when one hole per cell is present. In this case even at half-filling ($\delta = 0$) the quasiparticle effective bandwidth ($\sim tb_0$) is reduced but does not vanish. There is a continuous change of the renormalized band structure in going from $\delta < 0$ to $\delta > 0$. On the other hand, if $\varepsilon > \varepsilon_c$ the system becomes insulating at half-filling. This transition is marked by the continuous vanishing of b_0 when $\delta \to 0$ ($b_0^2 \approx \delta$), so that, when δ is exactly zero, the quasiparticle band has a vanishing spectral weight and a vanishing bandwidth ($b_0 = 0$). When $\varepsilon > \varepsilon_c$, if the doping is due to electrons ($\delta < 0$) the constraint can be satisfied without the d level approaching too much the p one ($\lambda < \varepsilon/2$), whereas, as soon as the doping becomes positive (hole doping) the constraint can be satisfied only if a sufficient p character is present in the lowest band to accomodate the added holes, so that the renormalized d level must be closer to the p level ($\lambda > \varepsilon/2$). In going from electron to hole doping there is, therefore, a discontinuous change of the quasiparticle band position inside the bare charge transfer gap $\varepsilon_p^0 - \varepsilon_d^0$. This shift as confirmed both from numerical calculations [33, 34] and from recent PES experiments [35], physically corresponds to a discontinuous jump in the chemical potential in adding or subtracting electrons around half-filling.

In particular a renormalized quasiparticle band is present close to the p levels at a distance of the order of t^2/ε. This phenomenon is nothing but a rephrasing in the Fermi liquid context of the Zhang-Rice picture, where the p-d hybridization and the related superexchange interaction originate the splitting down of singlet states from the oxygen band by an energy of the order of t^2/ε. The fact that the Fermi level for physically reasonable doping always falls into the lowest band opens the way to a possible reduction of this model to an effective single band model.

Another particularly appealing feature of the slave boson formalism is that in a very natural way it can describe an apparent dicotomy occurring in the superconducting oxides. Recent ARPES and ARIPES experiment in various classes of high temperature superconductors [36] have identified a large Fermi surface in rather good agreement with band structure calculations and satisfying the Luttinger theorem with a number of enclosed states corresponding to a filling $n = 1 + \delta$. On the other hand, transport experiments shows a dependence on a number of carriers of the order of the doping δ only. In particular Infrared (IR) reflectivity experiments [37] in metallic samples show a Drude peak with intensity increasing with doping. The very same behaviour can be obtained in a straightforward mean field slave boson treatment. In fact, within this scheme, the quasiparticles give rise to the formation of a band whose form is determined by the renormalized tight-binding model (3). The form of this band is consistent with the Luttinger theorem, the quasiparticles forming a Fermi liquid with $n = 1 + \delta$ states inside the Fermi surface. On the other hand, the mean field form of the physical current operator in the η direction, in the discretized version of a tight-binding model

$$J^{\eta}(j) = ib_0 \frac{et_{pd}}{2\hbar V} \sum_{\sigma} \left(d^{\dagger}_{j\sigma} p^{\eta}{}_{j+\frac{\eta}{2}\sigma} + d^{\dagger}_{j\sigma} p^{\eta}{}_{j-\frac{\eta}{2}\sigma} + p^{\eta}{}^{\dagger}_{j+\frac{\eta}{2}\sigma} d_{j\sigma} + p^{\eta}{}^{\dagger}_{j-\frac{\eta}{2}\sigma} d_{j\sigma} \right) \tag{10}$$

show a dependence on b_0 which is then reflected in the frequency dependent conductivity

$$\sigma(\omega) \propto \lim_{q \to 0} \frac{1}{\omega} \mathrm{Im}\{\mathrm{FT}\,[\langle TJ(j,t)J(0,0)\rangle]\} \propto b_0^2 \tag{11}$$

Then, since the mean field solution for b_0^2 in the region of interest for the superconducting oxides $\varepsilon > \varepsilon_c$, approximately scales with the doping (at least for not too large δ), it is clear that the slave boson formalism very naturally accounts for the observed doping dependence of the coherent Drude peak in optical experiments. In the same way all the transport properties will be controlled by $b_0^2 \approx \delta$.

The above analysis has recently been extended beyond mean field [38] including the first order fluctuations of the Bose fields (within a controlled 1/N expansion where the the maximum spin value is formally extended to N). While the mean field coherent contribution appears only away form half-filling, the fluctuation corrections provide incoherent contributions to the optical and to the photoemission spectra also in the insulating phase, whereas at finite doping they contribute to the midgap states.

All these features will be maintained in a more realistic five band model including the $d_{3z^2-r^2}$ orbital of copper and the p_z orbital of the apical oxygen, which will be discussed in section 6.

3 The extended tJ model

The presence of antiferromagnetic order in the insulating phase of the superconducting oxides suggests that magnetic effects could play a relevant role in the physics of these systems even in the conducting phase, possibly also providing a mechanism for the pairing, since at low doping the superconducting phase is close to the antiferromagnetic one. The model (1) can by itself generate a magnetic superexchange coupling among the spins of the holes on copper ions. Such a term, however, readily appears only in a perturbative approach on t/ε at the order t^4/ε^3 (when $U_d = \infty$). In our slave boson approach it can only appear beyond mean field due to a double exchange of fluctuations of the bosonic field b_i [39]. Since, however, for the sake of simplicity we restrict ourselves to a simple mean field approach, the effect of such a superexchange mechanism can be taken into account by the explicit inclusion of a term like

$$H_J = J\sum_{ij} \mathbf{S}_i \cdot \mathbf{S}_j, \tag{12}$$

with $\mathbf{S}_i = d^\dagger_{i\sigma}\sigma_{\sigma\sigma'}d_{i\sigma'}$ ($\sigma_{\sigma\sigma'}$ are Pauli matrices). The resulting model is the so called extended tJ model [10],[11]. Neglecting a term proportional to the densities which can effectively be included in a redefinition of the Hamiltonian parameters, one can rewrite (12) as

$$H_J = \frac{J}{2}\sum_{ij\sigma\sigma'} d^\dagger_{i\sigma}d_{j\sigma}d^\dagger_{j\sigma'}d_{i\sigma'}. \tag{13}$$

In order to deal with the Hamiltonian term (13), we have to introduce a decoupling scheme which allows for a local magnetic order without spin long range order. This is due to the fact that in real systems the doping rapidly spoils long range AFM order, while local magnetic correlations still play a role. Thus, in the context of high temperature superconductivity theories, terms of the form (13) are customarily decoupled via non local (bond) complex field, which, at mean field level assumes the form

$$\langle\Delta_{ij}\rangle \equiv \frac{J}{2}\sum_{\sigma}\langle d^\dagger_{i\sigma}d_{j\sigma}\rangle \tag{14}$$

so that the magnetic part (13) of the Hamiltonian becomes in mean field

$$H_J = \sum_{ij\sigma}\langle\Delta_{ij}\rangle d^\dagger_{j\sigma}d_{i\sigma} + \frac{2}{J}\sum_{ij}|\langle\Delta_{ij}\rangle|^2 \tag{15}$$

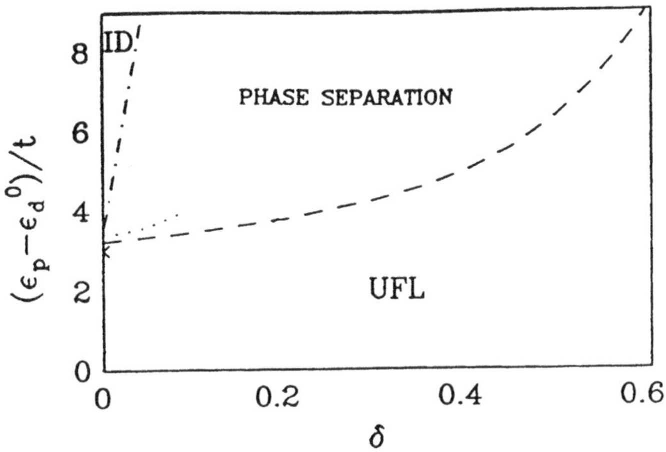

Figure 1: *Phase diagram at J=0.2 for various values of* $(\varepsilon_p - \varepsilon_d^0)/t$ *and* δ. $t = 1/\sqrt{2}$ *is fixed whereas* $(\varepsilon_p^0 - \varepsilon_d^0)$ *varies. UFL: Fermi liquid with uniform* Δ *parameters; ID Insulating Dimer. Above the dashed curve there is the phase separation region (the dotted curve shows the phase separation region when J=0.05). The cross indicates the point above which* $b_0 = 0$ *at half-filling. From ref. [11].*

In particular it is possible to mimic the AFM insulating phase when $b_0 = 0$ by means of the dimer phase $\langle \Delta_{ii+x} \rangle = J/2$, $\langle \Delta_{ii\pm y} \rangle = \langle \Delta_{ii-x} \rangle = 0$. The FL phase with short range AFM correlations can be represented by $b_0 \neq 0$ and $\langle \Delta_{ii\pm x} \rangle = \langle \Delta_{ii\pm y} \rangle = \Delta$ (uniform phase).

The extended tJ model in the simplified case of $t_p = 0$ with the possibility of uniform and dimer order has been investigated in ref. [11], where the phase diagram of Fig. 1 was obtained for $t = 1/\sqrt{2}$ and J=0.2 [2]. A phase separation region is present in the phase diagram above the dashed curve. A system with ε/t and δ above this curve phase separates producing regions with many holes and uniform symmetry (or metallic dimerized in a small region at ε/t not too large) and regions at half-filling with perfect dimerization of the copper spins.

The physical origin of the phase separation can ultimately be traced back to the competition between the short range magnetic interaction and the kinetic energy gain due to the hybridization. The orbital mixing tends to delocalize the holes reducing their

[2] Once the energy unit is suitably rescaled to give a physically reasonable value of the hopping $t_{phys} = 1.3eV$, J becomes $J_{phys} \approx 0.36eV$. This value is rather large compared to the estimated J in real oxides, but it has been chosen to make visible the effects of magnetism in the model

kinetic energy. At the same time, however, it reduces the number of holes on copper ($n_d < 1$) making more difficult the formation of a perfect dimer order (requiring $n_d = 1$) which minimizes the magnetic potential energy. The system prefers to create metallic regions at high doping where the hybridization is strong, and insulating regions at half-filling where the magnetic order is preserved.

It is worth noting that the phase separation region starts at $\delta = 0$ close to ε_c/t. In this region where the band tends to become flat ($b_0 \to 0$), the kinetic energy, which usually plays a stabilizing role, is strongly suppressed so that the kinetic-magnetic energy competition is particularly strong.

The presence of phase separation is a general feature of models with short range magnetic interactions. Numerical diagonalization of finite clusters [28], perturbative 1/N expansion [29] and slave boson-1/N calculations [32] all show that this phenomenon also occurs in the single band tJ model. In this latter case, however, contrary to the extended tJ model, a phase separation is always found at small doping (at least for J large enough [33]) because the single band tJ model corresponds to the $\varepsilon \gg t$ limit of the extended tJ model. Within the extended model, instead, also the $\varepsilon \sim t$ region is accessible, where the kinetic energy is large enough to stabilize the system.

In principle long range Coulomb forces should prevent the phase separation forbidding the formation of regions rich in charged holes. However, it is also possible that the confinement of positive charges in these regions is allowed by a related confinement of negatively charged ions (e.g. oxygen O^{-2} ions) compensating the charge imbalance. The experimental observation of phase separation in superconducting oxides [27] indicates that this phenomenon may indeed occur.

4 The Kondo-lattice model

A very simple physical idea is at the origin of the Kondo-lattice model (KLM) for HT_cSC. Starting from the low doping regime the mobile holes due to doping disturb the antiferromagnetically correlated substrate formed by holes residing on copper sites and lead to a rapidly decreasing Neél critical temperature. Nevertheless the substrate can still maintain an antiferromagnetically correlated nature, possibly providing a pairing potential.

Then the basic ingredient of this model is the existence of two different species of particles in the system: mobile holes moving on a substrate of heavy (almost) localized holes. This situation is realized in mixed valence systems (heavy fermions) and, on the basis of the previous argument, could also be present in HT_cSC. The presence of the heavy particles is due to the strong localizing electron-electron interactions leading to a single particle occupation on the copper sites in superconducting oxides. This localized holes (almost) entirely consists of spin degrees of freedom coupled via a superexchange Heisenberg interaction J, which further favors the localized behaviour. On the other hand, the mobile holes are weakly interacting with each other, but are coupled with the spins of the substrate via a Kondo-like interaction J_K.

Some experimental evidences can also support a scenario consistent with the existence of two different species of carriers in the HT_cSC systems. In particular NMR relaxation time T_1 experiments show a Korringa like behaviour in oxygen sites, which instead is not obeyed on copper sites. However, their relation with the temperature dependence of a unique spin susceptibility [40] indicates a strong correlation among these holes.

A more direct support to this model would come from the direct and inverse Angle-Resolved Photoemission Spectroscopy (ARPES and ARIPES) experiments [41],[42], which show the possible existence of two separate bands crossing the Fermi level. The presence of two bands across the Fermi level is still doubtful and debated, but, if confirmed, would probably be a relevant fact in the understanding of these materials. In the present context, however, it is important to notice that, if two nearby bands are formed by states with different symmetry and if they have different curvatures leading to different effective masses of the carriers, a two fluid picture can naturally arise.

In the previous section we have on the other hand seen that the Cu-O hybridization present in the planes of the superconducting copper oxides gives rise to a single type of quasiparticles with mixed p-d character. This situation seems to be rather general and as we shall see in section 6 also occurs in more complicated models including a larger number of orbitals per unit cell, apparently making the applicability of the KLM rather questionable.

Quasiparticles, however, are found in a mean field analysis to live in a single band with mixed symmetry and could still sustain a two-fluid system in a local picture: magnetic interactions, even though they involve energies, which are typically much lower than the energies involved in band structure calculations can act in different ways on the carriers with different symmetries. Thus they can by themselves select and distinguish on a local basis the wavefunction components of the quasiparticles.

The model we shall now briefly describe consists therefore of singly occupied $d_{x^2-y^2}$ orbitals, providing the localized spin degrees of freedom S_i, and of a band $\varepsilon(k)$ of mobile holes arising from the direct p-p hybridization [14]

$$H = J \sum_{<i,j>} \vec{S}_i \vec{S}_j + \sum_{k,\sigma} \varepsilon(k) c^{\dagger}_{k,\sigma} c_{k,\sigma} + J_K \sum_{i,j} \vec{S}_j \vec{\sigma}_i. \tag{16}$$

σ_i are the spins of the mobile holes of the band $\varepsilon(k)$. To mimic in a simple way a local magnetic order without long range spin symmetry breaking already in mean field approximation a magnetic flux order [3] has been assumed for the substrate of S_i spins [13],[14].

The presence of J_K couples the mobile holes to the localized spins, which are consequently disturbed. However, at low doping and/or small J_K, the mobile holes only slightly perturb the spin substrate, which maintain its flux order (spin liquid flux phase(SL)). Increasing the doping and/or J_K, the mobile holes tend to form singlets with the spins of the substrate. In this way their motion inside the magnetic substrate is favored and they give rise to a single band of bound states in order to minimize their kinetic energy. A Fermi liquid (FL) results where the quasiparticles have a mixed ((d-spin)-(p-charge)) character. This regime is usually known as "Kondo-compensated regime" because the bound states are created by the Kondo interaction J_K. Clearly the Kondo compensation tends to disrupt the magnetic order of the substrate, which, at large enough doping and/or J_K, looses its flux order. A mean field decoupling of the Kondo term introduces a parameter V describing the Kondo-compensated regime and effectively acting as a mobile holes-spin substrate (p-d) hopping term.

Again, as in the case of the extended tJ model, the kinetic energy gain due to the motion (hybridization) of the charges competes with the magnetic potential energy.

In the absence of an anomalous order parameter for a superconducting phase, the phase diagram has the form reported in Fig. 2a, where $J_K = 1$, $J = 0.3$ and the oxygen-oxygen hopping $t_p = 0.1$. Similarly to the extended tJ model, a phase separation occurs

[3] In the flux phase the Δ_{ij} mean field parameter is of the form $\langle \Delta_{ii\pm x} \rangle = \Delta e^{i\phi}$; $\langle \Delta_{ii\pm y} \rangle = \Delta e^{-i\phi}$.

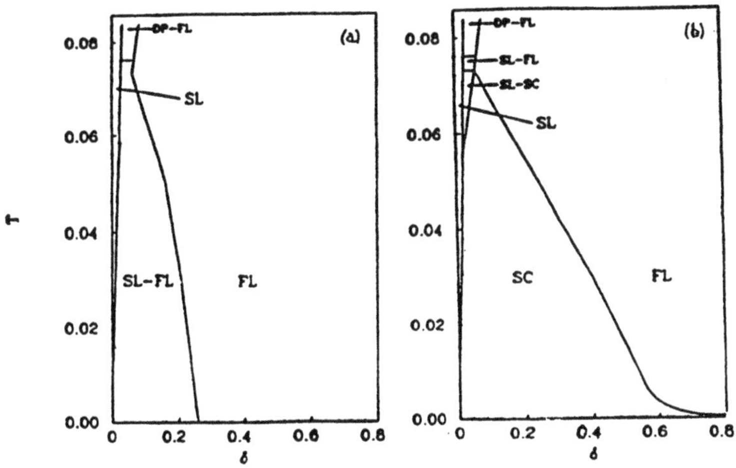

Figure 2: *Phase diagram temperature T vs. doping δ, at $J_K = 1$, $J = 0.3$ and $t_p = 0.1$. a) Only the non superconducting phases with non-anomalous decoupling parameters are considered. SL= spin liquid (flux); FL= Fermi liquid; DP= decoupled phase with flat bands (unphysical feature of the slave boson treatment at high temperature). SL-FL= region of phase separation between a SL and a FL phase; DP-FL= region of phase separation between a DP and a FL phase. b) An anomalous order parameter is introduced to describe a superconducting (SC) phase. From ref [14]*

between the Fermi liquid (FL) Kondo-compensated phase without flux magnetic order and the SL without p-d quasiparticle formation (i.e. without Kondo-compensation V=0). This phase separation is due to the above mentioned competition between the kinetic and magnetic energy terms.

The introduction of an anomalous order parameter describing the superconducting phase drastically changes this scenario. The formation of Cooper pairs, in fact, creates a state with lower energy just exploiting the competition effects. Whereas the motion of a single hole strongly disturbs the magnetic substrate, the coherent motion of two bound holes can be less annoying for the underlying spin system. The effective pairing interaction between the p-holes requires that the p-holes couple to the d-spins via a V different from zero. Thus the presence of the magnetic order in the substrate, not only provides a pairing mechanisms, but also allows the system to find a good compromise in the kinetic-magnetic energy dilemma. As a result, once a superconducting state is made possible by the introduction of a mean field anomalous order parameter, the phase diagram becomes the one reported in Fig. 2b. It is then apparent that the phase separation region has been enormously reduced by the presence of a large superconducting (SC) region.

In the second phase diagram it is also apparent that the critical temperature first increases and then decreases with doping. This behaviour reflects the fact that superconductivity connects the SL ($V = 0$) and the FL ($V \neq 0$) phases which are stable at low and large doping respectively. Due to J_K, at low doping the mobile holes interact with the substrate and, as soon as V is different from zero SC takes place. Then T_c grows with the number of carriers. Increasing the number of carriers, however, the substrate is increasingly disturbed and provides a weaker pairing potential leading to a decreasing critical temperature, eventually vanishing at large enough doping.

The following remark is now in order. Usually the phase separation is seen as a spurious source of instability requiring the introduction of long range forces to stabilize the system. The simple model we described in this paragraph, instead, indicate that superconductivity itself can provide a stabilizing mechanism with respect to phase separation, thereby making even more plausible the idea that the attractive forces giving rise to phase separation can be responsible for Cooper pair formation also. This suggestion will be further supported by the analysis of phase separation within the three band Hubbard model with nearest neighbor repulsion.

5 The three band Hubbard model with nearest neighbor repulsion

While in the previous section we have privileged the magnetic interaction in extending the three band Hubbard model, we now want to focus our interest on the effects of charge transfer (CT) excitations. We therefore extend the three band Hubbard model Hamiltonian (1) to include the nearest neighbor Coulombic repulsion

$$\tilde{H}_V = V \sum_{i,\sigma} \tilde{d}^\dagger_{i\sigma} \tilde{d}_{i\sigma} \sum_{\sigma',\alpha=x,y} \left(p^\dagger_{i\sigma'\alpha} p_{i\sigma'\alpha} + p^\dagger_{i\sigma'-\alpha} p_{i\sigma'-\alpha} \right) . \tag{17}$$

As a consequence of this repulsion the charge degrees of freedom will acquire the possibility of low energy excitations leading to important physical consequences. In particular the presence of charge excitations will affect the thermodynamic stability of the model leading to phase separation. As it will be apparent below, also in this case a connection seems to exist between superconductivity and phase separation.

This model has been extensively investigated in the weak coupling limit (small U and V) [15]-[21] where the following features have been detected: when the nearest neighbor repulsion V overcomes the Cu-O hybridization the system presents, by increasing the doping, i) a p-d charge-transfer (CT) instability leading to phase separation and ii) superconducting instabilities in the s-wave and d-wave channel, which are, however, very sensitive to the strength of U, posing the question if this result persists in the strong coupling regime.

The presence of a strong U repulsion in the real materials leads also to the existence of a metal-insulator transition, which could have relevant physical consequences.This latter point together with the sensitivity to U of the superconducting instabilities in the weak coupling analysis, have motivated a strong coupling investigation of the model (1)+(17) [22, 23].

By means of the slave boson technique, suitably generalized to allow the inclusion of fluctuations of the bose fields ($\frac{1}{N}$ expansion, where N is the maximum value of the spin index formally extended to assume large values), the model has been studied in the $U = \infty$ limit.

The mean field and the $\frac{1}{N}$ first order calculations have revealed the following behaviour of the model: i) the presence of the repulsion U introduces a coupling between the charge-transfer excitations describing the p-d charge fluctuations and the fluctuations of the total charge in the cell[4]; ii) as a consequence of i) a phase separation is present in a region of the $(\varepsilon_p^0 - \varepsilon_d^0)$–$\delta$ parameter space, in connection with the CT instability; iii) close to (but outside) the region where the system phase separates, superconducting instabilities occur both in the d-wave and s-wave channel[5]; iv) there is a critical value $V_c \approx 2t$: for $V < V_c$ the phase separation region is above the MIT point at zero doping. For $V > V_c$, the MIT is first order and the analysis indicates that it lays on the line of the phase separation.

The phase diagram in the case $V = 1.75t < V_c$ is shown in Fig. 3. The dashed curve indicated the points where the charge transfer fluctuations and the compressibility both

[4]This characteristic is also present in the weak coupling limit of the model.

[5]The presence of Cooper instabilities in the s-wave channel, renders this model particularly interesting, also in comparison with models with magnetic interactions, where the s-wave channel is usually stable and superconductivity of d-wave type only appears.

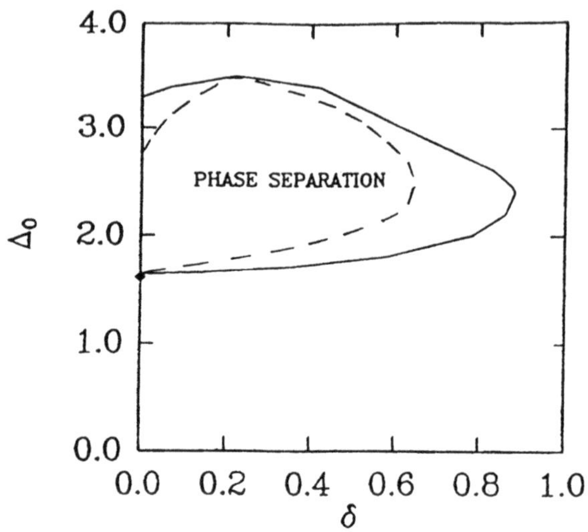

Figure 3: *Phase diagram $\Delta_0 \equiv \varepsilon_p^0 - \varepsilon_d^0$ as a function of doping for V=1.75t. The diamond indicates the metal-insulator transition point. The dashed curve indicates the points where the charge-transfer susceptibility and the compressibility diverge. From ref. [22]*

diverge. The continuous curve is the phase separation region obtained by means of a standard Maxwell construction. The fact that superconducting instabilities occur close to a phase separation region (point ii) and iii)) is a further indication of the possible relation between Cooper pair formation and the hole segregation occurring when the system phase separates.

In the presently considered model, where no magnetic interactions are present, the phase separation is due to the competition between the kinetic energy and the nearest neighbor repulsion. The former takes advantage of the hybridization, mixing p and d orbitals. On the other hand the nearest neighbor potential energy is larger when the charge tends to be distributed between nearby orbitals and, therefore, disfavors hybridization. In this situation the system prefers to create large regions with many holes where the kinetic energy wins leaving large regions at or close to half-filling where the potential energy term is minimized.

Also in this case, like in the extended tJ model, the phase separation region is close and above the metal-insulator transition, where the kinetic energy is depressed making more relevant the competition with the potential energy. Unfortunately the theoretical treatment of this model permits the detection of the instabilities, but makes much more difficult the determination of the energy of the superconducting phase. Thus it is not yet known if the allowance of a superconducting state lowers enough the energy in order to prevent phase separation similarly to what occurs in the Kondo-lattice model examined in the previous section.

In the context of the present extended Hubbard model, however, a qualitative argument can be put forward which makes plausible the common origin of superconductivity and phase separation. Close to the phase separation region, where superconducting instabilities occur, the compressibility $\frac{dn}{d\mu}$ is large (it diverges on the dashed curve of Fig. 3). Since in a Fermi liquid the compressibility is proportional to $\frac{m^*}{m}(1 - A_0^s)$ and since in the superconducting cuprates the effective mass enhancement is small ($m^* \sim 2 - 5m_e$), a large compressibility indicates a large and negative Landau scattering amplitude A_0^s. This fact shows the presence of strong attractive forces in the particle-hole channel (eventually leading to phase separation) which, however, can reasonably also contribute to the particle-particle channel leading to Cooper instabilities.

6 The five band Hubbard model: correlation between the hole symmetry and T_c

The models considered so far are extensions of the three band Hubbard model, which are obtained by the inclusion of magnetic (extended tJ and Kondo-Lattice models) or of nearest neighbor Coulombic terms in the Hamiltonian pointing towards general purely electronic mechanisms of pairing without any reference to special structures, which could differentiate among different cuprates. We therefore continue our journey in the land of purely electronic models of superconducting cuprates, by extending the three band Hubbard model with the inclusion of a larger number of orbitals, with the double aim of i) clarifying how complicated electronic structures together with strong correlation give rise to a suitable renormalized relatively simple band structure, and ii) of relating the different electronic structure of the various cuprates with their critical temperatures.

Within the context of the three band model, we already mentioned that Zhang and

Rice [9] proposed that the holes on $d_{x^2-y^2}$ orbitals form singlets with the holes on planar oxygen in the molecular orbital combination of local b_1 symmetry $\frac{1}{\sqrt{4}}(p_{x_1} - p_{y_2} - p_{x_3} + p_{y_4})$ by means of a hopping integral 2t ($t \sim$ 1.3eV) (the phases of the orbitals have been chosen with the same standard convention used in ref. [24]). This gives rise to a singlet band inside the charge-transfer gap; the equivalent result in the mean field slave boson approach is a quasiparticle band of predominantly b_1 symmetry crossing the Fermi level [7],[8] (see section 2).

Due to their similarity in the planar structure, the models entirely restricted to the CuO_2 planes hardly discriminate between the different high T_c compounds. In most of the actual crystals, pyramidal or octahedral structures are present with apical oxygen ions above or/and below the copper ions. Several experimental findings (including Raman spectroscopy [46], where a softening of the mode related to the motion of the apical oxygen was revealed when the temperature was lowered below T_c) indicate that also these out-of-plane oxygens can play a relevant role despite their small occupancy revealed by Electron Energy Loss Spectroscopy (EELS) [44] experiments. This conclusion is also supported by a recent analysis [47] of the isotope effect and by semiphenomenological analyses of the relation between crystal structure and the maximum critical temperature in the various high T_c cuprates. A clear correlation has been found between the maximum critical temperature attained within each class of compounds T_c^{max} and the Cu bond valence sum which is almost entirely due to variation in bonding of the copper ion to the apical oxygen [48]. Similarly a correlation [49] has been shown between T_c^{max} and the atomic level ε_p^z of the p_z orbitals of the apical oxygen.

As it turns out, the p_z orbitals hybridize rather effectively with the underlying $d_{3z^2-r^2}$ orbitals of copper and with the a_1 symmetry orbital combination of the p_x and p_y orbitals of the oxygen ions on the planes $\frac{1}{\sqrt{4}}(p_{x_1} + p_{y_2} - p_{x_3} - p_{y_4})$. Therefore the three band extended Hubbard model is further enlarged by including the $d_{3z^2-r^2}$ orbitals of copper [45],[24] The additional $d_{3z^2-r^2}$ orbital mixes with the a_1 symmetry orbital combination of the planar oxygen via an hopping term $\frac{2t}{\sqrt{3}}$. This gives rise [24] to a second band of predominantly a_1 symmetry slightly above the first, mainly b_1 symmetric, band, which also confers a sizable a_1 character to the occupied (hole) states via an overall hybridization of the two bands. Since the hybridization of the apical p_z orbital with all the planar states

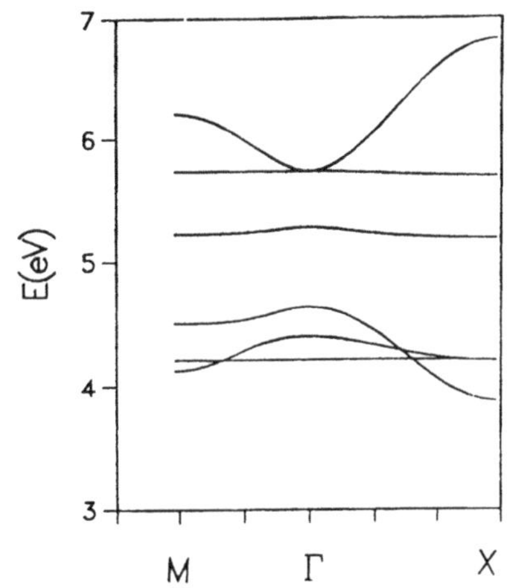

Figure 4: *Bandstructure for typical values of parameters: (in units of t)* $\varepsilon_p^0 - \varepsilon_d^0 = 5$, $t_p = 0.2$, *apex-plane hoppings* $t_{p_z d_z} = 0.6$, $t_{p_z p} = 0.2$, $\varepsilon_{p_z} = 5$ *and doping* $\delta = 0.3$. Γ, *X and M indicate the* $(0,0)$, (π,π) *and* $(\pm\pi, 0)$ *or* $(0, \pm\pi)$ *points in k-space respectively.*

with a_1 symmetry lowers the energy of the second lowest band (the one with mostly a_1 character), the presence if the apical oxygen tunes the mixing of the two bands and the a_1 contribution to the states occupied by holes. It must be emphasized, however, that the main effect of the apical oxygen is indirect: the p_z orbitals modify the occupancy of planar orbitals with a_1 symmetry while keeping their own occupancy rather small [25, 26] (less than a few percent, in agreement with the EELS [44] experiments). In addition polarized X-ray Absorption Spectroscopy (XAS) [43] and EELS [44] show a sizable (~10-15 per cent) occupancy of Cu $d_{3z^2-r^2}$ orbitals in all the investigated superconducting materials suggesting a possible role in high T_c superconductivity.

Extending the tight-binding Hubbard model by including the $d_{3z^2-r^2}$ orbital on copper and the p_z orbital on apical oxygen and by treating the infinite U repulsion on copper (for both the $d_{x^2-y^2}$ and the $d_{3z^2-r^2}$ orbitals) by means of the slave boson technique in mean field approximation, one obtains [25, 26] for typical values of the parameter a band structure of the form reported in Fig.4. The bare hopping parameters are chosen according to LDA calculations for La_2CuO_4, whereas the bare Cu(x^2-y^2)-O$(p_{x,y})$ atomic level difference is chosen in order to have a chemical potential jump in going from positive to negative doping of the order of 1.5eV. This jump, in fact, corresponds to the absorption gap in optical experiments [38] ranging from 1.2eV to 2eV.

The above analysis is a direct extension of the analysis of the three band Hubbard model carried out in section 2. All the considerations presented therein can be applied in the present context as well.

Two remarks are now in order: i) due to the bandwidth reduction introduced by the correlation within the slave boson technique, the width of the calculated bands is in rather good agreement with the ARPES and ARIPES experiments (e.g. ARPES experiments in Bi2212 compounds give a bandwidth $\sim$ 0.25eV, while we find a bandwidth $\sim$ 0.3eV, much closer to the experimental value than the LDA results, usually giving much broader bands); ii) two (rather narrow) bands are close to the Fermi level.

An important issue is whether there are actually two bands at the Fermi surface, and if so what the nature of the second band is. This is a delicate point also experimentally, because of the possibility that some of the broad features observed in ARPES do not originate from a quasiparticle band but should be attributed to an incoherent background.

According to our calculations, that two bands are crossing the Fermi level appears to be rather unlikely, or more precisely, that both could be planar Cu-O-like Fermi liquid quasiparticle bands: the situation where two such bands cross can in principle occur, but only for very strong apex-plane interaction, and that, as we shall see below, is expected to correspond to a low T_c or no superconductivity at all. Our results more likely suggest that only one Cu-O band crosses the Fermi level both in the ΓX and ΓM directions: the crossing Cu-O band would then be identified as the b_1-like lowest band (with an admixture of a_1 states), wit the non-crossing band a_1-like band (with an admixture of b_1 states) laying slightly above it.

The calculated band structure indicates that for the most reasonable parameters only one band crosses the Fermi level, having mixed b_1 and a_1 character. Moreover, as mentioned above, the amount of a_1 states is tuned by the apical oxygen. This leads to a simplification because, despite the presence of five bands in the model, only the total occupancy n_{a_1} of states with a_1 symmetry (i.e. p_z, $d_{3z^2-r^2}$ and the related planar combination of p_x and p_y) can be taken as a feature in the electronic structure that differentiates

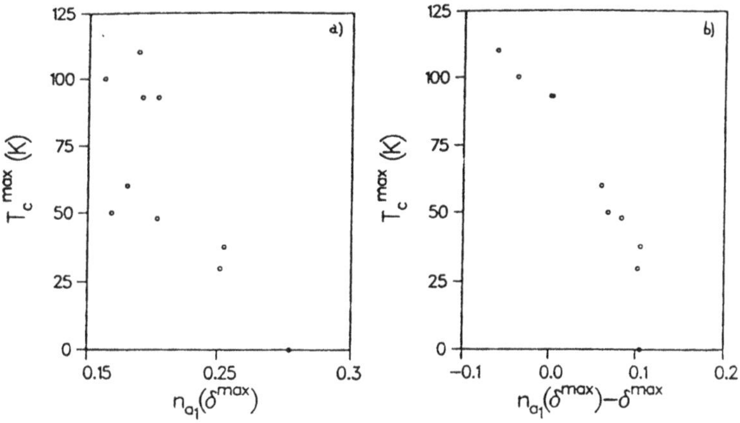

Figure 5: *Maximum critical temperature T_c^{max} of various of superconducting cuprates plotted (a) versus the density of holes with a_1 symmetry at the doping δ^{max} at which T_c^{max} is attained and (b) versus the corresponding excess density $n_{a_1}(\delta^{max}) - \delta^{max}$. The input parameters are obtained from ref. [49] for the compounds $La_2SrCu_2O_6$ ($T_c^{max} = 0K$), $La_{1.85}Sr_{0.15}CuO_4$ ($T_c^{max} = 38K$), $La_{1.85}Ba_{0.15}CuO_4$ ($T_c^{max} = 30K$), $(Ba_{0.67}Eu_{0.33})_2(Eu_{0.67}Ce_{0.33})_2Cu_3O_{8.78}$ ($T_c^{max} = 48K$), $Y_{0.8}Ca_{0.2}Ba_2Cu_3O_{6.11}$ ($T_c^{max} = 50K$), $YBa_2Cu_3O_{6.5}$ ($T_c^{max} = 60K$), $YBa_2Cu_3O_7$ ($T_c^{max} = 93K$), $Pb_{0.5}Tl_{0.5}Sr_2CaCu_2O_7$ ($T_c^{max} = 110K$), $TlBa_2CaCu_2O_7$ ($T_c^{max} = 100K$), $Bi_2Sr_2Ca_{0.9}Y_{0.1}Cu_2O_{8.24}$ ($T_c^{max} = 93K$). From ref. [25]*

among the various cuprate superconductors. We therefore consider ten sets of parameters calculated from the structural data of ref. [49]. They are representative of compounds with maximum critical temperature ranging from 0K to 110K. For each of them we have calculated n_{a_1} from our self-consistency equations *at the value* δ^{max} corresponding to T_c^{max}. No apparent correlation between T_c^{max} and $n_{a_1}(\delta^{max})$ is obtained Fig. 5a. However, having in mind a possible difference between the holes due to doping and the holes of the original substrate we also consider the excess of holes with local a_1 symmetry with respect to the doping itself $n_{a_1}(\delta^{max}) - \delta^{max}$. In Fig. 5b T_c^{max} is plotted versus this quantity and it is evident that a well defined trend exists: *the higher $n_{a_1}(\delta^{max}) - \delta^{max}$ is, the lower is T_c^{max}*. We made several tests: such a correlation shows up so clearly only for $n_{a_1}(\delta^{max}) - \delta^{max}$ or for the complementary quantity specifying the occupancy of states with local b_1 symmetry. We have also tested the robustness of this correlation by rescaling some of the parameters of each set by the same factor for all the compounds. Looking at the input parameters, we finally point out that the variation of $n_{a_1}(\delta^{max}) - \delta^{max}$ between compounds is largely due to the differences in $\varepsilon_p^z - \varepsilon_p$, i.e. in Madelung potential on the oxygen ions, rather than to differences in the hoppings between the apical and the planar orbitals.

The simple renormalized band structure resulting from the complex five band Hubbard model is therefore in reasonable agreement with the experiments as far as bandwidth, Fermi level crossing and relative occupations of states with different symmetries are concerned.

The established resulting correlation reflects once more the influence of the peculiar electronic structure of these cuprates in the possible pairing mechanism. Whatever will be the mediator of the pairing, a final theory of high T_c superconductivity must take into account the above relation.

7 Conclusions

Two distinct issues have been addressed in this paper. On one hand we pointed out that phase separation is commonly occurs in models with short range interaction, provided some mechanism (e.g. a strong local electron-electron repulsion) inhibits the stabilizing role of the kinetic energy. Moreover we stressed that phase separation and superconductivity can be related phenomena. In fact, both being due to attractive effective interactions, they take place in nearby regions of the phase diagram. In particular in the three band extended Hubbard model with nearest neighbor coulombic repulsion, superconductivity occurs in a region close to the phase separation and could be seen as a precursor effect of the latter.

In a completely different model, the Kondo lattice model, phase separation (almost) only occurs when the superconducting phase is artificially suppressed in the theoretical treatment. Otherwise the phase separation between a metallic FL phase and a magnetic phase is almost completely suppressed by the presence of a more stable SC phase. In this case SC not only preludes phase separation, but provides a valid alternative for the system with respect to phase separation.

The above investigation, therefore, establishes a strict relation between phase separation and SC irrespective of the nature (magnetic or nearest neighbor coulombic) of the short range interaction in the context of rather different simplified models.

Once again we would like to stress that the presence of long range Coulomb forces spoils a static phenomenon like phase separation, but could have a less dramatic effect on the dynamical effective interaction leading to pairing. This issue surely deserves further investigation. In any case the above investigated models could also retain their own validity because mobile ions like O^{--} (e.g. in oxygen rich $La_2CuO_{4+\delta}$ could phase separate together with the holes thereby compensating for the charge imbalance without any role being played by Coulomb forces. This latter occurrence has lately found some experimental confirmation [27]

All the above results and considerations seem to have a rather general validity, but just because of their generality, they are not terribly helpful in identifying the actual pairing mechanism in real cuprate materials. This is why in the last section we addressed the second main issue of this paper: i) what are the relevant features of the complicated copper oxides band structure and ii) under which respect these materials, which posses very similar CuO_2 planar structures, can be differentiated in order to explain the broad range of maximum critical temperature that they attain.

The analysis performed in Refs.[25, 26] and reported in section 6 indicates that i) a single band with mixed b_1 and a_1 symmetry crosses the Fermi level (if all atoms but the planar copper, the planar oxygen and the apical oxygen are disregarded); ii) the amount of holes with a_1 symmetry in this band is tuned by the position of the apical oxygen atomic level with respect to the planar orbitals atomic levels; iii) the excess of a_1 holes with respect to those introduced by doping (or, equivalently, the amount of holes with b_1 symmetry) is related to the maximum critical temperature T_c^{max} attained within each class of compounds.

Acknowledgments This work is a summary of papers produced in collaboration with N. Cancrini, S. Caprara, C. Castellani, L. F. Feiner, G. Kotliar and R. Raimondi. While the physical content is a result of the work of all these people, possible failures of this presentation must be ascribed to the authors only.

References

[1] J. G. Bednorz, and K. A. Müller, Z. Phys. B 64, 189 (1986).

[2] S. W. Tozer, et al., Phys. Rev. Lett. 59, 1768 (1987); S. Martin, et al., Phys. Rev. B 41 (Rap. Commun.), 846 (1990) and references therein.

[3] L. F. Mattheiss, Phys. Rev. Lett. 58, 1028 (1987); J. Yu, A. J. Freeman, and J.-H. Xu, Phys. Rev. Lett. 58, 1035 (1987); S Massidda, J. Yu, and A. J. Freeman, Physica C 152, 251 (1988); M. S. Hybertsen, and L. F. Mattheiss, Phys. Rev. Lett. 60, 1661 (1988); H. Krakauer, and W. E. Pickett, Phys. Rev. Lett. 60, 1665 (1988); A. K. McMahan, R. M. Martin, and S. Satpathy, Phys. Rev. B 38, 6650 (1988); M. S. Hybertsen, M. Schlüter, and N. E. Christensen, Phys. Rev. B 39, 9028 (1989).

[4] N. Nücker et al., Phys. Rev. B 39, 6619 (1989).

[5] For a general review see e.g. *Proceedings of the International M^2S HTSC Conference*, Stanford, USA, July 23-24 1989, Physica C 162-164, (1989); *Proceedings of the Anniversary Adriatico Research Conference and Workshop on "Strongly Correlated Electron Systems"*, ICTP, Trieste, Italy 19 June - 21 July 1989, Editors G. Baskaran, A. E. Ruckenstein, E. Tosatti, and Yu Lu, World Scientific.

[6] V. J. Emery, Phys. Rev. Lett. 58, 2794 (1987).

[7] G. Kotliar, P. A. Lee, and N. Read, Physica C 153-155, 538 (1988).

[8] M. Grilli, G. Kotliar, and A. J. Millis, Phys. Rev. B 42, 329 (1990).

[9] F. C. Zhang and T. M. Rice, Phys. Rev. B 37, 3759 (1988); T. M. Rice, and F. C. Zhang, Phys. Rev. B 3, 3754 (1988).

[10] C. Castellani, and G. Kotliar, Phys. Rev. B 39, 2876 (1989); E. Arrigoni, C. Castellani, and G. Strinati, Phys. Rev. B 41, 4838 (1990).

[11] C. Castellani, M. Grilli, and G. Kotliar, Phys. Rev. B 43, 8000 (1991).

[12] A. Aharony, R. J. Birgenau, A. Coniglio, M. A. Kastner and H. E. Stanley, Phys. Rev. Lett. 60, (1988) 1330; W.-M. Que, G. Kirczenow, Sol. St. Commun. 64, 1063

(1987); C. Castellani, C. Di Castro, and M. Grilli, Physica C 153-155, 1659 (1988); C. Castellani, C. Di Castro, and M. Grilli, International Journal of Modern Physics B, Vol. 1, No. 5 (1988) 659. H. Kamimura, S. Matsuno, and R. Saito, Sol. St. Commun. 67, (1988) 363; P. Prelovŝek, Phys. Lett. 126, (1987) 287.

[13] N. Andrei, P. Coleman, Phys. Rev. Lett. 62, 595 (1989).

[14] N. Cancrini, S. Caprara, C. Castellani, C. Di Castro, M. Grilli and R. Raimondi, Europhys. Lett. 14 (6), 597 (1991).

[15] C. M. Varma, S. Schmitt-Rink, and A. E. Ruckenstein, Solid State Comm. 62, 681 (1987);

[16] P.B. Littlewood, C. M. Varma, S. Schmitt-Rink, and E. Abrahams, Phys. Rev. B 39, 12371 (1989).

[17] R. Putz, G. Dopf, B. Ehlers, L. Lilly, A. Muramatsu, and W. Hanke, Phys. Rev. B 41, 853 (1990).

[18] Z. Tesanovic, A. R. Bishop, and R. L. Martin, Sol. State Commun. 68, 337 (1988).

[19] J. Bang, K. Quader, E. Abrahams and P.B. Littlewood, Phys. Rev. B 42, 4865 (1990).

[20] P. B. Littlewood, Phys. Rev. B 42, 10075 (1990).

[21] Y. Bang, G. Kotliar, C. Castellani, M. Grilli and R. Raimondi, Phys. Rev. B 43, 13724 (1991).

[22] M. Grilli, R. Raimondi, C. Castellani, C. Di Castro, and G. Kotliar, Phys. Rev. Lett. 67, 259 (1991).

[23] M. Grilli, R. Raimondi, C. Castellani, C. Di Castro, and G. Kotliar, International Journal of Modern Physics B Vol. 5, 309 (1991).

[24] M. Grilli, C. Castellani, and C. Di Castro, Phys. Rev. B 42, 6233 (1990).

[25] C. Di Castro, L. F. Feiner, and M. Grilli, Phys. Rev. Lett. 66, 3209 (1991).

[26] L. F. Feiner, M. Grilli, and C. Di Castro, to appear in Phys. Rev. B.

[27] J. D. Jorgensen et al., Phys. Rev. B 38, 11337 (1988); M. Hundley et al., Phys. Rev. B 41, 4062 (1990); J. D. Jorgensen et al., *Third International Symposium on Superconductivity*, Sendai, Japan, 6-9 November 1990, Springer-Verlag Tokyo.

[28] V. J. Emery, S. A. Kivelson, and H. Q. Lin, Phys. Rev. Lett. 64, 475 (1990).

[29] M. Marder, N. Papanicolau, and G. C. Psaltakis, Phys. Rev. B 41, 6920 (1990)

[30] C. Di Castro, and M. Grilli, *Proceedings of the 12th General Conference of the Condensed Matter Division* (EPS), Praha, Czechoslovakia, 6-9 April 1992.

[31] S. E. Barnes, J. Phys. F 6, 1375 (1976); P. Coleman, Phys. Rev. B 29, 3035 (1984); N. Read, and D. M. Newns, J. Phys. C 16, 3273 (1983); Read, J. Phys. C 18, 2651 (1985); J. Millis, and P. A. Lee, Phys. Rev. B 35, 3394 (1987); Kotliar and A.Ruckenstein, Phys.Rev. Lett. 57, (1986) 1362.

[32] M. Grilli, C. Castellani, and G. Kotliar, Phys. Rev. B, in print.

[33] E. Dagotto, A. Moreo, F. Ortolani, D. Poilblanc, J. Riera, and D. Scalapino, preprint.

[34] T. Tohyama, and S. Maekawa, Physica C 185-189, 1575 (1991).

[35] Z.-X. Shen et al., Phys. Rev. B 44, 12098 (1991).

[36] For a review see e.g. W. E. Pickett, H. Krakauer, R. E. Cohen, and D. J. Singh, Science 255, 46 (1992)

[37] G. A. Thomas et al., Phys. Rev. Lett. 61, 1313 (1988); see also S. Uchida et al, Phys. Rev. B 43, 7942 (1991).

[38] C. Castellani, G. Kotliar, R. Raimondi, M. Grilli, Z. Wang, and M. Rozenberg, preprint.

[39] A. Houghton, N. Read, and H. Won, Phys. Rev. B 37, 3782 (1988).

[40] A. J. Millis, H. Monien, and D. Pines, Phys. Rev. B 42, 167 (1990).

[41] T. Takahashi et al., Phys. Rev. B 39, 6636 (1989).

[42] T. Watanabe et al., preprint.

[43] A. Bianconi et al., *Proc. of the Int. M^2S HTSC Conference*, Stanford, USA, July 23-24 1989, Physica C 162-164, 209 (1990); A. Bianconi in *International Conference on Superconductivity - ICSC*, S. K. Joshi, C. N. R. Rao, and S. V. Subranyam editors, (Proc. Int. Conf. on Superconductivity, Bangalore, Jan. 10-14, 1990) World Scientific, Singapore, 448 (1990).

[44] H. Romberg et al., Phys. Rev. B 41, 2609 (1990).

[45] W. Weber, Z. Phys. B 70, 323 (1988);

[46] C. Thomsen, et al., Phys. Rev. B 37, (1988) 9860.

[47] K. A. Müller, Z. Phys. B 80, 193 (1990).

[48] D. M. de Leeuw, W. A. Groen, L. F. Feiner, and E. E. Havinga, Physica C 166, 133 (1990).

[49] Y. Ohta, T. Tohyama, and S. Maekawa, Physica C 166, 385 (1990), Y. Ohta, T. Tohyama, and S. Maekawa, Phys. Rev. B 43, 2968 (1991).

THE VIRTUAL EXCITON MECHANISM OF SUPERCONDUCTIVITY

M. CYROT

Laboratoire Louis Néel, CNRS, BP 166, 38042 Grenoble Cedex

ABSTRACT

We propose that the metallic state obtained by doping a Mott insulator is not described by a t-J model and has special properties near the transition.We consider that the reason of the Mott insulating state is the strong ionic potential. In the metallic state, this potential creates new low energy excitations where a hole sits on an ion and the electron is in a virtual bound state around it. These low energy excitations which are excitonic like,lead to properties different from a conventional metal. The ground state of this excitonic metal is superconducting.

The anomalous properties and superconductivity of the cuprates are observed closed to a metal non metal transition. It is widely believed that the undoped materials are Mott insulators. When doped, non standard metallic properties and superconductivity are found. However as soon as the doping is large enough, superconductivity disappears so the anomalous properties of the normal state. Thus superconductivity and anomalous properties are linked and restricted to a range of doping inbetween two critical concentrations c_m and c_M. Below c_m the material is antiferromagnetic. Above c_M the material is a metal described by the standard Fermi liquid theory and is no more superconducting. Following Anderson[1], many attempts have tried to describe this range of concentration using a t-J model or something similar. However practically all models based on a large value of the intra-atomic Coulomb correlation U lead to d-wave superconductivity. This is not confirmed by experiments. We believed that it is due to an inapropriate description of the doped Mott insulator. More precisely the charge excitations are completely absent in these models and this also prevents the understanding of a transition to a standard Fermi liquid. In this paper,we follow Mott's original suggestion. In this view, the insulating state is not driven by the large Hubbard U as always considered, but by the ion potential which is not enough screened if the electron density is too low. With this point of view, we reexamine[2] the description of a doped Mott insulator and we show that there exist metallic excitons of low energy i.e., of energy much smaller than the Fermi energy E_F. This lead us to call this range of doping excitonic metal. The anomalous properties of the low energy excitations are linked to the existence of the metallic excitons and leads to properties different from the standard Fermi liquid theory. We will describe their effects in the one electron density of states as measured for example by photoemission and tunnelling. Then we will describe the properties linked with electron-hole pairs as the frequency dependent conductivity and the Raman scattering. Finally we show that the excitonic mechanism of superconductivity can apply here and that the ground state of the excitonic metal is superconducting.

The basic physics behind the Mott[3] insulator is the following : we consider a crystalline array of hydrogen atoms with one electron per atom. It is always metallic in a simple band or Hartree-Fock model. Mott argued that the metallic state implies that the potential of each positive charge when screened by the electrons is not strong enough to trap an electron. Using a Thomas-Fermi potential for the ion potential, he showed that this is the case if

$$n^{1/3} a_H > 0.2 \tag{1}$$

where n is the density of electrons and a_H the Bohr radius.If this inequality is not fulfilled, i.e. for low electron density, the Hartree-Fock band is not stable. Each positive charge traps one electron and the system is an insulator. This criterium also means that the potential of an atom without electron is strong enough to create a bound state below the Hartree-Fock band. The insulating behavior is just due to the fact that we consider one electron per atom. We argue that when a Mott insulator is doped it is necessary to have a minimum doping in order to destroy this bound state. Indeed if the number of electrons per site is 1 + c, the minimum Mott doping c_M to destroy this bound state is

$$(1+c_M)^{1/3} a_H = 0.2 \, d \tag{2}$$

where d is the lattice constant. Thus when the material is doped with a concentration $c < c_M$, it is metallic because the number of electron is not an integer but on any atom there is always one bound electron in the ground state. This is not inconsistent with a

Fermi liquid theory. The discontinuity at the Fermi level is just smaller than one and states above k_F are partially filled.

A Gutzwiller many-body wave fonction with infinite U can describe the ground state of the Mott insulator as it prevents any site to be empty. However in that model no charge excitation is possible as it requires the infinite U energy. Our model differs from the standard Fermi liquid in the charge excitations. Usually to create a localized hole on a site required an energy of order E_F and is not considered. Here the situation is different. When a hole is created on a site, the new potential of this ion is large even when screened by the other electrons (we only consider screened potentials as screening occurs on a time scale given by the plasma frequency). It is large enough to create a bound state below the band. This bound state is empty and all the states are changed to scattering states. All these new scattering states have energy lower than the initial states. It decreases strongly the energy necessary to create a hole on a site which instead of being a high energy excitation of order E_F becomes a low energy excitation of energy ω_{th}. Indeed in order to create a bound state the potential has to be of order E_F except for special shapes of the density of states. Each state has an energy decrease of E_F/N so the many body state has an energy decrease of order E_F. ω_{th} the threshold energy is given by

$$\omega_{th} = \varepsilon_{HF} - \Delta E \tag{3}$$

where ε_{HF} is the Hartree-Fock energy needed to create a localized hole and ΔE is the change in the ground state energy arising from the transient potential. ω_{th} increases with doping as the ion potential becomes more effectively screened and the relaxation decreases. Thus we have low energy excitations with a hole on a given site and the electron in a virtual bound state around the hole. This is a metallic exciton. A Fermi liquid behaviour can exist only for very low energy and temperature $\omega << \omega_{th}$ $T << \omega_{th}$. As soon as this inequality is not fulfilled, the properties will deviate from a Fermi liquid behavior. In the following we want to describe these new properties.

We use a model similar to that introduced in the study of X-ray absorption in metal[4,5] and we consider the new Hamiltonian

$$H = \sum_k \varepsilon_k C^+_{k\sigma} C_{k\sigma} - \sum_{kk'q} V_{kk'} C^+_k C_{k'} C_{k-q} C^+_{k'-q} \tag{4}$$

where $V_{kk'}$ is the Coulomb interaction screened by the electrons, created by an ion when it has no electron. The Hamiltonien describes the appearence of a potential on a site i when no electron sits on that site. In a Hartree-Fock approach, the energy ε_k is calculated in a self consistant manner using a Hartree potential where the ion potential is nearly cancelled by the potential of the electron sitting on that site. Our Hamiltonien just recalls that if no electron sits on an ion, the potential of the ion is no more the Hartree potential but the bare potential which is large enough to create a bound state below the band.

In order to understand the properties of this excitonic metal, we first point out the similarities of this problem with the infrared catastrophe in the X-ray spectra of metals. As originally suggested by Mahan[4] the X-ray absorption band spectrum of metals displays a characteristic threshold singularity arising from the interaction between conduction electrons and the localized potential due to the hole created by an X-ray. Such an unusual behavior has its origin in the readjustment of the conduction electrons to the sudden change in the ionic core potential. The latter tends to excite electron-hole pairs out of the Fermi sea. Because of the large density of electron hole pairs the number of excited pairs is infinite, even though the total energy transfered to the electrons is finite. This is the infrared catastrophe. Here we describe a similar scenario. If a hole is created on a site the sudden new ionic potential perturbs in a non adiabatic way the distribution of electrons and creates a large number of electron-hole pairs. The appearence of a hole can be due either to temperature or to a frequency dependant pertubation. Thus at finite temperature or for a frequency dependant pertubation, the system will not react as a conventional Fermi liquid.

One of the main constraint for applying this physical scheme to our problem is that the disturbance has no recoil : it remains localized at a given site. We argue that the empty bound state has no recoil because the matrix element which permits the hole to move involves the overlap between all the scattering states which are different for the two sites. As it implies the overlapp of an infinite number of states, all smaller than one, it is exponentially small.This is Anderson's orthogonality theorem[6]. We have

$$t_{ef} = t\, N^{-K} \tag{5}$$

Where N is the number of electrons and K is related to the phase shift at the Fermi level due to the transient potential.However this is not valid for all potential strength. If we now take into account this life time, N has to be replaced by W/t_{ef} where W is the bandwidth[7]. The self consistent equation for t_{ef} is

$$t_{ef} = t\ (w/t_{ef})^K \tag{6}$$

This equation shows that $t_{ef} = 0$ when $K>1$ i.e. the hole is self trapped when the potential is large enough to create a bound state[7]. This is precisely the situation that we describe.

For a qualitative or semi quantitative description, we can rely on the known results obtained for this problem. However there are two obvious limitations.The first one is the existence of a large intraatomic Coulomb repulsion U. We cannot neglect the repulsion of two electrons on the same site which is always linked to a strong atomic potential. This leads to the magnetic properties and must be considered for a complete description. The second limitation is due to the fact that, contrary to the X-ray problem, holes can be created at temperature T of order of ω_{th}. The finite temperature problem is thus very different from the corresponding X-ray one.

We first consider the one particle density of states, as given for instance by photoemission. An anomalous contribution compared to the Fermi liquid one will appear at the threshold energy ω_{th}. As creating one localized hole produces an instantaneous potential and an infinite number of electron hole pairs , a delta function contribution is not obtained. Nozières and de Dominicis[5] have calculated it. It is related to the long time behavior for the Green function of the hole which is given by the phase shifts δ_l of electrons at the Fermi level due to the transient potential. The corresponding absorption is

$$I\ (\omega) = \frac{1}{(\omega-\omega_{th})^{1-\alpha}} \tag{7}$$

α is related to the square of the number of electrons n_l which must be bought in each channel near the excited core in order to achieve the new equilibrium ground state[8] . According to Friedel[9] an extra potential gives rise to an excess localized charge in the channel l equals to $\frac{\delta_l}{\pi}$. Since we have an empty bound state (for instance for $l = 0$) the excess localized charge becomes $\frac{\delta_0}{\pi} - 1$ in that channel and if we consider only that channel

$$\alpha = (\frac{\delta_0}{\pi} - 1)^2 \tag{8}$$

If lifetime of the hole is taken into account, the singularity is broadened[10]. At low temperatures, superposed to this absorption is that of a standard Fermi liquid if superconductivity is not considered.
The finite temperature properties at $kT \leq \omega_{th}$ can be only qualitatively described. Localized holes are created at those temperatures. This produces a large number of electron-hole pairs. However adiabaticity does not occur[11] . If a system is in the Fermi ground state and a potential is suddenly turned on and off , the probability that it finally returns to the ground state is always less than unity: i.e. appearance and disappearance of a hole let into the system a large number of electron-hole pairs of all energies. Thus qualitatively we have a finite density of electron-hole pairs for all ω contrary to a Fermi liquid where the density vanishes with ω. If we assume a constant density of electron-hole pairs as a function of energy, the life-time of an excitation becomes linear in ω instead of quadratic (the standard result for a Fermi liquid). In that case the system will behave qualitatively as a Marginal Fermi Liquid[12].However at temperatures of order ω_{th}, the hole mobility becomes important of order of T^{2K-1}, the recoil effect destroys this special effect and one recovers the finite temperature standard Fermi Liquid.

For the frequency dependent conductivity, an anomalous contribution is also obtained at $\omega_{th.}$ Combescot and Nozières[13] have shown that if the l channel is optically excited

$$\sigma(\omega) = \frac{w_l(\omega)}{(\omega-\omega_{th})^{1-\beta_l}} \tag{9}$$

where

$$\beta_l = (\frac{\delta_l}{\pi} - 2)^2 \tag{10}$$

If the hole is in an s state, we consider the phase shift of the partial p wave at the Fermi level as the matrix element connect l to l ± 1. β_l is the square of the number of electrons which must be brought in order to achieve the new equilibrium, knowing that the bound state is unoccupied and that one electron is provided by the creation of the hole. At low temperatures, this absorption superposes to the Drude's one if superconductivity is not considered. If we lower the temperature below the superconducting one, these features become more important because the Drude free carriers have condensed.

Another test would be the Raman scattering. This leads to additional informations as the coupling to particle hole excitations arise from different matrix elements than $\sigma(\omega)$. As shown by Doniach et al[14] the scattering cross section is

$$S(q,\omega) = \sum_{l} A_l(q)\, R_l(\omega) \tag{11}$$

with

$$R_l(\omega) = \frac{1}{(\omega-\omega_{th})^{1-\beta_l}} \tag{12}$$

As q increases, the predominant l term changes. Thus as the scattering angle is increased, the behavior of the scattering cross section as a function of energy should change from that predicted for the $l \pm 1$ phase shift to that predicted for the l phase shift.

We now comment on the possibility of an excitonic mechanism of superconductivity in the excitonic metal. Usually one needs another medium in which excitons exist, thus one has to consider multilayers. Here excitons are in the same medium. These excitons have low energy compared to E_F. Thus the direct Coulomb repulsion can be screened as in the phonon mechanism. Secondly these excitons exist at large q vector a condition stressed by Allender et al[13]. Third, the coupling between the exciton and the carriers is large as it implies the same electrons, contrary to others excitonic mechanisms. As shown by Ginzburg[(14)] the effective interaction is given by

$$V_{eff}(\omega) = \left\langle \frac{4\pi e^2}{q^2} \right\rangle \left[\int \frac{1}{\omega+\omega'} \left\langle \mathrm{Im}\, \frac{1}{\varepsilon(q,\omega')} \right\rangle .\, d\omega' + 1 \right] \tag{13}$$

The coupling is directly related to the scattering cross section at large value of q, which in the excitonic metal is large. In principle in a selfconsistent calculation this has to be calculated in the superconducting state as in the case of superfluidity in helium.

We end by discussing how this mechanism can be applied to the superconducting cuprates.Instead of a one band Mott insulator doped with electrons, we consider a charge tranfer insulator doped with holes. As usual we consider only Copper-Oxygen CuO_2 planes and the hybridized band made with the x^2-y^2 orbitals for the Copper and the $p\sigma$ orbitals of oxygen. We assume that creating a d^8 copper ion creates a potential large enough to extract one bound state below the band. In the ground state, the Copper ions are in the d^9 state. d^{10} states are prevented by the large U responsible of the insulating state. Holes in the plane are mainly in the oxygen p states. Adding more holes in the plane, permits to screen more effectively the d^8 copper ion and above a critical concentration of holes the bound state will disappears. Below this concentration metallic exciton will exist. This metallic exciton should be observed in photoemission or in frequency dependent conductivity or in Raman scattering. A d^8 copper ion will create a potential and a phase shift for the d and p character of the itinerant states. In the frequency dependent conductivity the p character of these states will be observed. In the Raman scattering as a function of increasing q one goes from the p character to the d character. This last experiment, if possible to make, would be the best test of a virtual exciton mechanism of superconductivity in the superconducting cuprates.

Acknowledgements :

We want to thank Prof. K.A. Muller who gave us the opportunity to write this paper and Prof. G. Benedek for a critical reading of the manuscript.

REFERENCES

1. P.W. Anderson, *Science* 235 (1987) 1196.
2. M. Cyrot, *Modern Physics Letters* B 6 (1992) 383 .
3. N.F. Mott, *Metal Insulator transitions*, Taylor et Francis LTD, London (1974).
4. Mahan, *Phys. Rev*. 163 (1967) 612 .
5. P. Nozières and C. de Dominicis, *Phys. Rev*. 178 (1969) 1097 .
6. P.W. Anderson, *Phys. Rev.Lett.* 18 (1967) 1049 .
7. K.Yamada, A.Sakurai and M.Takeshige, *Progress of Theoretical Physics* 70 (1983) 73 .
8. J.J. Hopfield, *Solid State Comment* 2 (1969) 41.
9. J. Friedel, *Phil. Mag* 63 (1952) 153 .
10. S. Doniach and M. Sunjic, *J. Phys*. C 3 (1970) 285.
11. E. Muller-Hartmann, T.V. Ramakristinan and G. Toulouse, *Phys. Rev*. B 3 (1971) 1102 .
12. C.M. Varma, P.B. Littlewood, S.Schmitt-Rinck, E.Abrahams and A.E. Ruckenstein, *Phys. Rev. Lett*. 63 (1989) 1996.
13. M. Combescot and P. Nozières, *Journal de Physique* 32 (1971) 913 .
14. S. Doniach, P.M. Platzman and J.T. Yue, *Phys. Rev*. B 4 (1971) 3345.
15. D. Allender, J. Bray and J. Bardeen, *Phys. Rev*. B 7 (1973) 1020.
16. V.L. Ginzburg and D.A. Kirshritz, *High-Tc superconductors*, Consultants Bureau New York (1982).

LINEAR ARRAYS OF NON HOMOGENEOUS Cu SITES IN THE CuO_2 PLANE: A NEW SCENARIO FOR PAIRING MECHANISMS IN A CORRUGATED-IRON-LIKE PLANE

A. Bianconi

Università dell'Aquila, v. San Sisto 20, 67100 L'Aquila, Italy
and INFM Research Unit of University of Rome, Department of Physics,
P. A. Moro 2, 00185 Roma, Italy,

ABSTRACT

Experimental results obtained by using x-ray absorption spectroscopy show that the configurations of Cu sites in the CuO_2 plane of Bi 2:2:1:2 high T_c superconductors are not homogeneous. Different Cu sites are characterized by short 2.3 Å and long 2.45 Å Cu-O(apical) distances. The linear arrays of different Cu sites forming domains with a corrugated-iron like shape is proposed to be a key characteristic of superconducting domains in the CuO_2 plane. The wavelength of the modulation is close to the superconducting coherence length. The ordering of the distorted Cu sites is suggested to be evidence for ordering of polarons driven by the pseudo Jahn Teller electron lattice interaction.
The Cu L_3 XAS experiments on Bi 2:2:1:2 system indicate that for $\delta = 19\%$ electronic states added by doping 4 ± 2 % have the a_1 symmetry (i.e. with Cu $\underline{3d}_{3z^2-r^2}$, the combination of O(planar) $2p_{x,y}$ orbitals with a_1 symmetry $\underline{L}(a_1)$, and O(apical) $2p_z$ orbital character) and $15\pm2\%$ have the b_1 symmetry ($\underline{3d}_{x^2-y^2}$ and the combination of O(planar) $2p_{x,y}$ orbitals with b_1 symmetry $\underline{L}(b_1)$). This new scenario supports the pairing mechanisms for high T_c superconductivity in the presence of two components: 1) the more delocalized component with b_1 symmetry and 2) the more localized component, with partially a_1 symmetry associated with different parts of the Fermi surface.

1. Introduction

So far the two dimensional CuO_2 plane where pairing takes place in high T_c cuprate superconductors[1] has been considered to be homogeneous, i.e. the Cu site structure configuration (the Cu square plane in electron doped systems, the square pyramid in YBaCuO and Bi 2:2:1:2 systems and the elongated bipyramid in doped La_2CuO_4) is assumed to be the same at all Cu sites. In this work we report that the Cu site structure is found to be not homogeneous in Bi

2:2:1:2 by EXAFS and diffraction works[2] and it is modulated with a period close to the superconducting coherence length. The different Cu sites are correlated both with ordering of polarons in the CuO_2 plane and with ordering of dopants: interstitial oxygen, defects, substituted ions. The presence of these ordered domains is difficult to be detected because the size of the ordered domains can be so small 10-30 Å to escape to be detected by electron, x-ray and neutron diffraction experiments and the system could appear to be a solid solution on a large scale. Only where ordered domains with size larger than 100 Å are formed they can be detected by diffraction and only if large samples giving intense superstructure spots are obtained the x-ray and neutron diffraction methods are able to solve the structure of the different Cu sites contributing to the superstructure. In high T_c superconductors the short coherence length 10-30 Å implies that the pairing could take place in small ordered domains and their ordering could be a key requirement for the pairing mechanism

X-ray absorption experiments by using synchrotron radiation probe the Cu site structure without requiring long range order. The extended x-ray absorption fine structure (EXAFS) probes the Cu-O interatomic distances with a 0.02 Å accuracy and with a measuring time of 10^{-16} sec. Therefore it provides a distribution of interatomic distances due to both static and dynamic contributions. EXAFS investigations[3] have found the presence of two Cu-O(apical) distances, 0.13 Å apart, in $YBa_2Cu_3O_{6.5}$, $YBa_2Cu_3O_7$, and about 0.2 Å apart in $TlBa_2Ca_3Cu_3O_{11}$ while diffraction data have reported a single Cu-O(apical) average distance. Anomalies on the variation of the Cu site structure at T_c were found by EXAFS and XANES.[3,4-6] The pulsed neutron diffraction experiments[7] probing the local pair distribution function without requirements for long range order have found evidence for a split position of the apical oxygen and changes in the local arrangement of oxygen atoms in in the CuO_2 plane across T_c. These results are not in agreement with x-ray and neutron diffraction data which probe the average atomic coordinates over domains of the order of 100-200 Å size.

In few superconducting systems Bi 2:2:1:2[8], $YBa_2Cu_3O_{6.5}$, $La_2CuO_{4.09}$[9] electron diffraction experiments show the presence of a superstructure due to ordering of dopants over domains larger than 100 Å, on the contrary in other systems, like LaSrCuO $YBa_2Cu_3O_7$, no superstructure has been detected indicating that if domains are present their size is not larger than 20-30 Å. However where the incommensurate superstructure is observed by diffraction methods the intensity of the superstructure spots are usually so weak that it is hard to extract the coordinates of the atoms of the different Cu sites contributing to the superstructure. This is the reason why only few diffraction experiments have revealed the different Cu site configurations by solving the superstructure.

We have recently investigated the local Cu site structure in Bi 2:2:1:2 system[2] by EXAFS in order to solve the controversy between EXAFS and diffraction methods. We have selected the Bi 2:2:1:2 system where the superstructure in neutron an x-ray diffraction has been solved by several groups[10-12]. In the Bi 2:2:1:2 system the Cu site structure modulation is stabilized by the mismatch between the BiO rock salt layers and the perovskite layers[10-12]. The crystals show the $\lambda a \times 1b$ incommensurate superstructure where the wavelength of the supestructure λ is found to be about 4.75. In superconducting samples where the hole doping has been controlled by Y to Ca substitution[13] and in samples prepared in Ar atmosphere with different oxygen content[14] the period λ has been found to change in the range 4.6 - 5 correlated with doping and the critical temperature. On the contrary λ is found to be around 4 in the insulating phase.

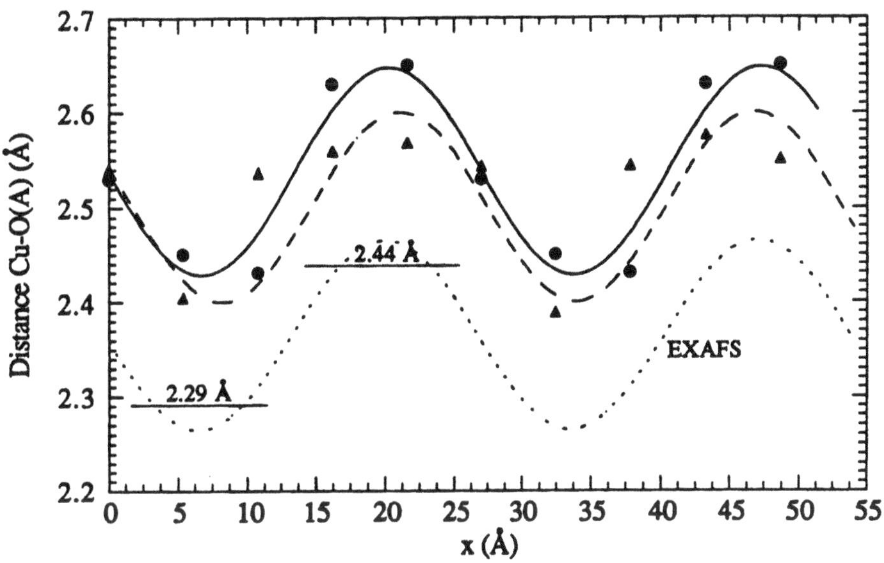

FIG. 1 The modulation of the Cu-O(apical) bond distances along the (100) direction due to the 5 x 1 superstructure obtained by x-ray diffraction ref. 10 (dots), and to the 4.75 x 1 superstructure solved by neutron diffraction ref. 11 (triangles) and as obtained by our EXAFS work, ref. 2.

2. Non homogeneous Cu sites

a) The modulated Cu apical oxygen distance.

Two average distances for the Cu-apical oxygen bond (Cu-O(apical)) have been found by EXAFS analysis: a short bond 2.29 Å with a probability of 60±10% and a long bond 2.44 Å with a probability of 40 ± 10% as shown in Fig. 1. The separation between the two positions is 0.15 ± 0.04 Å which is similar to that found in $YBa_2Cu_3O_7$ [3]. Because each Cu site is coordinated by only one apical oxygen two interpretations are possible: first, the apical oxygen moves in a double well potential, determined by a structural instability[3]; second, the Cu sites are not homogeneous i.e. there are two sets of different Cu sites with long and short distances. We have found that this second interpretation is in agreement with the three diffraction experiments[10-12] that solved the superstructure in Bi 2:2:1:2 crystals.

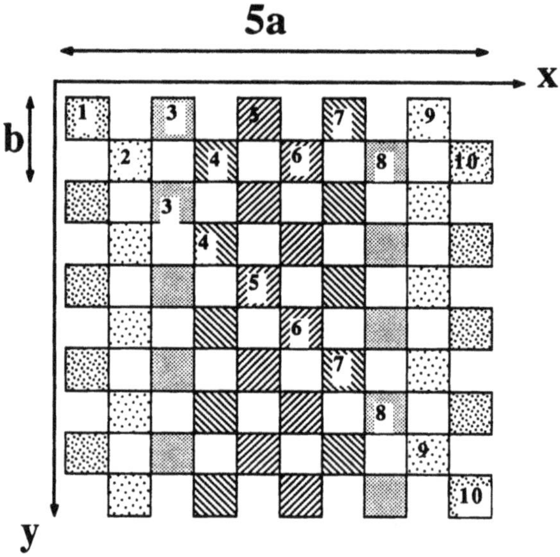

FIG. 2. Pictorial view of the CuO_2 plane with a 5x1 superstructure where the different CuO_4 square planes characterized by short and long Cu-O(apical) bonds are indicated.

In Fig. 1 we report the Cu-O(apical) (or Cu-O(A)) distances for different Cu sites along the (100) direction as found by different authors. The Cu-O(apical) distance has been found to be modulated over 5 or 4.75 crystalline unit cells i.e. with a period of 26-27 Å as it is shown in Fig. 1 where the Cu-O(apical) distances reported by ref. 10 and 11 are plotted and they have been fitted with a formula $d = d_0 + A\cos[2\pi/\lambda(x+\phi)]$. All authors agree on the amplitude of modulation A = 0.1 Å but they give different average values d_0 = 2.53 Å at 290K, $\lambda = 5a$[10], d_0 = 2.47 Å at 90K, $\lambda = 4.75a$ [11]. At the present status of EXAFS data analysis it is not possible to solve the five different distances expected for the 5x1 superstructure, therefore we have assumed only two distances in the EXAFS analysis and the effective Debye Waller factors take account of the distribution of the Cu-O distances. Therefore the two distances found by the present EXAFS work indicate the presence of the distribution of Cu-O(apical) distances over a range of 0.15 Å in agreement with diffraction data but around a shorter average Cu-O(apical) bond length of 2.37 Å. The joint interpretation of EXAFS and diffraction data clarifies that the short and long Cu-O(apical) distances are not due to the apex oxygen instability between two positions but it is due to the presence of different Cu site structure configurations in the CuO_2 plane. In Fig. 2 we present a pictorial view of the CuO_2 plane with a 5x1 superstructure with the linear arrays of different Cu square pyramids. In Fig. 3 a pictorial view of the modulation of the Cu-O(apical) distance in two dimensions is presented. Because the Cu-O(apical) distance modulates the local electronic structure of the Cu site, the CuO_2 plane looks more like a corrugated-iron foil, where the different Cu sites are aligned along the linear grooves, than like a flat layer as it was considered until now.

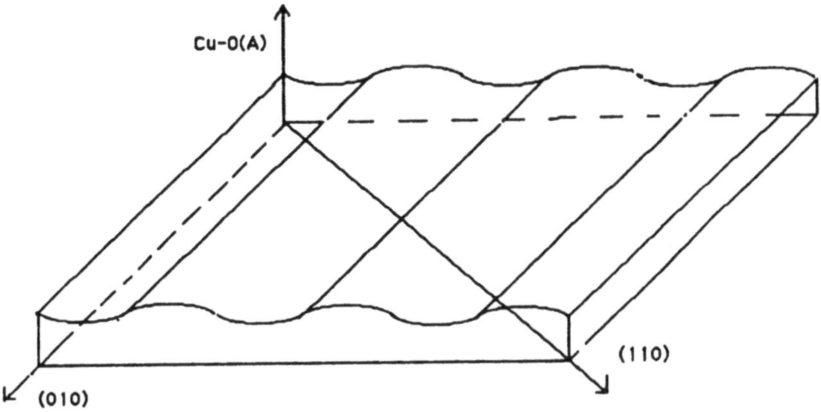

FIG. 3. Pictorial view of the CuO_2 plane where the Cu-O(apical) distance is modulated only in one direction, giving a corrugated iron like surface for apical oxygens which implies that also the electronic structure of the CuO_2 plane is modulated like a corrugated-iron foil.

In order to investigate if these results are unique for Bi 2:2:1:2 or indicate a common feature for the high Tc superconductors we have analyzed the case of the $YBa_2Cu_3O_{6.5}$. In fact also in this case many authors have reported the presence of a 2x1 superstructure due to linear and empty chains Cu(1)-O in the basal plane in the ortho II phase[15]. We have analyzed the average Cu(2)-O(apical) distance reported by Cava et al.[16]. In Fig.4 the fit of Cu(2)-O(apical) distance with two different distances in the range of oxygen concentrations 6.35-6.75 indicate that it is possible to assign the average Cu(2)-O(apical) distance measured by the diffraction experiments that have not solved the superstructure to the presence of two distances 2.27 Å and 2.45 Å where the probability of the short distance is taken to be given by the relative presence of four coordinated Cu(1) ions in the basal plane[17] increasing with oxygen doping. This result[18] is in agreement with EXAFS data showing the presence of two Cu(2)-O(apical) distances 0.13 Å apart. This result is in fully agreement with the results of Burlet et al.[19] who succeeded to solve the 2x1 superstructure in $YBa_2Cu_3O_{6.5}$ by measuring neutron diffraction data for a very large crystal. The results of Burlet et al. show the presence of sites with long 2.42 Å and short 2.32 Å Cu (apical) distances which reconciles the EXAFS and diffraction data for the presence of two different Cu-O(apical) distances within the experimental errors of 0.04 Å.

It is therefore possible to conclude that for the two superconducting crystals, Bi 2:2:1:2 and $YBa_2Cu_3O_{6.5}$, where the superstructure was solved by neutron diffraction and the local structure was investigated by EXAFS the two methods converge showing the presence of linear arrays of Cu sites with short and long Cu-O(apical) distances. In conclusion the linear arrays of different Cu sites configurations make the symmetry of the CuO_2 plane more like that of a corrugated iron foil as it is shown in Fig.3 than like a flat homogeneous two dimensional layer.

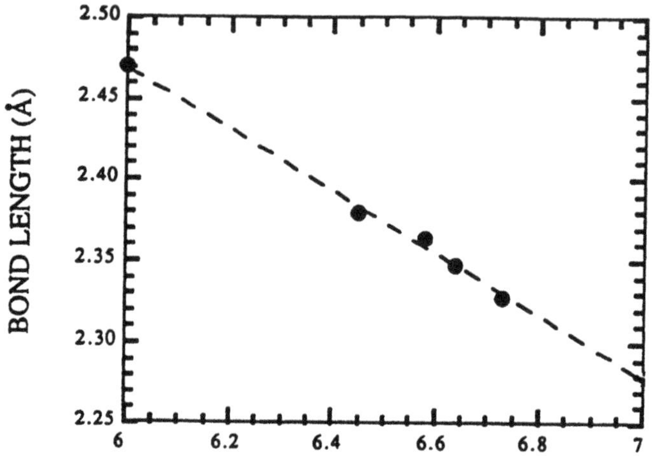

FIG. 4 The average Cu(2)-O(apical) bond distances measured by Cava et al.[16] in $YBa_2Cu_3O_{6+x}$ fitted with the combination of two Cu(2)-O(apical) distances 2.27 Å and 2.4 Å where the relative weight of the short 2.27 Å distance is taken to be proportional to the number of fourfold coordinated Cu1 ions in the basal plane[17].

b) The distortion of the CuO_4 square plane

Here we want to point out that the structural investigations indicate that the distortion of the CuO_4 square plane is a key requirement for superconductivity. From the EXAFS investigation of Bi 2:2:1:2 we have found two short 1.88 Å and two long 1.95 Å Cu-O(planar) distances. This result is in qualitative agreement with all diffraction data[10-12] and in quantitative agreement with neutron diffraction data at 90K and 55K on a crystal with a = 5.397 Å, b = 5.401 Å and c = 30.716 Å where the average Cu-O(planar) distances have been found to be 1.88 and 1.94 Å. Therefore this result confirms the distortion of the CuO_4 square planes in Bi 2:2:1:2 also if the **a** and **b** axis are very close.

The present results on the distortion of the Cu square plane on Bi 2:2:1:2 makes this system similar to other superconducting systems. In fact it seems that many crystallograpic investigations of superconductors indicate that the orthorhombic distortion is a key requirement for high T_c superconductivity. The tetragonal to orthorhombic transition has been found at the insulating to metal transition in $YBa_2Cu_3O_{6+x}$, around the oxygen concentration of 6.3[17]. At low temperature in the metallic superconducting phase $La_{1.85}Sr_{0.16}CuO_4$ shows the orthorhombic distortion. On the other hand the orthorhombic to tetragonal transition in $La_{2-x}Sr_xCuO_4$.[20] at low temperature above x=0.21 Sr doping has been found to be correlated with the transition from the superconducting phase to the normal metal phase. Moreover it is well established that the crystallographic transition from the low temperature orthorhombic phase to the tetragonal phase at

$x=0.12$ in $La_{2-x}Ba_xCuO_4$ suppresses the superconductivity.

c) The Cu displacement from the oxygen plane.

A characteristic feature of the superconducting phase of $YBa_2Cu_3O_{6+x}$ for $x > 0.3$ is the buckling (or dimpling) angle $\beta \sim 8^o$ degrees formed by O(planar)-Cu-O(planar) due to the displacement of the Cu ion from the coordinated oxygens O(planar) plane. By comparison of the Cu-O(planar) distances found by EXAFS and the value of the a axis we can deduce an average buckling (or dimpling) angle of $< \beta > = 4^o$ degrees in Bi 2:2:1:2. This result is in agreement with the diffraction data showing that the buckling (or dimpling) angle is modulated in different Cu sites between 0° and 8° and the displacement of the Cu ion from the plane of the four oxygens can be as large as $h \sim 0.25$ Å. The Cu displacement is modulated with the superstructure: it is close to zero at the Cu sites with long Cu-O(apical) distance and it is larger $h \sim 0.25$ Å for the Cu sites with the shortest Cu-O(apical) distance.

d) The electron - lattice interaction and the pseudo Jahn Teller coupling.

The crystalline structures of parent Cu(II) insulating compounds of the high T_c superconductors exhbit the Cu site structure configurations typical of the Cu(II) Jahn-Teller ions with elongated CuO_6 octahedra, as in La_2CuO_4, square pyramids, as in $Bi_2YSr_2Cu_2O_{8+\delta}$ and $YBa_2Cu_3O_6$, and square planes, as in Nd_2CuO_4. The Jahn Teller effect characteristic of the Cu(II) ions, removes the degeneracy of the upper E_g states $3d_{x^2-y^2}$, ($m_l=2$) and $3d_{3z^2-r^2}$, ($m_l=0$) in the octahedral O_h symmetry by pushing up the energy of the $3d_{3z^2-r^2}$, $m_l=0$, and lowering the energy of the $3d_{x^2-y^2}$, $m_l=2$, by reducing the O_h symmetry with an elongation of the Cu-O bond in the z direction, or in the extreme cases by pushing away one or two oxygen ions forming Cu sites with a square pyramid or a square plane coordination. Therefore in the divalent cuprate perovskite the single hole per unit cell is stabilized in the in the Cu 3d derived states with the component of orbital momentum $3d_{x^2-y^2}$, $m_l=2$.

The distortions of the CuO_4 square planes in doped superconductors can be interpreted as a pseudo Jahn-Teller type distortions[21,22] mixing the $m_l=2$ with $m_l=0,1$ states, therefore the transition from the tetragonal insulating phase to the orthorhombic superconducting phase can be correlated with the mixing of the $3d_{x^2-y^2}$ with $3d_{3z^2-r^2}$ (and $3d_{xz}$, $3d_{yz}$) hole states. In Fig. 5 the distortions of the Cu square plane mixing the $3d_{x^2-y^2}$ with $3d_{3z^2-r^2}$ hole states are shown. Therefore the distortions of the Cu square plane induced by doping can be classified as due to the increase of the pseudo JT distortions associated with the $3d^9\underline{L}$ states formed by doping.

The variation of the Cu-O(apical) distance induces the variation of the energy splitting Δ_{JT} between the $3d_{x^2-y^2}$ and $3d_{3z^2-r^2}$ states (called also the Jahn Teller splitting) and therefore by decreasing the Cu-O(apical) distance the mixing between the $3d_{x^2-y^2}$ with $3d_{3z^2-r^2}$ hole states in the CuO_2 planes increases.

The dimpling angle β enters in the expression for the coupling of the electronic states with the local lattice deformations in the theory of the pseudo-Jahn Teller effect[21]. In fact the electron lattice interaction of the pseudo JT type is given by $\lambda = g(Q)\ f(\Delta_{JT})\ h(\beta)$ where the configuration

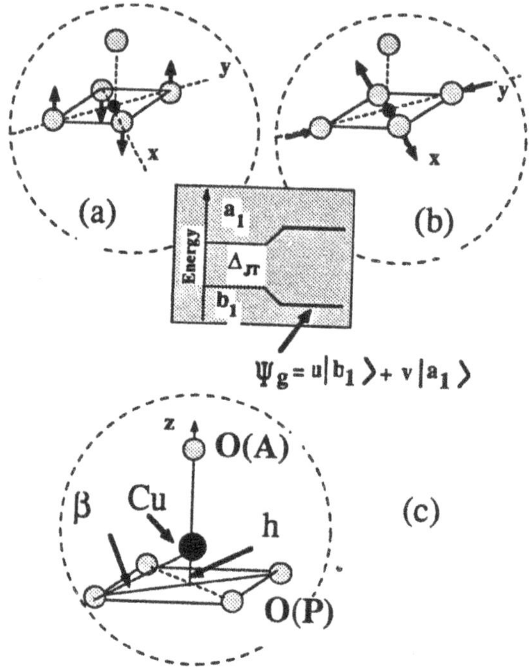

FIG. 5 The square plane distortions mixing the a_1 and b_1 states separated by the energy Δ_{JT} according to the pseudo Jahn Teller (JT) effect: a) The out of phase vibrations of in plane oxygens O(P) and b) the rhombic distortion of the square plane. The energy Δ_{JT} is a function of the Cu-O(apical) distance and the electron lattice coupling constant is function of the displacement h of the Cu ion from the oxygens plane or the dimpling angle β, panel c.

parameter Q is a measure of the distortion of the Cu square plane, for example for a rhombic distortion $Q= 2(d_1-d_2)/(d_1+d_2)$ where d_1 and d_2 are the Cu-O(planar) distances. The orthorombic crystal structure in superconductors can be associated with the stabilized JT distortion of the Cu square plane at low temperature driven by g(Q). The variation of the dimpling angle β or the Cu displacement from the oxygen plane in different Cu sites gives a modulation of the electron lattice interaction, via $h(\beta)$, and therefore it could indicate the dynamic coupling of the electronic states with lattice modes with wavevector $k\neq0$. The Cu-O(apical) distance modulates the electron lattice coupling λ via the induced variation of the energy splitting Δ_{JT} which is maximum for infinite Cu-O(apical) distance and it decreases by shortening the Cu-O(apical) distance. From the experimental determination of the Cu site configurations in the cuprate superconductors we think that it is possible to formulate the hypothesis that the Cu site distortions indicate the formations of polarons involving the dimpling of the in plane oxygens and the variation of the Cu-O(apical) distance. The formation of polarons has been indicated by photoinduced infrared absorption results[23]

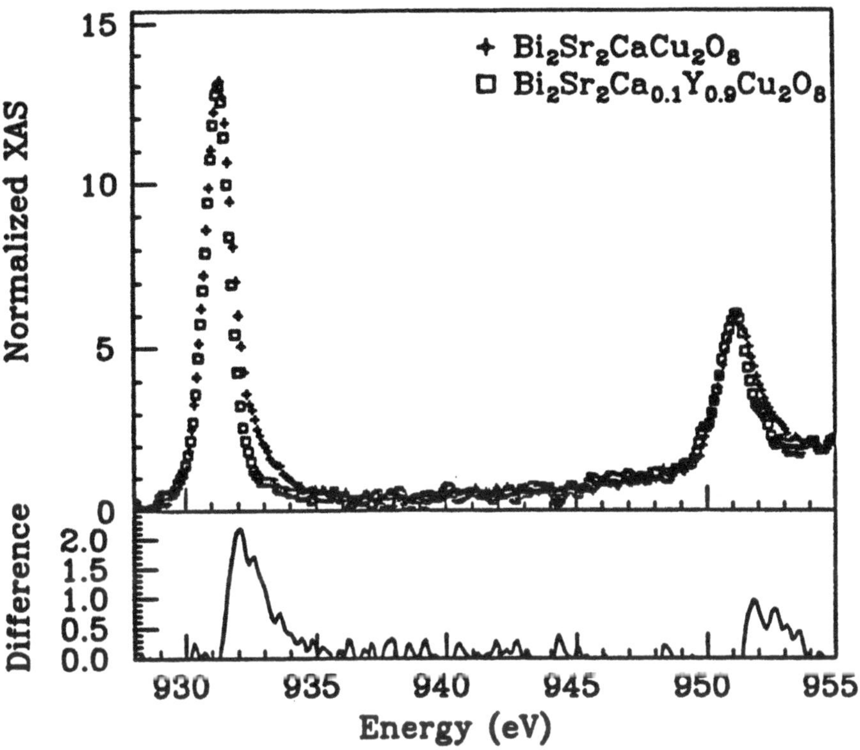

FIG. 6 The **E//ab** polarized Cu L3 x-ray absorption spectra (XAS) of $Bi_2Sr_2CaCu_2O_8$ and that of the insulating $Bi_2Sr_2Ca_{0.1}Y_{0.9}Cu_2O_8$ crystal and their difference, lower panel.

The superstructure can be razionalized as evidence for ordering of polarons along one direction in the CuO_2 plane. The modulation of the Cu displacement and of the Cu-O(apical) distance giving the superstructure indicates different Cu sites where the electron lattice coupling is large (large β angle and short Cu-O(apical) distance) and other Cu sites where the electron lattice coupling is small (low β angle and long Cu-O(apical) distance). The electronic states in the CuO_2 plane appears therefore formed by two components: the first component is associated with the linear arrays of Cu sites with the long Cu-O(apical) distance along the 010 direction (see Fig.2) where the electron lattice coupling is weak and the second component with large electron lattice interaction that can be associated with the pseudo Jahn Teller polarons. The local density calculations show that a large electron phonon interaction is expected along the (110) direction, i.e. along the Cu-O-Cu lines, and it is expected to be enanched by the dimpling angle β.[24] The relevance for superconductivity of the vibration modes mixing the $3d_{x^2-y^2}$ with $3d_{3z^2-r^2}$ hole states is

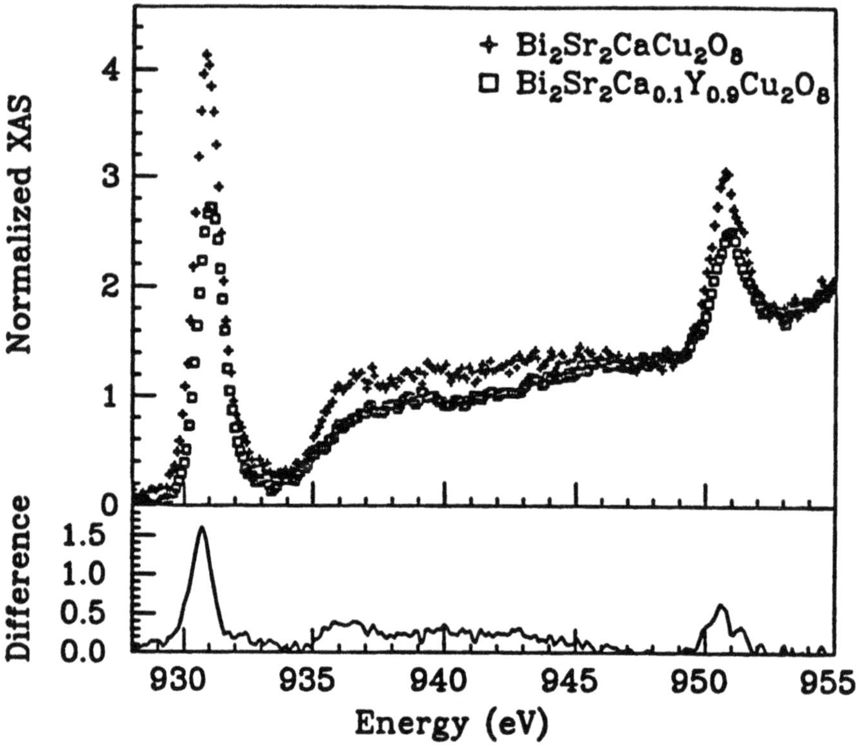

FIG. 7 The **E//c** polarized Cu L_3 x-ray absorption spectra (XAS) of $Bi_2Sr_2CaCu_2O_8$ and of the insulating $Bi_2Sr_2Ca_{0.1}Y_{0.9}Cu_2O_8$ crystal and their difference, lower panel.

demonstrated by the Raman results[25]. The mode at 335 cm^{-1} corresponding to out-of-phase vertical vibration (with B_{1g} symmetry) of two oxygen atoms in the CuO_2 plane, that becomes soft at T_c has the symmetry of the out of phase vibration ofte the O(P ions along the c axis shown in Fig. 5, mixing the $3d_{x^2-y^2}$ with $3d_{3z^2-r^2}$ states. The asymmetric Fano lineshape of this Raman line is an indication of the interaction of this mode with electronic transitions.

3. The modulation of the electronic structure

The single hole per Cu site in the insulating parent system is frozen in the $3d_{x^2-y^2}$ symmetry[26]. In high T_c superconductors the doping introduces additional δ hole states per Cu ion in the insulating system resulting in $1+\delta$ holes per Cu ion. The $3d^9\underline{L}$ character of additional hole states δ was found by x- ray absorption [27] and it is now well established. These states have a non

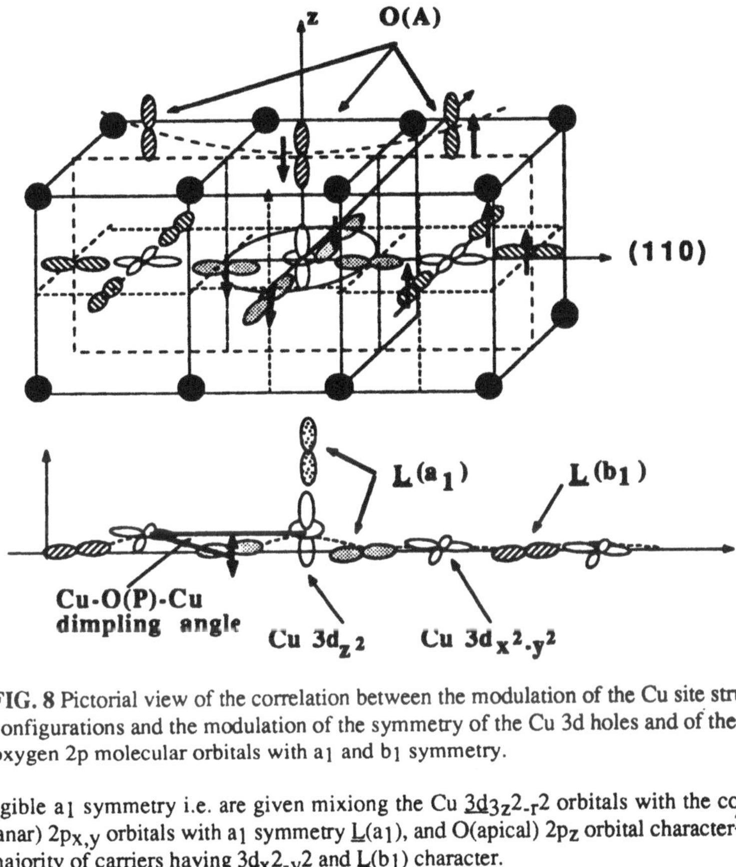

FIG. 8 Pictorial view of the correlation between the modulation of the Cu site structure configurations and the modulation of the symmetry of the Cu 3d holes and of the oxygen 2p molecular orbitals with a_1 and b_1 symmetry.

negligible a_1 symmetry i.e. are given mixiong the Cu $\underline{3d}_{3z^2-r^2}$ orbitals with the combination of O(planar) $2p_{x,y}$ orbitals with a_1 symmetry $\underline{L}(a_1)$, and O(apical) $2p_z$ orbital character[27-28] beyond the majority of carriers having $\underline{3d}_{x^2-y^2}$ and $\underline{L}(b_1)$ character.

The presence of some Cu 3d holes with a_1 symmetry in metallic systems was associated with the formation of pseudo Jahn Teller (JT) polarons characterized by short Cu-O(apical) distance and distortions of the CuO_4 square plane [28] and their relevance for some pairing mechanisms was discussed by Askenazi [29].

Theoretical calculations [30,31] show that the energy separation between the b_1 and a_1 states is modulated by the Cu-O(apical) distance therefore at the sites with short Cu-O(apical) bond the Cu 3d $m_l = 0$ component of the electronic carriers, or with a_1 molecular symmetry, is expected to increase. The stability of a Jhan Teller (called also anti-JT) polaron associated with a_1 states in LaSrCuO system was discussed recently [32.24].

We have measured the variation of the polarized Cu L_3 XAS absorption line of a Bi 2:2:1:2 superconducting single crystal compared with the Yttrium doped insulating system as it was previously reported for the LaSrCuO[33] superconductor and for a single layer Bi 2:2:0:1 system[34].

The Cu L_3 XAS spectra of the insulating system probes the single hole per Cu ion therefore the difference spectra probe the δ additional states $3d^9\underline{L}$. In the Bi 2:2:1:2 single crystal we have found a total of $\delta = .19 \pm .2$ $d^9\underline{L}$ states from the difference of the unpolarized spectra. The difference between the polarized **E//ab** spectra of the metal and of the insulating system, reported in Fig. 6, shows that $0.15 \pm .02$ added holes have the $3d_{x^2-y^2}\,\underline{L}(b_1)$ character. In Fig. 7 the difference spectrum between the **E//c** XAS spectra of the metal and the insulating system show the presence of the other $0.04 \pm .02$ states with $\underline{3d_{3z^2-r^2}}\,\underline{L}(a_1)$ character.

The final states due to the transition Cu $2p^6\,\underline{3d_{3z^2-r^2}}\,\underline{L}(a_1)$ -> Cu $2p^5\,3d^{10}\,\underline{L}(a_1)$ in **E//c** difference XAS spectra are at lower energy, at about 931.1 eV in the lower panel of Fig. 7, than the final states due to the Cu $2p^6\,\underline{3d_{x^2-y^2}}\,\underline{L}(b_1)$ -> Cu $2p^5\,3d^{10}\underline{L}(b_1)$ that appear in the **E//ab** XAS difference spectra, at about 932 eV in the lower panel of Fig. 6. Because the absorption line due to the Cu 2p -> 3d transition is a bound excitonic state, the higher energy position of the $2p^5 3d^{10}\underline{L}(b_1)$ final states indicates mobile carriers with b_1 symmetry screening the core hole. On the contrary the fact that the line due to the Cu $2p^5\,3d^{10}\,\underline{L}(a_1)$ final state is at the same energy as that of the absorption line in the insulating system indicates that the states with a_1 symmetry are more localized.

Therefore it is possible to associate the more delocalized holes with b_1 symmetry with the linear arrays along the b axis in Bi2212 with the Cu site configurations with long Cu-O(apical) distances. The more localized states, induced by doping, with partially a_1 symmetry can be associated with the electronic states with the modulated Cu site configurations along the Cu-O-Cu-O direction with the short and long Cu-O(apical) distances A pictorial view of the proposed correlation between the variation of the local symmetry of the electronic carriers and the variation of the Cu site structure is shown in Fig. 8.

4. Conclusions

The present work supports the formation of pseudo Jahn Teller polarons driven by the electronic states, induced by doping, with partially a_1 symmetry characterized by Cu sites with the short Cu-O(apical) distance and Cu displacement from the in plane oxygens. These Cu sites are distributed in the CuO_2 plane with a regular arrangement of linear arrays in the (010) direction separated by linear arrays of sites with long Cu-O(apical) distance.

This new scenario suggests pairing mechanisms for high T_c superconductivity in the presence of two components of the electronic structure: the more delocalized component formed by $\underline{3d_{x^2-y^2}}\,\underline{L}(b_1)$ states and the more localized states with partially a_1 symmetry. A possible pairing mechanism in this scenario see the pairing of the delocalized carriers with b_1 symmetry mediated by excitations in the more localized a_1 component. Therefore the system can be descrided as formed by an infinite number of parallel interacting chains of itinerant states separated by linear chains of localized states.

It is interesting to remark that the vawelength of the superstructure λ is close to the coherence length $\xi_{ab} = 26 \pm 5$ Å (values ranging from 21 Å to 31 Å are reported in the literature[35]) in Bi 2:2:1:2. Because $1/\xi_{ab} \sim 0.2\ a^{-1}$ gives the width of the momentum distribution involved in the pairing it is possible that the excitation exchanged in the pairing has the wavevector close to 1/λ.

The regular array of chains of polarons, separated by λ, seems to be correlated with the wavevector of the carriers at the Fermi level $k_F \sim 2\pi/\lambda$. More extensive investigations on the relation between the superstructure and the shape of the Fermi surface as function of doping will clarify if the structural instabilty giving the supestructure is driven by electronic structure.

The $k_F \cdot \xi_{ab} \sim 2\pi$ relation that is valid for high T_c superconductors seems to be related with the ordering in the real space of the spacing λ between the chains of polarons.

An interesting aspect of the hypothesis is that the resulting excitation exchanged by the carriers is related to the electron-lattice interaction of the pseudo Jhan Teller type is that it will have both phononic, excitonic and magnetic characters.

I would like to thank S. Della Longa, A.M. Flank, P. Lagarde, I. Pettiti, M. Pompa, P. Porta S. Turtù, D. Udron, and A. Di Cicco for experimental help and G. Calestani, C. Castellani, F. De Martini, C. Di Castro, M. Grilli, and L. Pietronero for useful discussions.

References

1. J.G. Bednorz and K.A. Müller *Rev. Mod. Phys.* **60** (1988) 565.
2. A. Bianconi, S. Della Longa, M. Missori, I. Pettiti, and M. Pompa in *Lattice Effects in High-Tc Superconductors,* edited by Y. Bar-Yam, World Scientific Pub., Singapore, 1992 (Proc. of the Meeting in Santa Fe, New Mexico, January 13-15 1992) to be published
3. J. Munstre de Leon, S.D. Conradson, I. Batistic, and A.R. Bishop, *Phys. Rev. Lett.* **65** (1990) 1675; S.D. Conradson, and I.D. Raistrick,*Science* **243** (1989) 1340; S.D. Conradson, I.D. Raistrick, and A.R. Bishop *Science* **248** (1990) 1394.
4. J. Röhler, and U. Murek in *High Tc superconductors: Electronic Structure* edited by A, Bianconi and A. Marcelli Pergamon Press (1989) pag. 271 .
5. H. Murayama, T. Ishii, N. Bamba, H. Maeda, A. Koizumi, Y. Yoshikawa, and H. Yamazaki *Physica C* **160** (1989) 524.
6. J. Röhler, A. Larish, and R. Shäfer *Physica C* 1992, to be published and J. Röhler in *Lattice Effects in High - T_c Superconductors* edited by Y. Bar-Yam, World Scientific Pub., Singapore, 1992.
7. B.H. Toby, T. Egami, J.D. Jorgensen, and M.A. Subramanian*Phys. Rev. Lett.* **64** (1990) 2414.; T. Egami et al. in *Electronic Structure and Mechanisms of High Tempearture Supersonductivity*, edited by J. Askenazi, and G. Vezzoli (Plenum, New York, 1992); S.J.L. Billinge, and T. Egami Proc. of the Meeting on *Lattice Effects in High-T_c Superconductors* Santa Fe, New Mexico, 1992 edited by Y. Bar-Yam, World Scientific Pub., Singapore, 1992, to be published
8. E.A. Hewat, M. Dupuy, P. Bordet, J.J. Capponi, C. Chaillout, J.L. Hodeau, and M. Marezio *Nature* **333** (1988) 53.
9. J. -C Grenier in *Phase Separation in Cuprate Superconductors edited by K.A. Müller* World Scientific Pub., Singapore, 1992
10. G. Calestani, C. Rizzoli, M.G. Francesconi and G.D. Andreetti *Physica C* **161** (1989) 598.
11. A.I. Beskrovnyi, M. Dlouhà, Z. Jiràk and S. Vratislav, *Physica C* **171** (1990) 19; and A.I. Beskrovnyi, M. Dlouhà, Z. Jiràk, S. Vratislav and E. Pollert, *Physica C* **166** (1990) 79.
12. V. Petricek, Y. Gao, P. Lee, and P. Coppens *Phys, Rev. B* **42** (1990) 387.
13 Y. Inoue, Y. Shichi, F. Munakata and M. Yamanaka, *Phys. Rev. B* (1989) **40**, 7307.
14. A.Q. Pham, M. Hervieu, A. Fuertes, and B. Raveau *Physica C* to be published
15. D. de Fontaine, G. Ceder, and M. Asta *Nature* (London) **343** (1990) 544; G. Ceder, M.

Asta, W.C. Carter, M. Kraitchman, D. de Fontaine, M.E. Mann, and M. Sluiter *Phys. Rev. B* **41** (1990) 8698

16. R.J. Cava, A.W. Hewat, E.A. Hewat, B. Batlogg, M. Marezio, K.M. Rabe, J.J. Krajewski, W.E. Peck Jr. and L.W. Rupp Jr. *Physica C* **175** (1990) 419
17. P.G. Radaelli, C.U. Segre, D.G. Hinks, and J.D. Jorgensen *Phys. Rev. B* **45** (1992) 4923.
18. S. Pagliuca, thesis University of Rome "La Sapienza" *to be published*
19. P. Burlet in *Phase Separation in Cuprate Superconductors edited by K.A. Müller* World Scientific Publ., Singapore, 1992
20. H. Takagi R.J. Cava, M. Marezio, B. Batlogg, J.J. Krajewski, W.F. Peck, Jr., P. Bordet, and D.E. Cox to be published; and in Proceedings of the workshop *Phase Separation in Cuprate Superconductors* edited by K.A. Müller, World Scientific Pub., Singapore, 1992
21. M. Bacci *Chemical Physics* **104** (1986)191; *Jpn. J. Appl. Phys.* **27** (1988) L1699
22. K.I. Kugel and D.I. Khomskii *Sov. Phys. Usp.* **25**(1982) 321 and *Sol. State Commun.* **13** (1973)763; M.D. Kaplan and D.I. Khomskii *Pis'ma Zh Eksp. Teor. Fiz.* **47** (1988) 631
23. C. Taliani, R. Zamboni, G. Ruani, F.C. Matacotta, and K.I. Pokhodnya *Sol. Stat Commun.* **66** (1988) 487.
24 I.I. Masin, O.K.Andersen, A.I. Liechtenstein, O. Jepsen, V.P. Antropov, S.N. Rasheev, V.I. Anisimov, J. Zanen,R.O> Rodriguez, and M. Methfessel, Proc. of the Meeting on *Lattice Effects in High-T_C Superconductors* Santa Fe, New Mexico, 1992 edited by Y. Bar-Yam, World Scientific Pub., Singapore, 1992, to be published
25. T. Ruf, C. Thomsen, R. Liu, and M. Cardona *Phys. Rev. B* **38** (1988) 11985; L. Genzel, A. Wittlin, M. Bauer, M. Cardona, E. Schönherr, and A. Simon *Phys. Rev. B* **40** (1989) 2170.
26. C. Li, M. Pompa, S. Della Longa, and A. Bianconi *Physica C* **178** (1991) 421; M. Pompa, C. Li, A. Bianconi, A. Congiu-Castellano, S. Della Longa, A. -M. Flank, P. Lagarde, D. Udron *Physica C* **184** (1991) 51.
27. A. Bianconi, A. Congiu Castellano, M. De Santis, P. Rudolf, P. Lagarde and A.M. Flank *Sol. State Commun.* **63**, 1009 (1987); A. Bianconi, P. Castrucci, A. Fabrizi, M. Pompa, A.-M. Flank, P. Lagarde, H. Katayama-Yoshida and G. Calestani *Physica C* **162-164** (1989) 209; A. Bianconi, C. Li, M. Pompa, A. Congiu-Castellano, S. Della Longa, D. Udron, A.-M. Flank, and P. Lagarde *Phys. Rev. B* **44** (1991) 10126; and A. Bianconi, C. Li, S. Della Longa, and M. Pompa *Phys. Rev. B* **45** (1992) 4989.
28. A. Bianconi, A.M. Flank, P. Lagarde, C. Li, M. Pompa, and D. Udron in *Electronic Structure and Mechanisms of High Tempearture Supersonductivity*, edited by J. Askenaazi, and G. Vezzoli (Plenum, New York, 1991).
29. J. Askenazi in *Electronic Structure and Mechanisms of High Tempearture Supersonductivity*, edited by J. Askenaazi, and G. Vezzoli (Plenum, New York, 1991)
30. M. Grilli, C. Castellani and C. Di Castro *Phys. Rev. B* **42**, 6233 (1990); C. Di Castro, L.F. Feiner and M. Grilli *Phys. Rev. Lett.* **66** (1991) 3209.
31. Y. Seino, A. Kotani and A. Bianconi *Jour. Phys. Soc. of Japan* **59** (1990) 815.
32. V.I. Anisimov, M.A.Korotin, J.Zanen, and O.K.Andersen *Phys. Rev. Lett.* **68** (1992) 345.
33. M. Pompa, S. Turtù, A. Bianconi, F. Campanella, A. M. Flank, P. Lagarde, C. Li, I. Pettiti, and D. Udron *Physica C* **185-189** (1991) 1061.
34. M. Pompa, P. Castrucci, C. Li, D. Udron, A. M. Flank, P. Lagarde, H. Katayama-Yoshida, S. Della Longa, and A. Bianconi *Physica C* **184** (1991) 102.
35. K. Kadowaki, *Physica C* **165-189** (1991) 1811; B. Batlogg, T.T.M. Palstra, L.F. Schneemeyer, R.B. van Dover, and R.J. Cava *Physica C* **153-155** (1988) 1062.

Phase Separation, Structure and Superconductivity in Oxygen-Annealed $La_2CuO_{4+\delta}$

P.C. HAMMEL, E.T. AHRENS, A.P. REYES, J.D. THOMPSON, D.E. MACLAUGHLIN,[†] Z. FISK, P.C. CANFIELD
Condensed Matter and Thermal Physics Group, Los Alamos National Laboratory, Los Alamos, New Mexico 87545, USA

and

J.E. SCHIRBER
Sandia National Laboratories, Albuquerque, New Mexico, 87185, USA

ABSTRACT

We report the measurements of the excess oxygen content in oxygen-annealed $La_2CuO_{4+\delta}$. Below the phase separation temperature, T_{PS} several features of the phase diagram for this material are unusual. At T_{PS} the oxygen content changes discontinuously in both the oxygen-poor and oxygen-rich phases; below T_{PS}, these change very little. We propose that the phase separation temperature is determined by the excess oxygen content of the oxygen-poor (antiferromagnetic) phase which in turn determines the Néel temperature. Phase separation occurs when the resulting sublattice magnetization is approximately 1/2 the zero temperature value. Bulk magnetization measurements show that the superconducting T_c is dependent on the cooling rate in a narrow temperature range well separated from T_{PS}. Nuclear quadrupole resonance (NQR) data provide evidence for the role of temperature dependent structural changes in this effect. The role structural changes may play in determining the normal and superconducting state properties is discussed.

1. Introduction

1.1 *Phase separation*

The properties of the superconducting cuprates continue to present a challenging puzzle. Evidence has begun to accumulate that the density of doped holes in the planar cuprates may be inhomogeneous. The time and distance scales describing this inhomogeneity vary but the phenomenon appears to be present in several of the superconducting cuprates. Several groups[1] have pointed out that there is a general tendency in the two-dimensional CuO_2 lattice toward phase separation of holes. In most cases, however, this tendency is frustrated by long range Coulomb interactions which prevent macroscopic phase separation. The result in this case is fluctuations of hole density occurring over short time and distance scales.[2] In this regard oxygen-annealed[3] $La_2CuO_{4+\delta}$ is a uniquely interesting system: it is the only system in which the mobility of the chemical dopant (oxygen) is sufficiently high at the temperatures at which phase separation is taking place that this frustration is absent. This allows one to study macroscopic phase separation over long time scales. We have made a variety of measurements which allow a deeper understanding of certain aspects of the system. Understanding long-range, static phase separation is probably quite important to furthering our knowledge of the more generally occurring, but less experimentally accessible systems displaying dynamical hole fluctuations. To achieve a full understanding of some of the fundamental features of the cuprates it may be necessary to incorporate the consequences of phase separation into models and theories.

We present very sensitive measurements of the temperature dependence of the doping of the antiferromagnetic phase δ_{AF} and estimates of the doping of the metallic phase δ_M. We find several unexpected features of the phase separation. These unusual aspects provide insight into the mechanism and dynamics of phase separation. In particular we observe that: i) there is an abrupt change in the doping (of both phases) at the phase separation temperature, T_{PS} ; ii) below $0.9T_{PS}$ the temperature dependence of δ_{AF} is very weak; iii) δ_{AF} remains finite at zero temperature; and iv) there is a close relationship between T_{PS} and the characteristic magnetic ordering temperature of the antiferromagnetic phase which suggests that the mechanism of phase separation and the development of long range antiferromagnetic order in the oxygen-poor phase are related.

1.2 *Structure*

In addition to information concerning the doping of the two phases, our NQR and NMR measurements reveal the existence of unexpected structural features of the metallic phase of this material, primarily related to the position of apical (LaO layer) oxygens and thus octahedral tilts. These features are observed by means of the quadrupole interaction which is very sensitive to the position of oxygen ions in the LaO layer as well as charge transfer between LaO layers and the CuO_2 planes. We also observe that the superconducting transition temperature T_c in the metallic phase is substantially increased with no change in transition width when the sample is slowly cooled. It is particularly interesting that it is only necessary to cool slowly through a narrow temperature range and that this temperature is well removed from both T_{PS} and T_c indicating that although it may involve motion of oxygen or doped holes it is not directly related to the phase separation process which occurs at T_{PS} and is nearly complete by $T = 0.9\,T_{PS}$. Abrupt changes are observed in the NQR spectrum in this same narrow temperature region indicating that the structural changes we observe are intimately connected with the ultimate development of superconductivity.

2. Experiment

The samples were produced by annealing stoichiometric $La_2CuO_{4+\delta}$ in a 3 kbar oxygen atmosphere at 575 °C for 12 hours as previously discussed.[4] The NMR and NQR measurements were made using a conventional phase coherent pulsed NMR spectrometer. NQR measurements were performed in zero applied field; spectra were obtained both by Fourier transform of the spin echo signal and by varying spectrometer frequency and plotting the intensity in a narrow bandwidth as a function of frequency. ^{63}Cu, ^{17}O, and ^{139}La measurements were all performed; we will emphasize the ^{139}La results here.

The ^{139}La (I=7/2) spectrum[5] is determined by the local electric field gradient (EFG) and the local hyperfine field arising from the ordered copper moments:

$$\mathcal{H} = \mathcal{H}_Q + \mathcal{H}_Z$$

$$\mathcal{H}_Q = \frac{h\nu_Q}{6}\left[3I_z^2 + \eta(I_x^2 - I_y^2) - I(I+1)\right]$$

$$\mathcal{H}_Z = -h\nu_Z I_{z'}$$

The fundamental parameter determining the nature of the zero-field spectrum is, ν_Q, the quadrupole frequency. This is determined by the electric field gradient (EFG), $V_{ij} = \partial^2 V/\partial x_i \partial x_j$ where V is the electrostatic potential at the nuclear site. In its principal axis frame, the EFG is diagonal; we choose the orientation of this frame such that the largest component of the EFG lies along the z-axis. The hyperfine field lies along the z'-axis which makes a polar angle θ with respect to the z-axis. The Zeeman frequency ν_Z, is given by $\nu_Z = \gamma H_{hf}$ where γ is the gyromagnetic ratio and H_{hf} the hyperfine field arising from the ordered copper moments. The magnitudes of ν_Q and η are given by

$$\nu_Q = A\sqrt{V_{zz}^2 + \frac{1}{3}\left(V_{xx} - V_{yy}\right)^2}$$

$$\eta = \frac{V_{xx} - V_{yy}}{V_{zz}}$$

Here A is a constant which includes the magnitude of the nuclear quadrupole moment and the asymmetry parameter η indicates degree of departure from axial symmetry. In the case of axial symmetry ($\eta = 0$) we have $\nu_Q = AV_{zz}$. The z component of the electric field gradient at the origin due to a collection of point charges of magnitude qe at the points $r_i = (x_i, y_i, z_i)$ is given by:

$$V_{zz} = eq\sum_i \frac{3z_i^2 - r_i^2}{r_i^5}.$$

Thus the quadrupole frequency in the present case will be primarily determined by the location and charge of oxygen ions in the LaO layer. Because the quadrupole interaction is proportional to m^2, it leaves $\pm m$ states degenerate: in the absence of a magnetic field there is a single NQR line corresponding to both the $\pm m \leftrightarrow \pm(m-1)$, $\Delta m = 1$ transitions. The Zeeman interaction splits these states; the magnitude of the frequency difference for the $m \leftrightarrow (m-1)$ and the $-m \leftrightarrow -(m-1)$ transitions is $\nu_Z \cos\theta$ to first order in ν_Z/ν_Q.

3. Phase Separation

3.1 *Properties of nearly stoichiometric lanthanum cuprate*

For very small values of δ, $La_2CuO_{4+\delta}$ is an antiferromagnetic insulator. Johnston et al.[6] observe a maximum in the Néel temperature T_N at some value of $\delta = \delta_0$. Increasing oxygen content suppresses T_N, at first slowly and then more rapidly with increasing δ: we estimate these results with the expression $T_N(\delta) = T_{N,max} - a(\delta - \delta_0)^2$. Employing the technique of perturbed $\gamma\gamma$ angular correlations (PAC), Saylor et al.[7] demonstrated that the maximum Nèel temperature, now believed to be $T_{N,max} \approx 328\,K$ is achieved for $|\delta_0| < 0.16$; they conclude $\delta_0 = 0$.

Zero field NQR techniques have been employed by several groups[8-11] to determine the parameters which describe the EFG and internal field in these nearly stoichiometric materials. Of the five parameters ν_Q, ν_Z, η, θ, and ϕ (if $\eta \neq 0$) which determine the transition frequencies, the first four have been determined to high accuracy. This work has clearly established that the internal field detected at the La site is due to the long range antiferromagnetic order of the Cu moments in the CuO_2 planes and has shown that $\nu_Z(T/T_N)$ has a universal form regardless of the value of T_N. Surprisingly, the onset of sublattice magnetization has been shown to be first order.[10,11] The observed discontinuity in ν_Z at the ordering temperature is $\nu_Z(T_N)/\nu_Z(0) \cong 0.34$. The value of η is very small (~0.015) and only weakly temperature dependent, θ also varies only very slightly with temperature in nearly stoichiometric $La_2CuO_{4+\delta}$. The angle ϕ is difficult to determine due to the small value of η.

3.2 *Properties of oxygen-annealed lanthanum cuprate*

As the excess oxygen content δ is increased a concentration is reached at which the system phase separates[3] at surprisingly low temperatures (250-310 K) into an oxygen-poor insulating antiferromagnetic phase and a oxygen rich metallic phase. The metallic phase is superconducting with T_c ~ 26-38K. Neutron diffraction measurements[12,13] performed on samples having δ~0.03 have determined[12] the tetragonal-to-orthorhombic transition temperature, (430 K), as well as the (closely related) crystal structures of the two phases. The oxygen-poor phase has the Bmab structure[12] observed in nearly stoichiometric $La_2CuO_{4+\delta}$. The location of the excess oxygen in the metallic state has been determined.[13] Some disagreement exists, however, concerning the structure of the metallic phase. The uncertainty focuses on the location of the apical oxygen: some measurements find it to be directly above the copper,[12,14] whereas others find it offset[13] (tilted octahedra). Our NMR measurements, which are quite sensitive to the square of the tilt angle, are able to show that the oxygen octahedra are not tilted just above or below T_{PS}. Jorgensen[12] suggests that his neutron diffraction results are consistent with tilted octahedra if the tilts are incoherent throughout the lattice. As we will discuss below, this interpretation is inconsistent with our NMR results. For the present case, we are unable to distinguish an orthorhombic from a tetragonal unit cell.

By means of μSR measurements, Ansaldo[15] observed an internal magnetic field and demonstrated that one of the phases was magnetic. This strengthened the assertion that one of the phases was indeed an oxygen-poor antiferromagnetic insulator. The phase separation has visible consequences for the transport and thermodynamic properties of the materials.[4,16] The temperature dependence of the resistivity indicates transitions which are broad and slightly hysteretic.

We experimentally determine T_{PS} by four different magnetic resonance methods. The first two detect the onset of the existence of the antiferromagnetic phase through the appearance of lines in the ^{139}La NMR or NQR spectra which are known to be due to this phase (based on studies of nearly stoichiometric $La_2CuO_{4+\delta}$). The third relies on the sensitivity of the planar spin susceptibility to the magnitude of hole doping: the sudden shift in metallic phase doping at T_{PS} causes a discontinuous shift in the ^{17}O Knight shift (see Fig. 3 below). Finally, phase separation produces a narrow peak in the oxygen relaxation rate. All of these measurements produce consistent results independent of cooling history or rate of cooling. Our measurements show[17] the oxygen-poor phase to appear quite suddenly although the relative fraction of this phase may continue to change below T_{PS} down to 220 K.

3.3 *Temperature dependence of the excess oxygen content*

NQR is a very powerful probe of phase separated $La_2CuO_{4+\delta}$ because of its ability to distinguish signals from the two phases of the material. We will focus first on the oxygen-poor, antiferromagnetic phase. Below T_{PS} the long range antiferromagnetic order in the oxygen-poor phase causes an internal magnetic hyperfine field at the La site. By splitting the single NQR line, the hyperfine field in the oxygen-poor phase distinguishes that La NQR signal from the metallic phase signal. The magnitude of this splitting is proportional to the hyperfine field. Zero field NQR is the most sensitive method of determining the temperature dependence of the order parameter in the antiferromagnetic phase. In general δ_{AF}, and therefore T_N will be temperature dependent. We define a temperature dependent parameter T_N^* which determines ν_Z at any given temperature. We assume the same dependence of ν_Z on T/T_N^* as is observed in stoichiometric $La_2CuO_{4+\delta}$. We employ the measured $\nu_Z(T)$ in the oxygen-poor phase to extract $T_N^*(T)$, and hence $\delta_{AF}(T)$ (through the known dependence of T_N on δ). We further define the parameter T_N^0 and δ_{AF}^0 to be the values of $T_N^*(T)$ and $\delta_{AF}(T)$ as $T \to 0$. In the phase separated material, the disappearance of long range order is the result of phase separation occurring at $T_{PS} < T_N^0$ so that the antiferromagnetic phase transition cannot be observed at T_N^*.

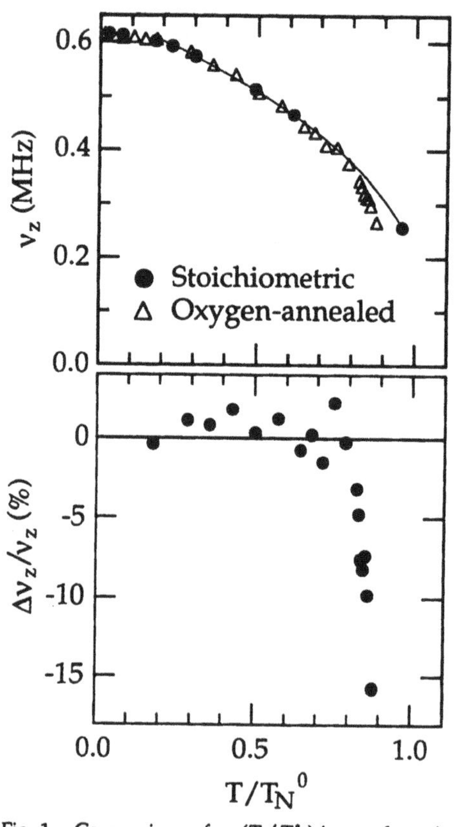

Fig. 1 Comparison of $\nu_Z(T/T_N^*)$ in nearly stoichiometric ($T_N^0 = T_N = 318$ K) [ref. 10] and oxygen-annealed ($T_N^0 = 279$ K) $La_2CuO_{4+\delta}$ (top panel). The lower panel shows the difference in ν_Z for the two samples. Below $0.8T_N^0$ the difference is slight.

By performing exact diagonalization of the nuclear Hamiltonian we have obtained ν_Q, ν_Z, η and θ at several temperatures from measurements of the frequencies ν_m where $h\nu_m = E_m - E_{m-1}$ and $m = 7/2$, $5/2, -5/2$, and $-3/2$. The values of the parameters θ and η are found to be in close agreement with the corresponding values in stoichiometric $La_2CuO_{4+\delta}$. The agreement of $\nu_Q(T)$ for the two samples is very good: they agree to within less than 0.1%.

We infer $\delta_{AF}(T)$ from $\nu_Z(T)$ by inverting functional fits to data from stoichiometric $La_2CuO_{4+\delta}$ for $\nu_Z(T)$ [ref. 11] and $T_N(\delta)$:[8,9]

$$v_Z(t) = 0.65(1-t)^{0.39}; \quad t \equiv \frac{T}{1.054T_N^*} \tag{1}$$
$$T_N(\delta) = 328 - 3.55\times10^5\delta^2$$

There is intuitive significance to T_N^* since δ_{AF} has very little temperature dependence below $0.9T_{PS}$. As a consequence, in this temperature range $v_Z(T)$ is virtually indistinguishable from what one would observe in stoichiometric $La_2CuO_{4+\delta}$ having $T_N = T_N^0$. The temperature dependence of v_Z is compared with the corresponding quantity obtained in stoichiometric $La_2CuO_{4+\delta}$ in Fig. 1. We have plotted $v_Z(T/T_N^0)$ where T_N^0 =279 K is obtained from the data below $0.9T_{PS}$ using Eq. 1. The plot of the difference in v_Z for the oxygen-annealed and the stoichiometric sample demonstrates that the data below $0.9T_{PS}$ is well described by a single value of $T_N^* = T_N^0$. In Fig. 2 we show the derived temperature dependence of δ_{AF}. The nearly constant value of δ_{AF} below $0.9T_{PS}$ is the expected result of the fact that, for a single determination of T_N^*, $v_Z(t)$ is very similar to that observed in nearly stoichiometric material.

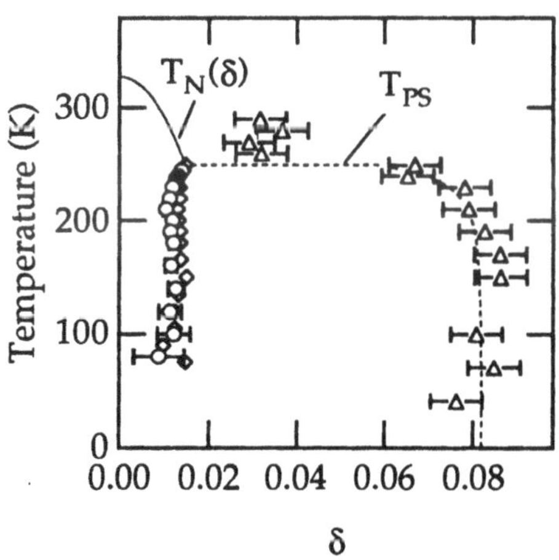

Fig. 2 The temperature dependence of the excess oxygen content is shown. Below T_{PS} =247 ±2 K two phases are formed. The oxygen-poor phase is described by the left set of data. Results from two samples (circles and diamonds) are shown. The second phase having higher doping is metallic (triangles) and, at low temperatures, superconducting. The region bounded by the two curves and below T_{PS} is inaccessible. The value of δ above T_{PS} is obtained from Ref [13].

The temperature dependence of δ_{AF} is shown in Fig. 2 along with the results from the metallic phase. We postpone discussion of the metallic phase results until later and focus on the properties of the oxygen-poor (low δ) phase. Its behavior is unexpected and it differs from earlier proposals.[18] Four features in particular are unusual. i) A discontinuous change in δ_{AF} occurs at T_{PS}, ii) except for a slight rounding near T_{PS}, δ_{AF} is essentially constant once phase separation occurs, iii) δ_{AF} is not tending toward zero at $T = 0$, and iv) the close relationship between phase separation temperature and the characteristic magnetic ordering temperature in the oxygen-poor phase.

Of these the first is unambiguous because below T_{PS}, the oxygen-poor phase is antiferromagnetic with a substantial sublattice magnetization. This demands a considerably smaller value of δ_{AF} than the value of δ (=0.03) which prevails above T_{PS}. As we will discuss below, the doping of the metallic phase also changes discontinuously at T_{PS}. We have observed no hysteresis within the resolution of our measurements (phase separation occurs between two temperature points separated by 5 K) indicating that this is not the result of supercooling.

We now examine the possibility that items ii) and iii) above can be understood to be the result of a sudden decrease in oxygen mobility immediately below T_{PS}. At this point, we do not have direct information concerning the temperature dependence of oxygen mobility in these compounds. As a consequence, we cannot rule out this possibility. However there are indications that this may not be the cause of the near constancy of δ_{AF}. Neutron diffraction linewidths[12] indicate that the domain sizes for the two phases are quite large: ~1000Å. This means that the mobility of the excess oxygen is exceptionally large at T_{PS}. It seems unlikely that oxygen mobility drops so dramatically immediately below T_{PS} that it could completely prevent further changes in oxygen concentration.

Nuclear spin-lattice relaxation measurements have the potential to determine oxygen mobility. These measurements have not yet been completed in the oxygen-annealed compounds, however they have been made in stoichiometric material.[11] ^{139}La nuclear spin-lattice relaxation is known to be due to fluctuations of the EFG as well as magnetic fluctuations. It is reasonable to expect that the highly mobile excess oxygen would have a considerable effect on ^{139}La nuclear spin lattice relaxation rate because the interstitial oxygen significantly alters a La-O bond distance for eight La nuclei[13] in its vicinity Thus for δ=0.01, 4% of La nuclei will experience a strongly altered EFG. Rapid motion of this interstitial should have a significant effect on ^{139}La nuclear spin lattice relaxation. If oxygen mobility is decreasing rapidly, the average length of time (correlation time τ_c) a ^{139}La nucleus experiences an altered EFG will increase. At the temperature where oxygen mobility becomes so slow that $1/\tau_c$ becomes comparable to the nuclear Larmor frequency one might expect a feature in the ^{139}La nuclear spin lattice relaxation rate. The relaxation rate observed in stoichiometric $La_2CuO_{4+\delta}$ is smoothly and monotonically decreasing below 280 K down to low temperatures where magnetic relaxation becomes dominant. This monotonic decrease suggests that there is no sudden loss of oxygen mobility. Rough determinations of this relaxation rate in the oxygen-annealed samples indicate no strong changes in this temperature range. We believe this to be evidence against the proposition that oxygen mobility is suddenly lost at T~225 K~ $0.9T_{PS}$.

Based on entropy considerations alone one expects a finite δ_{AF} at finite temperature, but these considerations become less important as the temperature approaches zero. Our determination of δ_{AF} becomes insensitive at low temperature since $\nu_Z(T=0)$ is entirely independent of δ_{AF}. The trend of the data at higher temperatures indicates, however, that the system does prefer to maintain a non-zero δ_{AF}. This is consistent with the observation that for low levels of excess oxygen the three-dimensional antiferromagnetic ordering temperature is only slightly suppressed ($\Delta\delta$ of ~.014 suppresses T_N by less than 8%).[6] The rate of suppression of T_N becomes increasingly rapid as δ is increased further. Our results suggest that, in the oxygen-annealed compound, the cost of expelling a low level of dopants from the antiferromagnetic phase is not compensated by its increased stabilization.

In order to gain some knowledge of the sample to sample variation of $\delta_{AF}(T)$ we studied two different samples. The crystals were grown in different growths and high pressure oxygen annealed in separate runs. As is evident in Fig. 2, the temperature dependence of δ_{AF} for these two samples is very similar as is the phase separation temperature. We note however that samples having a range of phase separation temperatures have been reported.[12,17] We will discuss the relationship between δ_{AF} and T_{PS} and possible conclusions concerning the sample dependence of δ_{AF} below.

The data in Fig. 2 support the following description of phase separation. A single homogeneous phase will persist with decreasing temperature down to a temperature sufficiently below T_N^0 that phase separation will produce an antiferromagnetic phase with sublattice magnetization greater than some critical value which is approximately 1/2 the low temperature value. Thus T_{PS} is determined by the equation $\nu_Z(T_{PS}) \approx (1/2)\nu_Z(T=0)$. Upon cooling through T_{PS}, phase separation occurs producing a discontinuous change in the excess oxygen content of both the antiferromagnetic and metallic phases. Cooling below T_{PS}, δ_{AF} continues to decrease slowly until $T \approx 0.9T_{PS}$; below this temperature δ_{AF} maintains the constant value δ_{AF}^0.

From this point of view, δ_{AF}^0 is the crucial physical parameter because it determines T_N^0. δ_{AF} varies just below T_{PS}, causing T_N^* to decrease by ~7% on cooling from T_{PS} to $0.9T_{PS}$. The relationship between T_N^0 and T_{PS} is very similar for the two samples: $T_{PS}/T_N^0 \cong 0.9$. Thus, in our samples, phase separation always occurs at a temperature about 90% of T_N^0 and the values of T_N^0 are the same within the uncertainty. Values of T_{PS} [12] greater than T_N^0~280 K require the unexpected result that δ_{AF} would have to increase with decreasing temperature below T_{PS} or that δ_{AF}^0 is a sample dependent quantity. A complete understanding of this question requires that $T_N^*(T)$ be determined in those samples having exceptionally high values of T_{PS}.

The Néel transition in nearly stoichiometric $La_2CuO_{4+\delta}$ is strongly first order.[10,11] The development of the sublattice magnetization in oxygen-annealed materials is also discontinuous as a consequence of phase separation. The similar

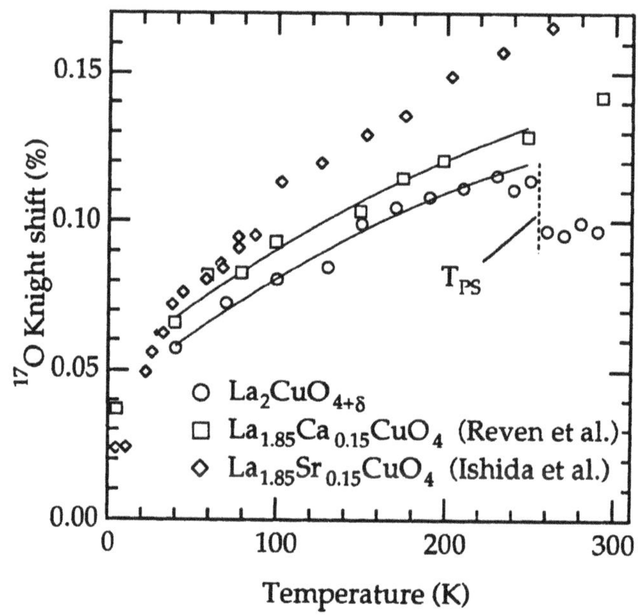

Fig. 3 Comparison of the planar oxygen Knight shift in lanthanum cuprate measured with the applied field perpendicular to the CuO_2 planes. Three sets of data are presented: Reven et al. [ref. 20], Ishida et al. [ref. 21] and this work. The planar spin susceptibility is sensitive to the level of doping. The temperature dependence of oxygen-annealed $La_2CuO_{4+\delta}$ is very similar to that of the Ca doped sample except very near T_{PS}. This allows us to estimate the temperature dependence of δ_M in the oxygen-annealed compound.

behaviors of $\nu_Z(T)$ in this respect suggests that phase separation may also occur in the nearly stoichiometric materials. This would produce such a small volume of metallic phase as to be very difficult to observe. The observation[19] of superconductivity (T_c~40 K) in a very small volume of these stoichiometric samples supports the idea of a tiny fraction of metallic phase. In this picture, $T_N^*(T \to 0)$ in these samples would be higher in less doped samples. Thus we would have to conclude that $\delta_{AF}(T \to 0)$ is a sample dependent quantity. The discontinuous change in the sublattice magnetization in nearly stoichiometric materials[10,11] is 25% smaller ($\sim 0.34\, \nu_Z(T=0)$) than is observed in oxygen-annealed $La_2CuO_{4+\delta}$ ($\sim 0.45\, \nu_Z(T=0)$).

3.4 *Metallic phase*

We turn now to a discussion of the properties of the metallic phase. In particular we will discuss i) the temperature dependence of the excess oxygen content δ_M; ii) the observation that slowly cooling the sample through a narrow temperature range centered around 190 K increases the superconducting transition temperature by ~4 K (14%); and iii) aspects of the (temperature dependent) structure of the metallic phase.

We compared the temperature dependence of the sublattice magnetization in the oxygen-poor phase to that in stoichiometric $La_2CuO_{4+\delta}$ to determine the temperature dependence of δ_{AF}. In the metallic phase we employ a similar strategy relying on the known dependence of the spin susceptibility of the CuO_2 planes on the doping. By comparing the temperature dependence of the susceptibility in the metallic phase to that of a compound for which the doping is constant we infer the temperature dependence of the doping and thus δ_M. This technique has the attractive virtue of being exclusively sensitive to the doping of the CuO_2 planes rather than the average doping of the metallic phase.

In Fig. 3 we show the temperature dependence of the total ^{17}O Knight shift (applied field perpendicular to the CuO_2 planes) from oxygen-annealed $La_2CuO_{4+\delta}$ and two metal-cation doped compounds: $La_{1.85}M_{0.15}CuO_4$, where M=Sr[21] or Ca[20]. Because the metal cations are not mobile below room temperature, the average doping is temperature independent; we assume here that planar doping is also constant (ie, that there is no temperature dependence below room temperature to the charge transfer from the LaO intergrowth layers to the CuO_2 planes).

The nominal dopings of the metal cation doped compounds are identical yet the Knight shifts are not. The most probable explanation is that the actual dopings differ slightly (indicating the sensitivity of Knight shift to doping). Except for a small offset, the temperature dependence of the spin susceptibility of $La_{1.85}Ca_{0.15}CuO_4$ is very similar to that of oxygen-annealed $La_2CuO_{4+\delta}$. From this we conclude that throughout the temperature range where the temperature dependences are the same δ_M is not changing. To obtain a quantitative estimate of δ_M we rely on recent measurements of Chou et al.[22] They show that the temperature dependent susceptibilities of $La_{1.85}Sr_{0.15}CuO_4$ and $La_2CuO_{4.11}$ are very similar, and they conclude from this that the doping of these compounds is the same. We estimate the temperature dependence by making the admittedly crude assumption that spin susceptibility varies linearly with doping. If we assume that the smaller spin susceptibility in the Ca doped material in Fig. 3 indicates a slight sub-stoichiometry relative to the nominal doping, then we can associate the Knight shift in this material with $\delta_M = 0.10$. For $T>T_{PS}$, $\delta = 0.03$.[13] Then a linear interpolation provides the results shown for the right hand branch in Fig. 2. As the doping decreases, the spin susceptibility seems to becomes less temperature dependent at the higher temperatures. Thus our scheme may overestimate the temperature dependence of δ_M just below T_{PS}.

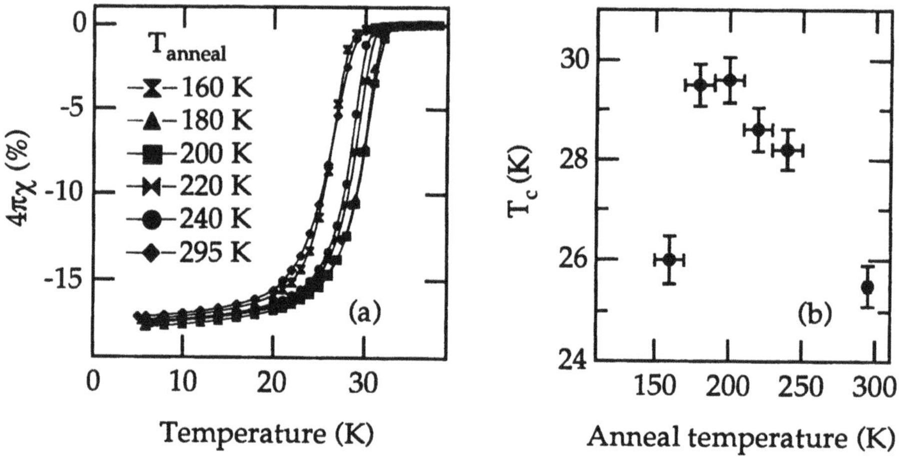

Fig. 4 Oxygen-annealed $La_2CuO_{4+\delta}$ was slowly cooled through 20 K intervals centered about several temperatures. The various curves in panel (a) show the zero field cooled susceptibility for the different temperature ranges through which the sample was slow cooled. The horizontal error bars in panel (b) indicate the width of this temperature range; the vertical error bars indicate the temperature width for the susceptibility fall from 40% to 60 % of the full $4\pi\chi$. T_c is peaked for an annealing temperature of ~200 K.

4. Structure and Cooling Rate Dependence of T_c

4.1 *Dependence of T_c on cooling rate*

The unexpected nature of the phase diagram of oxygen-annealed $La_2CuO_{4+\delta}$ motivated us to look for some underlying process which could be playing a role in phase separation. This led us to undertake an examination of the temperature dependence of the local structure of the metallic phase insofar as it can be studied by magnetic resonance techniques. Our NMR studies[17] indicate that the structure (octahedral tilt) of the metallic phase becomes temperature dependent just below T_{PS}. We will discuss a study of the effect of cooling rate on superconducting T_c and local structure in the LaO layer which was undertaken to understand these features.

As a first step we employed bulk magnetization to study the dependence of the superconducting transition temperature on the rate of cooling (such as have been observed earlier[23]) above T_c. We found T_c to be quite sensitive to this rate. T_c was shifted by more than 4 K, but there was little change in the width of the transition and the total shielding fraction. Furthermore, we found it necessary to cool slowly in a narrow range of temperature only--one well separated from both T_{PS} and T_c.

These measurements are shown in Fig. 4(a) which displays a family of curves representing the ZFC susceptibility obtained after the sample was slowly cooled through various 20 K temperature intervals centered about the temperatures listed. In Fig. 4(b) the dependence of T_c on the center temperature in that interval is shown. This curve is peaked when the sample is slowly cooled through a band of temperatures near 200 K. The striking feature is that annealing at temperatures higher than 210 K actually produces a lower T_c if one then passes rapidly through the critical temperature band. This strongly suggests that a change occurs around 200 K which requires some time to proceed to the optimum condition. There are at least two possibilities: there may be a change in the equilibrium structure or the excess interstitial oxygen may abruptly freeze into its preferred lattice position. In Section 3.3 we discussed our reasons for believing that oxygen motion is not lost at T~200 K, however, we are not able to rule out the possibility of a freezeout of oxygen mobility at the present time.

4.2 *Metallic phase structure*

The NQR spectrum is extremely sensitive to small changes in the positions of the oxygen ions in the LaO layer. We have previously studied[24] the NQR spectrum obtained from this compound; here we pay particular attention to gaining a microscopic understanding of the changes which the cooling rate studies indicate are occurring in the vicinity of 200 K. We find that there is a clear change in the spectrum which occurs between 190 K and 200 K.

In Fig. 5(a) we show the NQR spectra in the frequency range near the antiferromagnetic peaks for $T = 190\ K$ and $T = 200\ K$. There is a marked decrease in the intensity of the sharp central peak associated with the metallic, non-magnetic phase. The decrease in this intensity below 200 K is accompanied by the development of a peak at lower frequency (~18 MHz). This peak is evident in Fig. 5(b).

It is clear that local, microstructural features can have an important influence on superconductivity and T_c. Neutron scattering results[25] in $YBa_2Cu_3O_{6.5}$ show that chain oxygens are ordered, supporting the view[26] that such order is necessary to facilitate charge transfer to the CuO_2 planes. Annealing experiments[26] in oxygen deficient $YBa_2Cu_3O_{7-y}$ support this view; here the ordering is thought to be occurring on the local or mesoscopic scale. Recent experiments on $La_{2-x-y}Nd_yM_xCuO_4$ materials[27,28] have shown that changes in the pattern and degree of tilting of the oxygen octahedra can dramatically suppress superconductivity. The substantial mismatch between the Cu-O and La-O bond-lengths leaves the CuO_2 planes under compression. One of the mechanisms for relieving this compression is a rotation of the oxygen octahedra, such as occurs at the tetragonal-to-orthorhombic transition. Transfer of a hole from the La-O to the CuO_2 layers also reduces this compression.[29] Evidence from x-ray[30] and neutron scattering[31] experiments for various local distortions has been accumulating. Thus one can imagine that the changes we observe in the local oxygen structure in oxygen-annealed $La_2CuO_{4+\delta}$ reflect the response of the system to an increasing (with decreasing temperature) bond-length mismatch. This would alter the charge transfer to the planes and thus the superconducting T_c.

The NQR data are not sufficient to completely determine the local structure in the vicinity of the ^{139}La probe. However, the evolution of the NQR spectra which we observe is consistent with a change in the degree of the octahedral tilt or the axis about which that tilt occurs. A distribution of Cu-O(apical) distances[30,32] would also account for these results. Such changes could explain both the NQR results and the clear relationship to superconducting properties if the structural changes altered the charge transfer from the LaO layer to the planes. The large width of the NQR line at low temperature is indicative of substantial disorder in this structure or a period for the structural modification which exceeds the lattice parameter thus effectively producing a number of inequivalent La sites. Direct evidence for changes in the tilt of the EFG comes from NMR measurements.

In contrast to NQR measurements, the shift of the NMR resonance[33] (in an

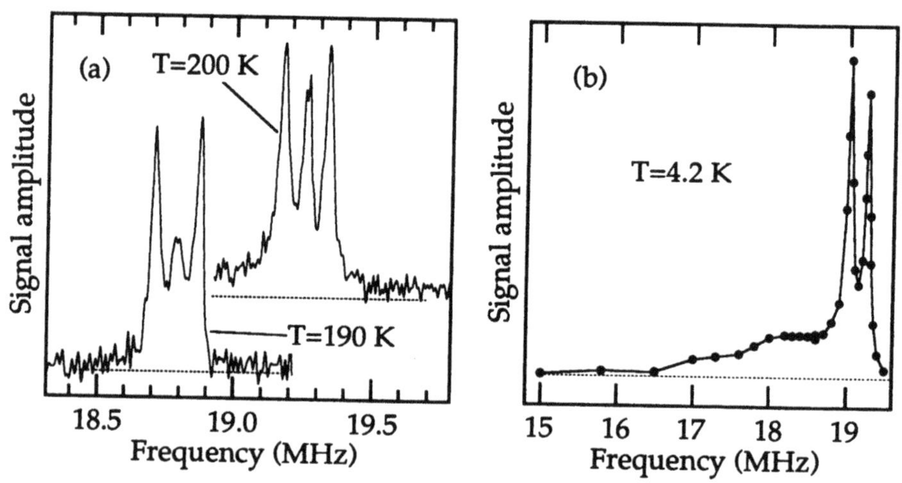

Fig. 5 ^{139}La NQR spectra taken at various temperatures. (a): Fourier transform spectra taken in the vicinity of the critical annealing temperature determined in cooling rate studies. The central peak is due to the metallic phase; the outer peaks originate in the antiferromagnetic phase. The intensity of the sharp metallic peak decreases considerably between 200 and 190 K. The upper spectrum (200 K) has been offset by 0.5 MHz for clarity. (b): A low temperature frequency swept spectrum: the metallic intensity has almost entirely moved into a broad, downshifted peak.

applied field large compared to ν_Q/γ) is sensitive the angle θ between the applied magnetic field and the direction of the dominant EFG (defined to be along the z-axis of the principal axis coordinate frame of the EFG), V_{ZZ} through the second order quadrupole shift,

$$\Delta\nu: \ \Delta\nu = \frac{15}{16}\frac{\nu_Q^2}{\nu_Z}\left(1-\cos^2\theta\right)\left(1-9\cos^2\theta\right)$$

In our NMR measurements with the field aligned along the axis perpendicular to the CuO_2 planes, we are thus able to directly and sensitively measure the deviation of the direction of V_{ZZ} from the c-axis which occurs when the octahedra rotate. The change in the orientation of V_{ZZ} is directly related to the tipping angle of the oxygen octahedra away from the axis perpendicular to the planes (in particular, untilted octahedra will produce an untilted V_{ZZ}). We find that the octahedra are untilted for temperatures just below T_{PS} , but that when the sample is cooled to 40 K a tilt of ~10° has developed. For small tipping angles,

$$\Delta\nu(\theta) \propto -\frac{15}{4}\frac{\nu_Q^2}{\nu_Z}\theta^2.$$

Under certain conditions such as spatially incoherent octahedral tilts or sufficiently rapid temporal fluctuations for which $\langle\theta\rangle = 0$, a probe such as x-ray or neutron diffraction will give a result of $\theta = 0$. However an NMR measurement is sensitive to $\langle 1-\cos^2\theta\rangle$. Thus, a vanishing shift cannot be understood as resulting from a spatial or temporal average of tilted octahedra--it demonstrates that the individual octahedra are untilted.

5. Discussion

5.1 *Excess oxygen content below the phase separation temperature*

The unusual temperature dependence of the excess oxygen content (Fig. 2) in oxygen-annealed $La_2CuO_{4+\delta}$ is provocative. A variety of behaviors in the related $La_{2-x-y}Nd_yM_xCuO_4$ compounds[27,28] are observed having to do with temperature dependent tilts of the octahedra. These structural effects are known to have a strong effect on superconductivity. One can expect on general grounds having to do with the bond-length mismatch[29] between the LaO and the CuO_2 layers that these changes will also be related to local charge transfer between these two substructures in the lattice. The peculiar nature of the phase separation diagram leads one to consider the possibility that the macroscopic phase separation observed in oxygen-annealed $La_2CuO_{4+\delta}$ may be influenced by some underlying structural or local charge-transfer process which remains unrecognized.

The clear relationship between macroscopic phase separation and the onset of three-dimensional long range antiferromagnetic order in the oxygen-poor phase suggests the possibility that phase separation is stabilized by the magnetic order which develops in a portion of the sample if most of its excess oxygen is expelled. The onset of sub-lattice magnetization in nearly stoichiometric $La_2CuO_{4+\delta}$ is discontinuous as it is in the oxygen-annealed compounds. One way of understanding this discontinuity in the nearly stoichiometric compounds is to postulate that phase separation is occurring here as well. This would form a very small volume of metallic (oxygen-rich) phase which would be hard to detect. Its existence is supported, however, by the observation of trace superconductivity in stoichiometric materials.

5.2 *Cooling rate dependence of the superconducting transition temperature*

The peak in the superconducting T_c which is achieved with slow cooling through the temperature range from 210 K to 180 K provides clear evidence for the onset of some process in this temperature region. The increase in T_c with annealing time demonstrates that optimum completion of this process is advantageous for superconductivity. The fact that the distribution of T_c's and the shielding remains constant suggests that the change is homogeneous on the scale of the superconducting coherence length. This particular value of temperature is striking in that it is well removed from both T_{PS} and T_c. This indicates that although it may involve motion of oxygen or doped holes it is not directly related to the phase separation process which occurs at T_{PS} and is nearly complete by the time $T = 0.9T_{PS}$

5.3 *Temperature dependence of local structure as reflected in NQR and NMR*

We have observed a shift in spectral weight in the NQR spectrum which takes place between 200 and 190 K. With decreasing temperature the intensity of the line originating from a metallic phase having a structure very similar to that of the stoichiometric material (but having no magnetic order) decreases substantially. Simultaneously a peak due to metallic phase having an altered structure appears. At low temperatures, the integrated area of this peak is comparable to that of the two antiferromagnetic peaks. With decreasing temperature the EFG axis rotates away from the c-axis and there develops a considerable distribution in the degree of rotation ($0 < \theta < 15°$). One explanation for the observed spectra is that the change occurring at 195 K involves the rotation of the CuO_6 octahedra.[27,28] The broadened lines would suggest that this rotation is about an axis which may vary from unit cell to unit cell. A second and appealing possibility is suggested by the observation by x-ray absorption measurements[30] of a modulation of the Cu-O apical oxygen distance with a 27Å period along the a-axis in the Bi 2:2:1:2 compound. Such a modulation in the compound we are studying would result in a rotated EFG at the La site and modulation distance of several lattice parameters would produce a large number of inequivalent La sites thus accounting for the distribution of rotation angles and the broad NQR line.

6. Conclusions

We report the temperature dependence of the doping below the phase separation temperature in oxygen-annealed $La_2CuO_{4+\delta}$. Exploiting the extreme sensitivity of the antiferromagnetic ordering temperature in $La_2CuO_{4+\delta}$ to the value of δ, we have determined the temperature dependence of the doping of the oxygen-poor (antiferromagnetic) phase of oxygen-annealed $La_2CuO_{4+\delta}$. In a similar fashion the doping dependence of the planar susceptibility has been used to estimate the temperature dependence of the doping of the CuO_2 planes in the metallic phase. The results are contrary to expectations and suggest that phase separation is the result of other changes in the material. We have explored the possibility that structural changes associated with the tendency toward rotation of the oxygen octahedra may be involved. Such considerations are supported by the evolution of the NQR and NMR spectra which indicate that such structural changes are occurring. The temperature at which these changes set in coincides with an optimum annealing temperature for producing the highest superconducting transition temperature in the material.

At phase separation, the doping of both the oxygen-poor and oxygen-rich phase occurs. Once phase separation has occurred, the doping changes very little except for a small change just below T_{PS}. This feature suggests that the phase separation temperature is determined by the low temperature value of δ_{AF} through the dependence of T_N^* on δ_{AF}. Once the system has cooled sufficiently below the T_N^* which corresponds to $\delta_{AF}(T_{PS})$, a discontinuous decrease in the doping occurs in certain regions of the sample allowing a substantial sublattice magnetization (approximately 1/2 of the zero temperature value) to develop. In this view, the ability to develop an ordered antiferromagnet with a substantial magnetization is a driving force behind phase separation.

The recent demonstration that the appearance of sub-lattice magnetization in stoichiometric $La_2CuO_{4+\delta}$ is also discontinuous leaves open the possibility that even in these materials magnetic ordering is the result of phase separation. If this scenario is borne out, a reconsideration of the nature of the magnetic ordering may be in order.

We gratefully acknowledge enlightening discussions with R.H. Heffner. This work was performed under the auspices of the U.S. Department of Energy. One us (ETA) acknowledges the support of the University of California Institutional Collaborative Research (INCOR) Program in High Temperature Superconductivity and the Associated Western Universities (AWU). One of us

(D.E.M.) acknowledges support through the U.S. National Science Foundation, Grants no. DMR-8814783 and DMR-9114911.

† Permanent address: Department of Physics, University of California, Riverside, CA 92521.

1. V. J. Emery, this volume; C. DiCastro, this volume; V. J. Emery, S.A. Kivelson, and H. Q Lin, *Phys. Rev. Lett.* **64**, 475 (1990); M. Grilli, R. Raimondi, C. Castellani, and C. Di Castro, *Phys. Rev. Lett.* **67**, 259 (1991); M. Marder, N. Papanicolaou, and G. C. Psaltakis, *Phys. Rev. B* **41**, 6920 (1990).
2. A. J. Heeger, this volume; G. Yu, H. Lee, A. J. Heeger, N. Herron, and E. M. McCarron, *Phys. Rev. Lett.* **67**, (1991).
3. J.E. Schirber, B. Morosin, R.M. Merrill, D.E. Hlava, E.L. Venturini, J.F. Kwak, P.J. Nigrey, B.J. Baughman, and D.S. Ginley, *Physica C* **158**, 183 (1989).
4. M.F. Hundley, J.D. Thompson, S-W. Cheong, Z. Fisk, and J.E. Schirber, Phys Rev. B **41**, 4062 (1990).
5. For an review of nuclear quadrupole resonance (NQR) see: T. P. Das and E. L. Hahn, "Nuclear Quadrupole Resonance Spectroscopy", in *Solid State Phys.*, edited by F. Seitz and D. Turnbull (Academic Press, New York, 1958), suppl. 1, chap. I.
6. D. C. Johnston, S. K. Sinha, A. J. Jacobson, and J. M. Newsam, *Physica C* **153-155**, 572, (1988).
7. J. Saylor, L. Takacs, C. Hohenemser, J. I. Budnick, and B. Chamberland, *Phys. Rev. B* **40**, 6854 (1989).
8. H. Nishihara, H. Yasuoka, T. Shimuzu, T. Tsuda, T. Imai, S. Sasaki, S. Kanbe, K. Kishio, K. Kitazawa, and K. Fueki, J. Phys. Soc. Jpn. **56**, 4559 (1987).
9. I. Watanabe, Y. Nakamura, K. Kumagai, *Physica C* **152**, 262 (1988).
10. V.A. Borodin, V.D. Doroshev, Y.M. Ivanchenko, M.M. Savosta, A.E. Filippov, *JETP Lett.* **52**, 469 (1991).
11. D.E. MacLaughlin, J.P. Vithayathil, H.B. Brom, J.C.J.M. de Rooy, P.C. Hammel, P.C. Canfield, A.P. Reyes, J.D. Thompson, and S-W. Cheong, to be published.
12. J. D. Jorgensen B. Dabrowski, S. Pei, D. G. Hinks, L. Soderholm, B. Morosin, J. E. Schirber, E. L. Venturini, and D. S. Ginley, *Phys. Rev. B* **38**, 11337 (1988).
13. C. Chaillout, S-W. Cheong, Z. Fisk, M. S. Lehmann, M. Marezio, B. Morosin, and J. E. Schirber, *Physica C* **158**, 183 (1989).
14. J. Zhou, S. Sinha, and J. B. Goodenough *Phys. Rev. B* **39**, 12331 (1989).
15. E. J. Ansaldo, J.H. Brewer, T.M. Riseman, J.E. Schirber, E.L. Venturini, B. Morosin, and D.S. Ginley, and B. Sternlieb, *Phys. Rev. B* **40**, 2555 (1989).

16. J. Ryder, P.A. Midgley, R. Exley, R.J. Beynon, D.L. Yates, L. Afalfiz, and J.A. Wilson, *Physica C* **173**, 9 (1991).
17. P.C. Hammel, A.P. Reyes, Z. Fisk, M. Takigawa, J.D. Thompson, R.H. Heffner, S-W. Cheong and J. E. Schirber, *Phys. Rev. B* **42**, 6781 (1990).
18. B. Dabrowski, D.G. Hinks, J.D. Jorgensen, and D.R. Richards, *Mat. Res. Soc. Symp. Proc.*, **156**, 69 (1989).
19. M. Mehring, this volume; R.K. Kremer, E. Sigmund, V. Hizhnyakov, F. Hentsch, A. Simon, K.A. Müller, and M. Mehring, *Z. Phys. B-Condensed Matter*, **86**, 319 (1992).
20. L. Reven, J. Shore, S. Yang, T. Duncan, D. Schwartz, J. Chung, and E. Oldfield, *Phys. Rev. B* **43**, 466 (1991).
21. K. Ishida, Y. Kitaoka, G-Q. Zheng, and K. Asayama, *J. Phys. Soc. Jpn.* **60**, 3516 (1991).
22. F.C. Chou, J.H. Cho, and D.C. Johnston, to be published, Physica C.
23. See also Y. Oda, A. Sumiyama, T. Kohara, M. Yamada, K. Asayama, S. Kashiwai, and M. Motoyama, *Physica C* **185-189**, 917 (1991).
24. P.C. Hammel, E.T. Ahrens, A.P. Reyes, R.H. Heffner, P.C. Canfield, S-W. Cheong, Z. Fisk, and J.E. Schirber, *ibid*, 1095 (1991).
25. P. Burlet et al., this volume.
26. J.D. Jorgensen, D.G. Hinks, P.G. Radaelli, S. Pei, P. Lightfoot, B. Dabrowski, C.U. Segre, and B.A. Hunter, *Physica C* **185-189**, 185 (1991) and B.W. Veal et al. *Phys. Rev. B* **42**, 6305 (1990).
27. J.D. Axe, A.H. Moudden, D. Hohlwein, D.E. Cox, K.M. Mohanty, A.R. Moodenbough, and Y. Xu, *Phys. Rev. Lett.* **62**, 2751 (1989).
28. M.K. Crawford, W.E.Farneth, R.L. Harlow, E.M. McCarron, R. Miao, H. Chou, and Q. Huang, *Lattice Effects in High-T_C Superconductors*, ed. by Y. Bar-Yam, World Scientific Pub., Singapore, 1992.
29. J. B. Goodenough in *Superconductivity and its Applications*, AIP Conference Proceedings, p. 219, ed. by Y.H. Kao, P. Coppens, H. Kwok, Buffalo, NY 1990.
30. A. Bianconi, this volume and A. Bianconi, S. Della Longa, M. Missori, I. Pettiti, and M. Pompa, to be published in *Lattice Effects in High-T_C Superconductors*, ed. by Y. Bar-Yam, World Scientific Pub., Singapore, 1992.
31. S.J.L. Billinge and T. Egami, *ibid*.
32. J. Mustre de Leon, S.D. Conradson, I. Batitstic, A.R. Bishop, I.D. Raistrick, M.C. Aronson, and F.H. Garzon, *Phys. Rev. B* **45**, 2447 (1992).
33. M. H. Cohen and F. Reif, *Solid State Phys.* **5**, 321 (1957).

PHASE SEPARATION IN $La_{2-x}Sr_xCuO_4$ AND $YBa_2Cu_3O_x$ STUDIED BY MOSSBAUER SPECTROSCOPY

P. IMBERT, J.A. HODGES, G. JEHANNO, P. BONVILLE and C. GARCIN
Service de Physique de l'Etat Condensé,
Département de Recherche sur l'Etat Condensé, les Atomes et les Molécules,
Centre d'Etudes de Saclay, 91191 Gif-sur-Yvette Cedex, France

ABSTRACT

In a first part, we present a Mössbauer study made on ^{57}Fe impurities substituted for Cu in the series $La_{2-x}Sr_xCuO_4$ for $0 \leq x \leq 0.30$. In a second part, we present some results of a Mössbauer study made on ^{170}Yb impurities substituted for Y in the series $YBa_2Cu_3O_x$ for $6 \leq x \leq 7$. In both series, we have studied at low temperature the competition between magnetism and superconductivity on a microscopic scale and we have evidenced the coexistence of two fractions showing different local behaviours, over part of the explored x range. The first fraction corresponds to local Cu spin freezing, whereas the second fraction is attributed to local superconducting behaviour with no evidence of intrinsic Cu magnetic moments.

1. ^{57}Fe Doped $La_{2-x}Sr_xCuO_4$ Samples[1]

In the $La_{2-x}Sr_x(Cu_{0.995}\,{}^{57}Fe_{0.005})O_4$ series, the isomer shift value, as well as the saturated magnetic hyperfine field value measured at low temperature on the Mössbauer isotope ^{57}Fe, are characteristic of the high spin ($S = 5/2$) Fe^{3+} charge state. This ion, which is almost magnetically isotropic, is a suitable magnetic probe for both identifying local magnetic ordering in the CuO_2 planes and following its magnetic orientation. Throughout the series, we observed, at room temperature, a quadrupole doublet whose characteristics are consistent with a substitution of the $^{57}Fe^{3+}$ probe in the octahedral Cu site. When the Sr doping x is varied in the series, both the isomer shift ε and the quadrupole splitting Δ are independent of x in the semiconducting composition range ($0 \leq x \lesssim 0.06$). For larger x values, when the holes become mobile, both ε and Δ decrease quasi linearly when x increases, so reflecting the increasing concentration of delocalized carriers in the CuO_2 planes.

At 1.5 K, magnetic hyperfine spectra are observed on ^{57}Fe for all x values, but they look different depending on whether x is smaller or larger than about 0.16 (Figure 1). This change in the response of the probe actually reflects a crossover between two competing magnetic regimes: for $x < 0.16$, the dominant regime is that of a spontaneous Cu spin ordering (or freezing), whereas for $x > 0.16$ the dominant regime is that of iron-induced local magnetic ordering within a Cu sublattice which is not intrinsically magnetic. We assign respectively these two regimes to a semiconducting phase and to a superconducting (or normal metallic) phase. The composite character of the spectra for $x = 0.14$ and 0.16 reveals that both fractions coexist for intermediate compositions.

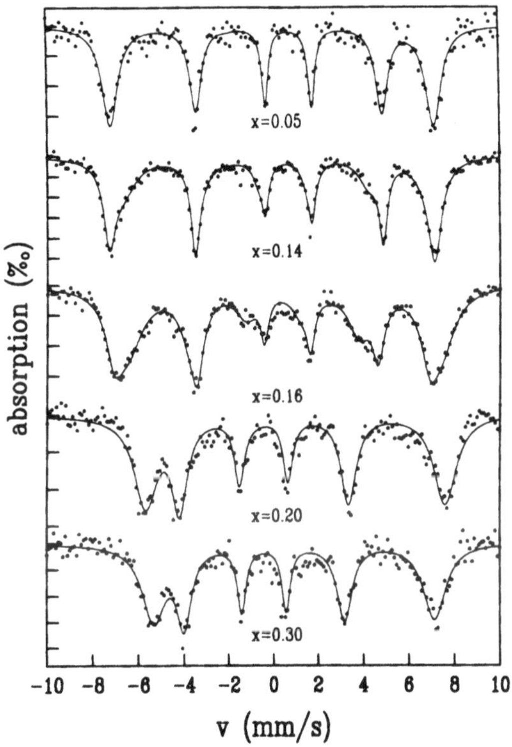

Figure 1: ^{57}Fe Mössbauer absorption spectra at 1.5 K in $La_{2-x}Sr_xCu_{0.995}Fe_{0.005}O_4$ for different x values. Each spectrum is fitted to a single hyperfine sextet except for $x = 0.14$ and 0.16 where two sextets are used.

In order to discuss these properties in more detail in connection with the phase diagram, we present in Figure 2 the x dependence of the following parameters: first, the local magnetic ordering temperature T_m experienced by the Mössbauer probe, and the superconducting critical temperature T_c obtained from the resistivity measurements; secondly, the saturated magnetic hyperfine field H_0 on the ^{57}Fe probe; thirdly, the angle $\theta = \left(\overline{OZ}, \overline{H}_0\right)$ between H_0 and the main axis OZ of the electric field gradient at the Fe site.

The semiconducting composition range ($0 < x < 0.06$) can be divided in two parts labelled a and b in Figure 2. The zone a ($0 < x < 0.02$), where T_m (or T_N)

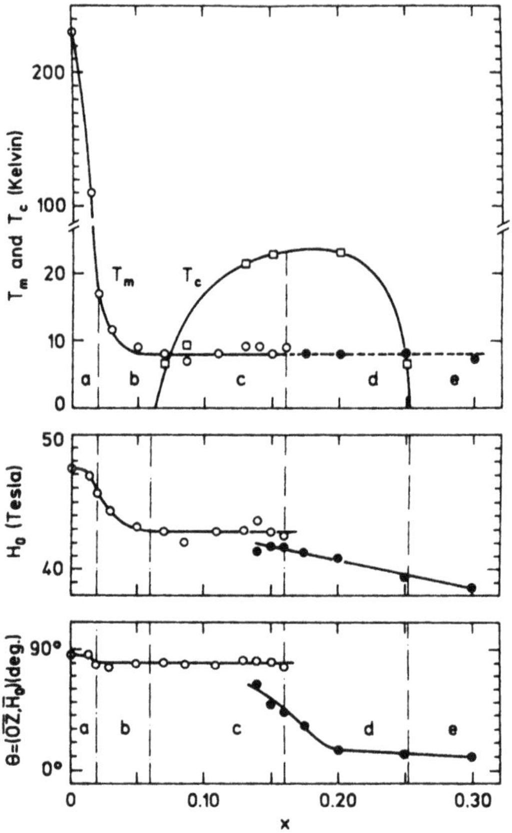

Figure 2: ^{57}Fe Mössbauer and resistivity measurements in $La_{2-x}Sr_xCu_{0.995}Fe_{0.005}O_4$ versus x.
Top: superconducting critical temperature T_c (squares) and magnetic freezing temperature T_m (circles). For $x < 0.16$, T_m is quasi independent of the Fe concentration. For $x > 0.16$, T_m is Fe concentration dependent.
Middle: saturated hyperfine magnetic field H_0.
Bottom: angle $\theta = \left(\overline{OZ}, \overline{H}_0\right)$ between H_0 and the main electric field gradient axis OZ.

In these figures, open circles and full circles are respectively measured in the phase with spontaneous Cu spin freezing (zones a, b, c) and in the phase with Fe-induced local spin freezing (zones c, d, e).

drops abruptly for increasing x, corresponds to 3D antiferromagnetic ordering. The zone b ($0.02 < x < 0.06$), where T_m decreases to slightly below 10 K, is assigned to a Cu spin glass phase. In both zones a and b the angle θ remains between 80° and 90° which is the value in La_2CuO_4, showing that the Cu magnetic moments remain essentially oriented within the basal plane $(\bar{a}, \bar{b})$.

The superconducting composition range ($0.06 < x < 0.25$) can be separated (about the value $x = 0.16$) in two roughly equivalent parts c and d, where the magnetic behaviours are quite different. In the first part (zone c, $0.06 < x < 0.16$), the local properties experienced by the ^{57}Fe probe look essentially similar to those observed within the preceding semiconducting zone b and we conclude that a dominant Cu spin glass fraction persists for these x values. In the remaining part of the superconducting region (zone d, $0.16 < x < 0.25$), the value of θ drops, showing that the magnetic moment of the Fe probe is no longer oriented in the basal plane but is more oriented towards the $\bar{c}$ direction. Besides for samples in zone d, T_m decreases towards zero as the Fe doping is reduced. We conclude that there is no spontaneous Cu magnetic ordering in the zone, but that a local spin freezing is induced in the superconducting phase by the Fe impurities. Finally, in the normal metallic composition range (zone e, $x > 0.25$), the local magnetic properties are the same as in zone d.

Near the high x value limit of zone c, the analysis of the magnetic hyperfine spectra clearly confirms the coexistence of two fractions: see the θ and H_0 measurements in both fractions near $x = 0.15$ in Figure 2. The fraction with $\theta \simeq 80°$ (open circles) corresponds to persisting semiconducting Cu spin glass regions, whereas the other fraction (black circles) corresponds to superconducting regions without spontaneous Cu magnetic ordering. The Cu spin glass fraction vanishes for $x \simeq 0.20$ (Figure 3).

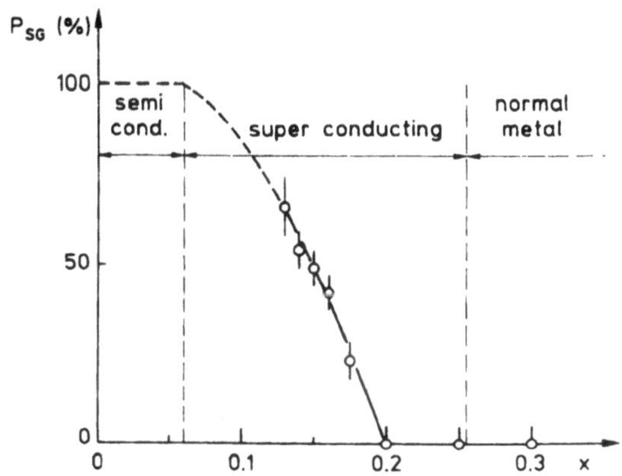

Figure 3: Relative volume $P_{SG}(\%)$ of the Cu spin glass fraction in $La_{2-x}Sr_x(Cu_{0.995}Fe_{0.005})O_4$ versus x from the ^{57}Fe Mössbauer measurements.

In order to obtain more information on the local perturbation induced by the Fe^{3+} impurities in the superconducting Cu sublattice, we have also performed Mössbauer *emission* spectroscopy measurements in $La_{2-x}Sr_xCuO_4$ samples for $x \geq 0.20$. In these experiments, extremely dilute $^{57}Fe^{3+}$ impurities are generated by the decay of the parent ^{57}Co diluted in the matrix. In the superconducting phase, these isolated Fe probes also experience a local molecular field directed near the $\bar{c}$ direction. This molecular field is assigned to the presence of neighbouring localized Cu^{2+} magnetic moments (with $g_c > g_a, g_b$) induced by the Fe impurities. Similar induced Cu^{2+} localized moments have also been detected near Zn impurities by EPR measurements in superconducting $La_{1.8}Sr_{0.2}CuO_4$ by Finkelstein et al.[2].

2. ^{170}Yb Doped $YBa_2Cu_3O_x$ Samples[3,4]

We have examined the magnetic properties of the Cu(2) sublattice in $YBa_2Cu_3O_x$ samples using $^{170}Yb^{3+}$ Mössbauer probes substituted for Y^{3+}. The substitution of a rare earth is non perturbative, as it does not introduce any charge unbalance in the system and it does not influence any superconducting properties. In these samples, the ground crystal field level of the Yb^{3+} ion is a very well isolated Kramers doublet with effective electronic spin $S' = 1/2$. At the temperatures of the measurements, the paramagnetic fluctuation rate of S' is well below the lower limit associated with the measurement technique ($2 \times 10^8 s^{-1}$) and the Mössbauer spectrum corresponds to the so-called slow paramagnetic relaxation lineshape. The Figure 4 shows some of these Mössbauer spectra, recorded at 1.4 K for various values of x.

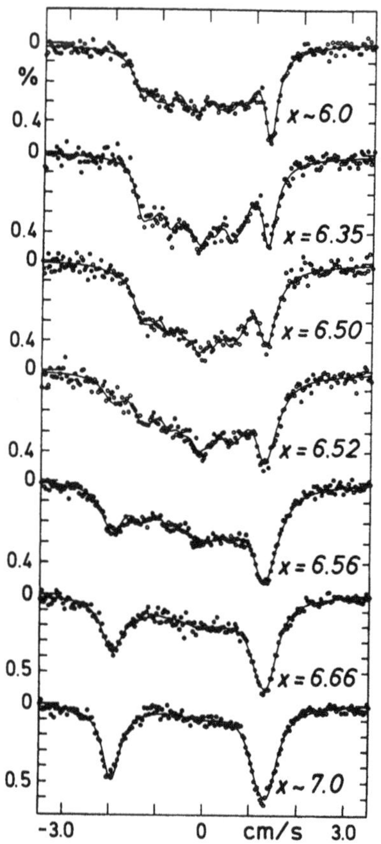

Figure 4: $^{170}Yb^{3+}$ Mössbauer absorption spectra at 1.4 K in $Y_{0.97}Yb_{0.03}Ba_2Cu_3O_x$ as a function of x. In the crossover region, two subspectra are visible: one evidencing a static Cu(2)-derived molecular field is due to the magnetic fraction, the other where no molecular field is visible is due to the superconducting fraction.

In the $x \sim 7.0$ superconducting sample, the $^{170}Yb^{3+}$ Mössbauer spectrum is made of two lines, corresponding to the two eigenstates of the electro-nuclear hamiltonian

$$\mathcal{H} = \bar{S}'.\tilde{A}.\bar{I} \tag{1}$$

in the excited nuclear level with spin $I = 2$ (in the nuclear ground level $I = 0$). The characteristics of this spectrum identify the $\tilde{A}$ tensor as being almost isotropic and close to that associated with a Yb^{3+} cubic Γ_7 ground state.

The lineshape obtained in the $x \sim 6.0$ antiferromagnetic compound is quite different. The hamiltonian

$$\mathcal{H} = \bar{S}'.\tilde{A}.\bar{I} - \beta \bar{H}_m.\tilde{g}.\bar{S}' \tag{2}$$

then includes an additional term due to the presence of a molecular field H_m of average value 1700 G acting on the Yb^{3+} electronic spin. This field results from superexchange and dipolar interactions with the magnetically ordered Cu(2) neighbours. For magneto-crystalline symmetry reasons, it was initially not expected that the sum of these interactions over the eight Cu neighbours would lead to a nonzero net result at the Yb site. The observation of a significant molecular field shows that some symmetry lowering is present at this site.

At intermediate x levels ranging from ~ 6.5 to about 6.7, the line fits show that both subspectra corresponding respectively to Eqs.(1) and (2) are simultaneously present. We interpret this result as indicating that some of the probes are in regions of the sample which show local magnetic behaviour (the relative volume of these regions is given in Figure 5) and some are in regions which show local superconducting behaviour.

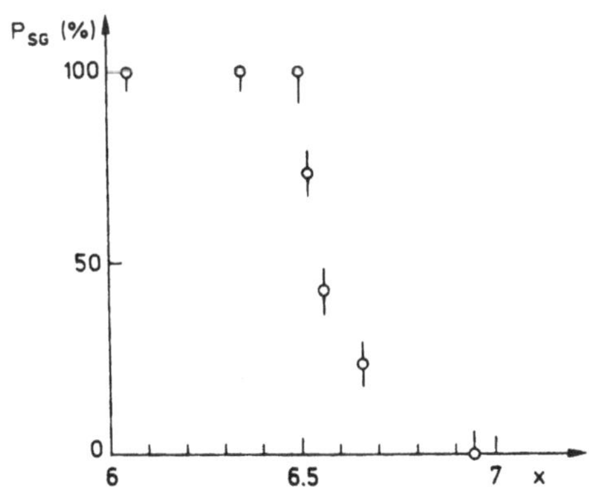

Figure 5: Relative volume $P_{SG}(\%)$ of the antiferromagnetic or spin glass like fraction in $(Y_{0.97}Yb_{0.03})Ba_2Cu_3O_x$ versus x, from the ^{170}Yb Mössbauer measurements.

In addition, we have observed the deformation of the $^{170}Yb^{3+}$ Mössbauer line-shape in $YBa_2Cu_3O_x$ as a function of the temperature T when the Cu derived molecular field fluctuates. In this way, we follow the fluctuation rates of the short range correlated Cu(2) moments as a function of x and T [4].

3. Concluding Remarks

Using low temperature Mössbauer spectroscopy on both ^{57}Fe in $La_{2-x}Sr_xCuO_4$ and ^{170}Yb in $YBa_2Cu_3O_x$, we have first identified the response of the probes in the two following limit situations:

i) at low or moderate carrier levels in the semiconducting phase, when a Cu magnetic ordering is present (antiferromagnetic or spin glass like ordering)

ii) at high carrier levels in the superconducting phase, when no intrinsic Cu magnetic ordering is present.

Then, exploring the semiconducting/superconducting transition for intermediate carrier levels, we have evidenced in both series the superimposition of the two types of signals over a rather large composition range. These results compare favourably with the results obtained by Weidinger et al.[5] using another local probe technique, i.e. muon spin rotation experiments.

In both series, the observed "phase separation" could result from a segregation of carriers into hole rich and hole poor regions, with a possible contribution from atomic composition heterogeneities. In the $YBa_2Cu_3O_x$ series, local modulations of the oxygen occupancy of the chains could drive the carrier segregation. In the $La_{2-x}Sr_xCuO_4$ series, X ray diffraction data shows that it is unlikely that there are sizeable fluctuations of the Sr concentration x on a macroscopic scale. However the presence of possible local fluctuations of x on a microscopic scale cannot be discarded.

4. References

1. P. Imbert, G. Jéhanno, P. Debray, C. Garcin and J.A. Hodges, to be published in *J. de Phys.* (Paris).
2. A.M. Finkelstein, V.E. Kataev, E.F. Kukovitskii and G.B. Teitelbaum, *Physica* **C168** (1990) 370.
3. J.A. Hodges, P. Bonville, P. Imbert and G. Jéhanno, *Physica* **C184** (1991) 259 (Part I).
4. J.A. Hodges, P. Bonville, P. Imbert, G. Jéhanno and P. Debray, *Physica* **C184** (1991) 270 (Part II).
5. A. Weidinger, Ch. Niedermayer, H. Glücker, A. Goldnik, G. Nowitzke, E. Recknagel, H. Eickenbusch, W. Paulus, R. Schöllhorn, J.I. Budnick, *Hyperfine Interactions* **63** (1990) 147.

PHASE DIAGRAM AND TRANSPORT STUDIES ON $La_{2-x}Sr_xCuO_4$

H. TAKAGI, R. J. CAVA, M. MAREZIO, B. BATLOGG, J. J. KRAJEWSKI
and W. F. PECK, JR.
AT&T Bell Laboratories, Murray Hill, NJ
07974, U.S.A.

P. BORDET
Laboratoire de Cristallographie, CNRS, Grenoble, France

D. E. COX
Brookhaven National Laboratory, Upton, NY 11973

ABSTRACT

The electronic and structural phase diagram of heavily doped $La_{2-x}Sr_xCuO_4$ is re-examined with high quality samples annealed at high temperatures. In contrast to recent reports of phase separation, we observed instead complete solid-solution at least up to x=0.3. Based on high resolution synchrotron x-ray diffraction and magnetization studies, we demonstrate that the superconducting phase is confined to a much narrower range than believed, and that the superconducting phase boundary in the overdoped region coincides with the structural phase boundary between orthorhombic and tetragonal phases. Systematic study of electrical resistivity up to 1000K as a function of doping is also presented, which demonstrates that the "T-linear" resistivity over wide temperature range is observed only in the narrow region for optimal superconductivity.

1. Introduction

In high-T_c cuprates, the electronic and structural phase diagram, most frequently as a function of hole (electron) concentration, has been recognized to have a critical importance in disclosing an essential constraint for high-T_c superconductivity. The control of hole concentration can be achieved by the chemical modification of charge reservoir layers, e.g. heterovalent ion substitution or oxygen absorption/desorption [1]. It is commonly assumed that the hole concentration is a well-defined quantity and can be changed continuously and homogeneously over a certain range. Chemical modification, however, inherently introduces a certain chemical inhomogeneity to the system, which, taken with the extremely short superconducting coherence length, often raise serious doubts about the validity of the assumption on the hole concentration.

In this context, we first present here our re-examination of the phase diagram in heavily doped $La_{2-x}Sr_xCuO_4$ ($x\geq0.15$) where superconductivity disappears without a metal-insulator transition.[2-7] The essence of this study has already been described in Ref. 8. In heavily doped $La_{2-x}Sr_xCuO_4$, it has been pointed out that the system is thermodynamically unstable against the phase separation, for x>0.2, and that the material is always a mixture of two phases with different Sr content [9]. We therefore first address the question of whether or not the phase (solid-solution) can exist. We demonstrate that the solid-solution can be formed at least up to x=0.3 in contrast to the previous report.

Then, we demonstrate the main result of the phase diagram study, revised superconducting and structural phase boundaries. We show that the superconducting phase is confined to a much narrower composition range than believed so far, up x~0.2, and that the disappearance of superconductivity occurs at the low temperature orthorhombic (LTO) to high temperature tetragonal (HTT) structural phase boundary.

Finally, we present systematic evolution of electrical resistivity up to 1000K as a function of doping, including the lightly doped region. The result indicates that the much discussed T-linear resistivity over a wide temperature range is observed only in a narrow composition region of optimal superconductivity near x=0.15. Outside the optimum composition for superconductivity, we found intriguing non-T-linear behaviors of resistivity. The details of this study has been described in Ref. 10.

2. Search for Phase Separation

The previous report for phase separation in the heavily doped $La_{2-x}Sr_xCuO_4$ was mainly based on the observation of the broadening in the x-ray or the neutron powder diffraction peaks.[9] However, we would like to emphasize here that in order to examine the possibility of thermodynamic instability with respect to the phase separation, it is not sufficient to check the phase purity of the materials prepared under specific synthetic conditions. One cannot distinguish whether an observation of an inhomogeneity originates from the presence of an equilibrium two-phase region, or from defective processing of the materials.

Our approach to the issue of chemical phase purity was to anneal the series of samples for a prolonged period. If a two-phase region exists in thermal equilibrium then prolonged annealing will enhance the tendency towards phase separation. If a single phase region exists in thermal equilibrium, then the samples will become more uniform during an extended anneal. Thus, we have prepared a series of sinterd pellets ($0.15 \leq x \leq 0.3$) and have divided the samples into two sets. We have annealed these for one month (720 hrs.), which is one to two orders of magnitude longer than the time scale for sintering and annealing in the usual sample preparation procedure. The two sets of samples were annealed at 850°C and 1000°C respectively, low enough to allow for the formation of a chemical miscibility gap, if one is to occur, and high enough to allow for interdiffusion of the cations. After the annealing, the samples were cooled to room temperature at the rate of 10°C/hr. to allow for maximum oxygen uptake. During these treatments, the samples were kept under a very pure oxygen gas flow, which was passed through a drying agent to remove any moisture.

We have extensively investigated the chemical homogeneity of the samples by two experimental probes: high resolution synchrotron x-ray powder diffraction, which was performed at the beam line X-7A of National Synchrotron Light Source at Brookhaven National Laboratory, and Meissner effect measurements performed in a SHE magnetometer. While the former gives us information on a length scale a few hundreds of Å or longer, the latter may be able to probe electronic properties on much smaller length scales, since the coherence length of high-T_c cuprates is estimated to be as short as a few tens of Å(ξ_{ab}~30Å).

We have not observed any indication of the phase separation in the result of the high resolution powder x-ray diffraction studies which are summarized in Fig. 1. In the inset of Fig. 1, the $(200)_T$ diffraction peaks shift systematically upon doping without any significant change in the linewidth, indicating the formation of the solid-solution. To check further for any hidden information on phase stability, we plot the composition dependence of the linewidth obtained by Lorenzian fitting (top) and the peak position (bottom). In the bottom panel, we again see the systematic variation of peak positions with composition. The linewidth (0.025-0.03 deg) are only slightly larger than the instrumental resolution measured with a highly crystalline Si standard (0.015-0.02 deg) and tend to decrease upon Sr-doping. Considering this decrease of the linewidth, and the saturation in lattice constant, i.e. peak position shown in the bottom panel, upon Sr doping, we conclude that most of the observed line broadening originates from a compositional distribution $\pm\Delta x$ about the average composition x. From the observed broadening and the x-dependence of the peak position, we roughly estimate a compositional distribution of $\Delta x \simeq 0.01$. This compositional distribution, however, is not due to an instability against phase separation. If this were the case, the lines would broaden in the phase separation composition region on extended annealing. We do not, however, observe any such line broadening after annealing over the whole composition range investigated [11]. We therefore conclude that the observed compositional distribution is due to the imperfect mixing inherent in chemical solid-solutions.

The highly sensitive Meissner measurements provide insights into homogeneity on a shorter length scale than the x-ray powder diffraction. Can we see any indication of phase separation on a microscopic scale which x-ray powder diffraction could not detect? We show the temperature dependence of the Meissner signal for representative samples. On a linear scale shown in the upper panel, one can see a quite systematic shift of what appears to be "T_c onset" with increasing Sr content. This "T_c onset" seems to disappear somewhere between x=0.25 and x=0.3. This is in good agreement with the previous reports [2-7]. If the same data are plotted on a logarithmic scale, however, the "T_c onset" is actually quite ambiguous for x>0.15, with a low magnetization tail starting from 36-39K in all samples. Since 39K corresponds to the optimized T_c for x=0.15 in this system, this is evidence for short range compositional distribution extending at least down to $\simeq$0.15 for all x. It should be noted here that we are discussing the homogeneity on a length scale smaller than that probed by the x-ray powder diffraction. We focus now on the tail structure. This is <u>not</u> an indication of the phase separation. If there is a tendency for phase separation into a high-T_c phase and a non-superconducting phase, long time annealing would <u>enhance</u> the tail structure associated with the optimal-T_c. However, comparing the as-prepared samples, indicated by the solid line, and the annealed samples, indicated by the broken line, in the lower panel of Fig. 2, it is clear that the annealing process <u>suppresses</u> the tail structure. This means that the solid-solution becomes more homogeneous with time, even on a microscopic scale. There is no evidence for phase separation. We therefore proceed to the re-examination of the electronic and structural phase diagram in this composition range.

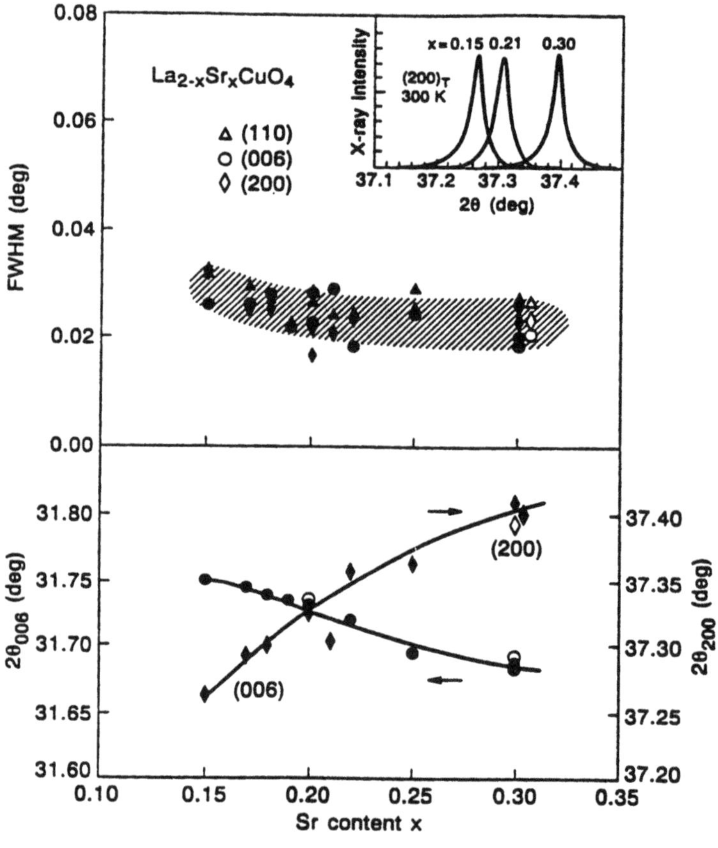

Fig. 1 The results of high resolution synchrotron x-ray powder diffraction measurements as a function of Sr content x. The full width at half maximum (top) and the peak position (bottom) of the $(110)_T$ (triangle), $(006)_T$ (circle) and $(200)_T$ (diamond) peaks obtained by least squares fits with $\lambda = 1.2$ Å [21] at 300K. symbols: open, as-prepared; closed, annealed at 850°C for one month; shaded, annealed at 1000°C for one month. The inset shows raw diffraction data for the $(200)_T$ reflection for samples annealed at 850°C for one month.

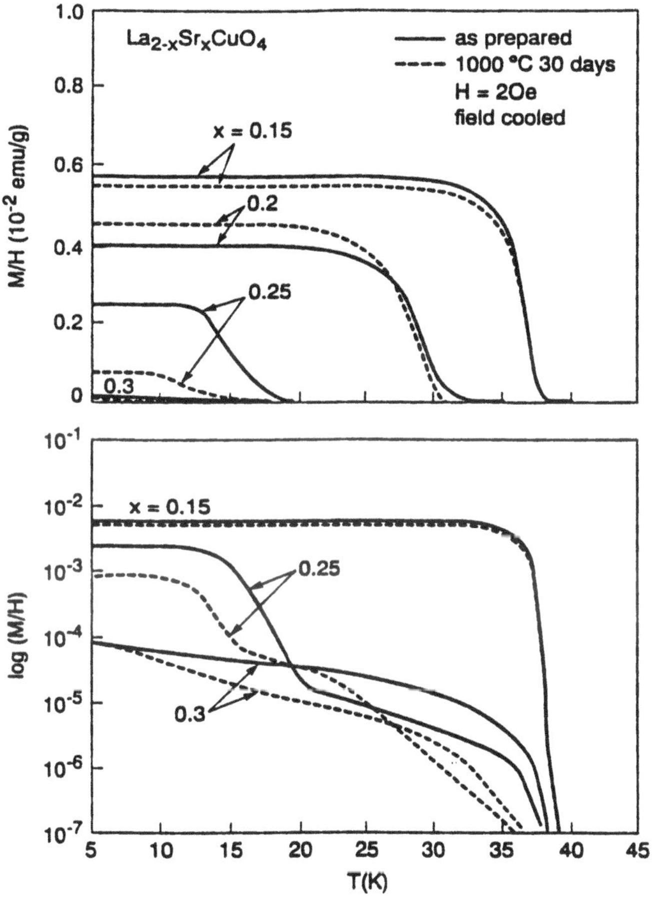

Fig. 2 Temperature dependence of the Meissner signal (field cooled, H = 2 Oe) for samples as prepared and annealed at 1000°C for one month.

3. Re-examination of the Phase Diagram

The superconducting phase boundary in the overdoped region has been previously believed to be near x = 0.3, since a "continuous" decrease in the "T_c onset" is observed up to this composition range [2-4]. This can be misleading, however, because T_c can be modified by the proximity effect in a microscopic mixture of a metal and a superconductor.

To locate the superconducting phase boundary, we summarize the magnetization data for our series of samples in Fig. 3. In the Meissner (field cooled) signal, there is a remarkable decrease of the signal near x = 0.2, although the "T_c onset" decreases smoothly through x = 0.2. Furthermore, this drop in the Meissner signal is more pronounced in the long-time annealed samples, as a consequence of the homogenization on annealing. We therefore conclude that the bulk superconducting phase is confined to a very narrow composition range with a maximum x of 0.2. The observation of only minor amounts of superconductivity above x = 0.2 by a specific heat measurement [8] and a muon spin rotation measurement [9] further confirms the present conclusion. One of the consequences of this conclusion is that quantitative arguments concerning T_c as a function of doping are not meaningful, as "T_c" is an ambiguous quantity due to the inhomogeneity inherent to a solid-solution system, particularly in non-optimal composition range away from x = 0.15.

The other key result of this study relates the disappearance of superconductivity near x=0.2 to a structural phase boundary. The structural phase boundary from high temperature tetragonal (HTT) to low temperature orthorhombic (LTO) structure had been reported to be just above x = 0.15, and is thus believed to have little to do with the superconducting phase boundary [10]. We have investigated the LTO/HTT structural phase transformation in detail by high resolution synchrotron powder x-ray diffraction. The correlation between superconducting and structural phase boundaries becomes evident in Fig. 3. While finite peak splitting due to the orthorhombic distortion at 30K can be seen up to x = 0.2, it disappears above x = 0.22. The LTO/HTT phase boundary is located just above x = 0.2, which is the same composition at which the Meissner signal shows a significant drop.

The combined structural and magnetic studies we have presented here clearly indicate that the disappearance of superconductivity on overdoping, which occurs near x=0.2, is associated with a structural phase transformation. The disappearance of superconductivity has heretofore been believed to be purely the consequence of electronic overdoping, which induces a modification of the electronic states [4-7]. The present observation shows that not only hole concentration, but also a delicate structural distortion is very likely to play a role in the disappearance of the superconductivity in heavily doped $La_{2-x}Sr_xCuO_4$. This is similar to the suppression of superconductivity in $La_{2-x}Ba_xCuO_4$ near x=1/8 [15].

The HTT/LTO structural transformation manifests itself only in the out of plane conductivity as a transition from metallic to semiconducting temperature dependence, not in the in-plane conductivity, indicating the modification of the coupling between planes [16]. This is probably because $La_{2-x}Sr_xCuO_4$ has a relatively low anisotropy among

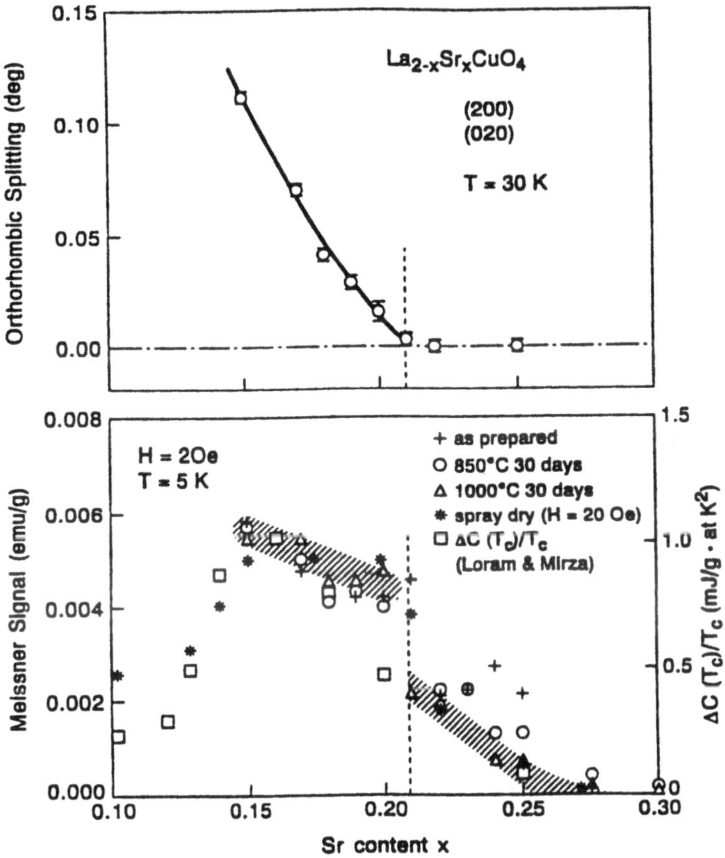

Fig. 3 Relationship between orthorhombicity and superconductivity. The orthorhombic splitting between $(200)_o$ and $(020)_o$ at 30K obtaind by synchrotron x-ray diffraction (top) and the Meissner signal at 5K in a field of H=2 Oe (bottom). The Meissner data from ref.4 and the specific heat jump at T_c from ref.12 are included in the bottom panel for comparison.

cuprates, and is located in a marginal region between a two-dimensional metal and an anisotropic three-dimensional metal [7]. Although the effect of two-dimensionality on the superconductivity is not yet understood, even a delicate structural distortion may influence superconductivity through a change in the effective layer coupling. On the other hand, one may argue that a specific structural environment is indispensable for the occurrence of high-T_c superconductivity. Although most high-T_c cuprates have an orthorhombic structure at low temperatures, a simple correlation between orthorhombic symmetry and superconductivity obviously does not hold true. We know of apparent exceptions: e.g. $La_{2-x}Sr_xCaCu_2O_6$ [18] and $Tl_2Ba_2Ca_2Cu_3O_{10}$ [19] which are tetragonal below T_c. However, since the local structure can be different from the average structure observed by conventional diffraction techniques, further studies to clarify a possible interplay between local symmetry and superconductivity will be interesting. Indeed, recent neutron pair distribution function analysis on $Tl_2Ba_2CaCu_2O_8$ has revealed short range buckling type distortion to occur near T_c [20].

4. Systematic Evolution of Temperature Dependent Resistivity up to 1000K as a function of Doping.

In most high-T_c superconducting cuprates showing optimal superconducting properties, the resitivity shows the "T-linear" behavior over a wide temperature range from just above T_c to near 1000K [21]. This T-linear resistivity has been often discussed in terms of a novel charge dynamics closely related to mechanism of high T_c superconductivity. In order to get further insight into the charge transport in CuO_2 planes, we have measured electrical resistivity up to high temperatures, not only for the optimal composition but also for the under doped and the over doped composition range.

In Fig.4, systematic evolution of electrical resistivity as a function of doping is presented. The point we would like to make in the figure is that the resistivity linear in T over a wide tempearture range is observed only in a narrow composition range near x=0.15, where we observe a substantial amount of Meissner signal and specific heat jump at T_c in Fig.3. In the lightly doped region below x=0.15, the resistivity shows a pronounced S-shaped temperature dependence. At low temperatures, the resistivity shows a distinct decrease of slope near 500K. The decrease of resistivity slope reminds us of the resistivity saturation which has been firmly established in A15 compounds [22]. In the heavily doped region above x=0.2, the resistivity appears almost insensitive to doping, in contrast to lower x, and shows a significant upturn even above room temperature. The temperature dependence of resistivity for x=0.3, for example, can be well described by a simple power law, $\rho = \rho_0 + AT^n$ where n ~ 1.5, over a whole measured temperature region up to 1000K. The super-linear behavior up to such a high temperature is surprising. In a conventional metal where electron-phonon scattering is dominant, the resistivity shows a T-linear behavior above 1/5 to 1/3 of the Debye tempearture [23]. One may argue the dominant electron-electron scattering, which gives rise to T^2 dependence. The observed temperature dependence, however, substantially weaker than T^2 [24]. Further study is required to disclose the microscopic origin of the anomalous super-linear behavior up to high temperature.

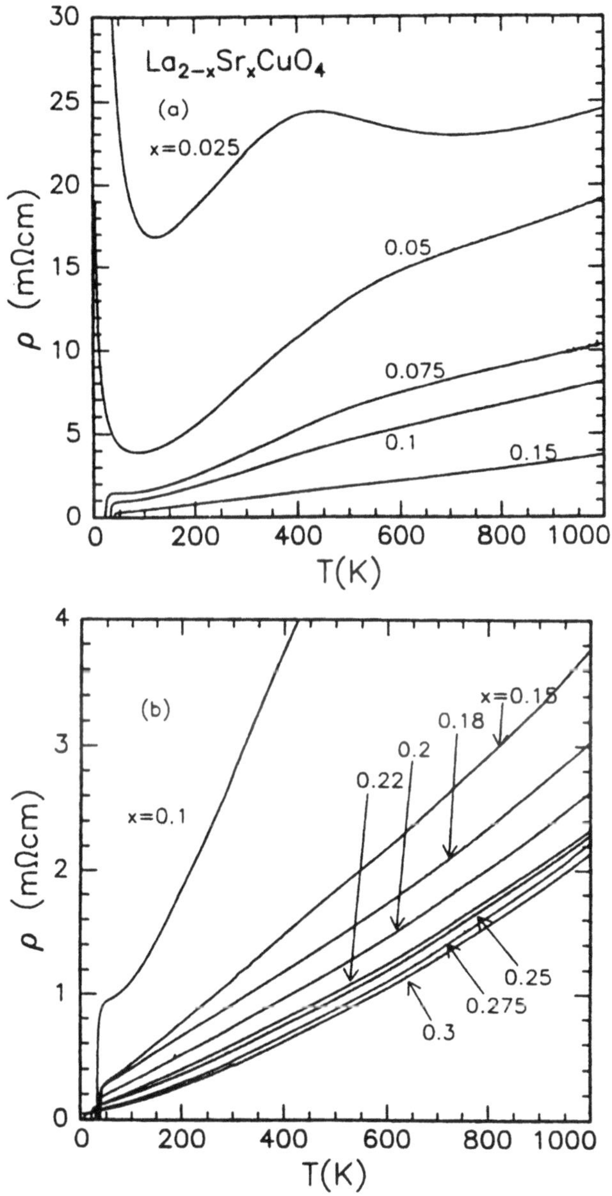

Fig. 4 Systematic evolution of temperature dependent resistivity as a function of doping. (a) $0 < x \leq 0.15$ and (b) $0.15 \leq x \leq 0.3$.

5. Summary and Conclusion

Through systematic studies of the structural and physical properties of $La_{2-x}Sr_xCuO_4$, we have shown the following:

1. The solid-solution in $La_{2-x}Sr_xCuO_4$ extends at least to x=0.3. We have not observed any tendency toward phase separation.
2. The bulk superconducting phase is confined to a narrow composition region up to x=0.2.
3. The superconducting phase boundary at x=0.2 coincides with the structural phase boundary between the low temperature orthrhombic and the high temperature tetragonal phases. Thus revised phase diagram reveals an intriguing relationship between structural distortion and the disappearance of superconductivity in overdoped $La_{2-x}Sr_xCuO_4$. Finally, we are beginning to get insight into the longstanding issues.
4. The much discussed T-linear resitivity over a wide temperature range is observed only in a narrow region for optimal superconductivity. In the lightly doped region, we observe significant S-shaped temperature depence. In the heavily doped region, the resistivity shows a super-linear temperature dependence up to near 1000K.

Acknowledgement

We would like to thank S-W. Cheong, J. Kwo, J. C. Phillips, M. Crawford, J. Jorgensen,Y. Maeno, Y.Tokura, E.Abrahams, B.Shastry, J. Zaanen and N.P.Ong for useful discussions. This work was partly supported by the Division of Material Science, U.S., Department of Energy under contract No. DE-AC02-76CH00016.

REFERENCES

1. R. J. Cava, *Science* **247** (1990) 656; Y. Tokura and T. Arima, *Jpn. J. Appl. Phys.* **29** (1990) 2388.
2. R. B. van Dover, R. J. Cava, B. Batlogg and E. A. Rietman, *Phys. Rev.* **B35,** (1987) 5337.
3. J. B. Torrance, Y. Tokura, A. I. Nazzal, A. Bezinge, T. C. Huang and S. S. P. Parkin, *Phys. Rev. Lett.* **61** (1988) 1127.
4. H. Takagi, T. Ido, S. Ishibashi, M. Uota, S. Uchida and Y. Tokura, *Phys. Rev.* **B40** (1989) 2254.
5. J. B. Torrance, A. Bezinge, A. I. Nazzal, T. C. Huang, S. S. P. Parkin, D. T. Keane, S. J. LaPlaca, P. M. Horn and G. A. Held, *Phys. Rev.* **B40** (1989) 8872.
6. Y. Ando, M. Sera, S. Yamagata, S. Kondoh, M. Onoda and M. Sato, *solid State Commun.* **70** (1989) 303.

7. M. Oda, T. Ohguro, N. Yamada and M. Ido, *J. Phys. Soc Jpn.* **58** (1989) 1137.

8. H. Takagi, R. J. Cava, M. Marezio, B. Batlogg, J. J. Krajewski, W. F. Peck, Jr., P. Bordet and D. E. Cox, submitted.

9. J. D. Jorgensen, P. Lightfoot, S. Pei, B. Dabrowski, D. R. Richards and D. G. Hinks, Advance in Superconductivity III, Eds. K. Kajimura and H. Hayakawa, Springer Verlag 1990, p. 337.

10. H.Takagi, B.Batlogg, H.L.Kao, J.Kwo, R.J.Cava, J.J.Krajewski and W.F.Peck, submitted.

11. Our preliminary measurement on samples with $x \geq 0.35$ showed definite broadening compared with $x < 0.35$. It is not clear at this stage whether it is an indication of the phase separation or a possible oxygen vacancy effect.

12. J. W. Loram and K. A. Mirza, *Electronic Properties of High-*T_c *Superconductors and Related Compounds,* Eds. H. Kuzmany, M. Mehring, J. Fink, Springer Verlag 1990, p. 92.

13. D. Harshman, G. Aeppli, B. Batlogg, G. P. Espinosa, R. J. Cava, A. S. Cooper, L. W. Rupp, Jr., E. J. Ansaldo and D. Ll. Williams, *Phys. Rev. Lett.* **63** (1989) 1187.

14. For example, Ref. 5: In contrast to this report, we should note that one of the early papers (R. M. Fleming, B. Batlogg, R. J. Cava and E. A. Rietman, Phys. Rev. **B35** (19887) 7191) suggested the correlation between the boundaries: See also T. Fujita, Y. Aoki, Y. Maeno, J. Sakurai, H. Fukuba and H. Fujii, *Jpn. J. Appl. Phys.* **26** (1987) L363; K. Sun, J. H. Cho, F. C. Chou, W. C. Lee, L. L. Miller, D. Johnston, Y. Hidaka and T. Murakami, *Phys. Rev.* **B43** (1991) 239.

15. A. R. Moudenbaugh, Y. Xu, M. Suenaga, T. J. Folkerts and R. N. Shelton, *Phys. Rev.* **B38** (1988) 4596: J. D. Axe, D. E. Cox, K. Mohanty, H. Moudden, A. R. Moudenbaugh and Y. Xu, *Phys. Rev. Lett.* **62** (1989) 2751.

16. S. Kambe, K. Kitazawa, M. Naito, A. Fukuoka, I. Tanaka and H. Kojima, *Physica* **C160** (1989) 35.

17. T. Ito, H.Takagi, S. Ishibashi and S. Uchida, *Nature* **350** (1991) 596.

18. R. J. Cava, A. Santoro, J. J. Krajewski, R. M. Fleming, J. V. Waszczak, W. F. Peck, Jr. and P. Marsh, *Physica* **C172** (1990) 138.

19. D. E. Cox, C. C. Torardi, M. A. Subramanian, J. Gopalakrishnan and A. W. Sleight, *Phys. Rev.* **B38** (1988) 6624.

20. B. H. Toby, T. Egami, J. D. Jorgensen and M. A. Subramanian, *Phys. Rev. Lett.* **64** (1990) 2414.

21. See for example, B.Batlogg, *Phisica* **B169** (1990) 7.

22. See for example, M.Gurvich, *Physica* **135B** (1985) 276.

23. See for example, W.E.Pickett, *J.Superconductivity* **4** (1991) 397.

24. See for example, C.C.Tsui,A.Gupta and G.Koren, *Phisica* **C161** (1989) 415.
25. Since the measurements were performed in two independent runs with slightly different x-ray wavelengths, the two groups of peak positions were adjusted to fit each other.

Static and Dynamic Transport Aspects of Phase Separation

B. Bucher and P. Wachter
Laboratorium für Festkörperphysik, ETH Zürich, 8093 Zürich, Switzerland

Abstract

We report on experimental evidence of different phases in the superconductor $YBa_2Cu_4O_8$: simple metallic CuO chains and highly correlated CuO_2 planes. $YBa_2Cu_4O_8$ is a genuine untwinned sample; hence, we were able to determine the anisotropy of the resistivity along the a - and b - direction. Along the b direction (chain - and plane conduction channels), a normal metallic temperature behavior of the chain dominates. For the a direction (only plane conduction channel), the resistivity reveals an unconventional behavior with a saturation at higher temperatures.

Further, we present results of the dynamical (optical) conductivity of the CuO_2 plane as a function of frequency and temperature. The frequency dependence is consistent with a model of ferromagnetic polarons in an antiferromagnetic matrix. The temperature behavior is in agreement with a real space condensation.

1 Introduction

From the beginning [1], the high T_c superconductors (HTSC) have been very challenging systems. As regards the solid state chemistry, the wide variability of their stoichiometry gives rise to homogeneity problems; the samples may show a chemical phase separation of different oxygenated regions. So, the proper preparation of homogeneous samples is crucial. Concerning the physics, the normal as well as the superconducting state show unusal features. In addition, as we will show at least for $YBa_2Cu_4O_8$ (124), one has to take into account the coexistence of different physical phases. The CuO chains behave as a normal Fermi liquid whereas the CuO_2 planes exhibit a strongly correlated behavior. These two subsystems are connected by the apex oxygen and it is an interesting task to resolve the role of the apex oxygen with respect to charge transfer, plane - chain interaction switch etc.

The oxides are prototypes of ionic crystals and, therefore, also for a polaronic nature of the charge carriers. On the other hand, the magnetic properties of the CuO_2 planes of the HTSC have been established to be crucial for superconductivity. Considering these two features for the HTSC, one is tempted to anticipate magnetic, polaronic charge carriers : magnetic polarons. The magnetic polarons could be ferromagnetic polarized regions in the short range ordered, antiferromagnetic CuO_2 plane. A physical manifestation of such magnetic polarons is a diffusive motion instead of a propagating one. The dynamical (optical) conductivity to be presented, indeed is consistent with the hypothesis of magnetic polarons.

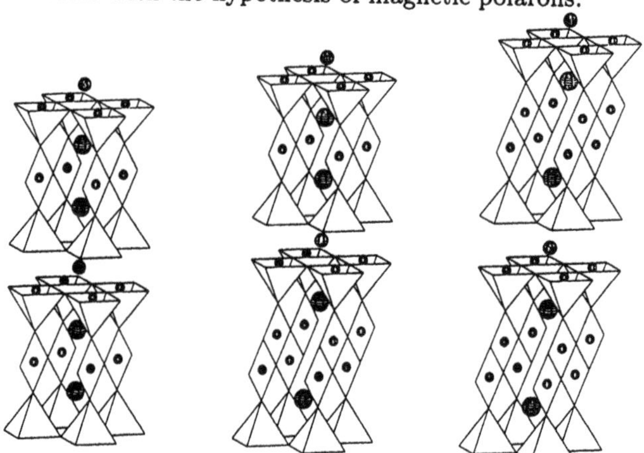

Figure 1: Crystal structure of the n = 0, 1, 2 members of the homologous series $Y_2Ba_4Cu_{6+n}O_{14+n}\,(n = 0,1,2,\ldots)$. The picture of the n=0 member shows two unit cells. For the n=1 structure, only a half of the unit cell is depicted. The n=2 member is represented by just one unit cell.

The well studied (123) superconductor has the disadvantage of being twinned. Thus, the "ab-plane" properties are rather a mixture of the CuO_2-plane and the CuO-chain (along the b-axis). Untwinned $YBa_2Cu_4O_8$ provides the opportunity to measure the polarized reflectivity along the **a**-direction i.e. the genuine response of the 2-dim. CuO_2-plane.

To our knowlege, $YBa_2Cu_4O_8$ is the only stoichiometric HTSC i.e. it is not possible to vary the oxygen content because of the high coordination of the oxygen inside the double chains [2]. This is one great advantage of the 124 system as we have not to worry about chemical phase separation, in addition to the genuine untwinned crystal structure.

2 Experimental

We present the static [3] and dynamic (optical) transport properties of $YBa_2Cu_4O_8$. In fig. 2, the dc resistivity is depicted for the current along the a - (plane) and b (plane & chain) directions, respectively. A strong anisotropy is evident. For the b - direction, one expects two conduction channels : the plane and the chain. The a - direction represents the intrinsic conductivity of the interesting CuO_2 - plane alone. Assuming a model of shunted conduction channels along the b - direction (plane and chain) and a single conduction channel for the a - direction (plane), one recognizes a dominant chain contribution along the b - axes.

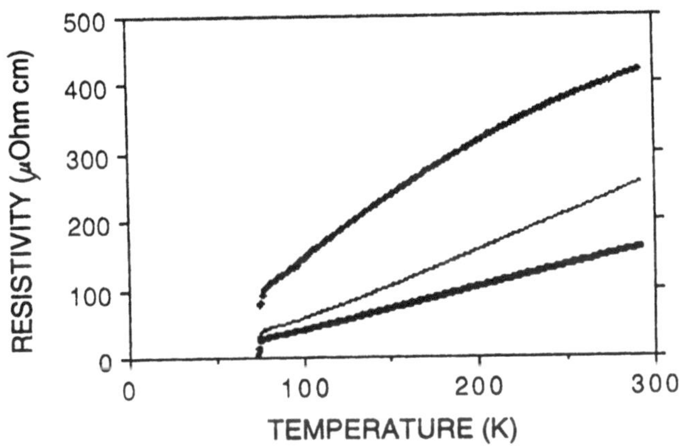

Figure 2: Resistivity of a $YBa_2Cu_4O_8$ single crystal for the two plane directions along the b - axes (lower curves) and the a - axes (upper curve). Also shown is the calculated resistivity of the chain alone, exploiting the model of shunted resistors along the b direction.

Obviously, the two conduction channels show a completly different temperature dependence. The chain reveals a Bloch - Grüneisen signature as expected for a normal metal whereas the plane shows an unusual saturation and quite a high value. The intersection of a linear extrapolation to zero temperature gives for both directions a negative value, indicating a good sample quality and, more important, a nonlinear resistivity has to be expected at low temperature. These results are in concordance with measurements of the spin - lattice relaxation time T_1 of $YBa_2Cu_4O_8$ by Zimmermann *et al.* [4]. They found a normal metallic relaxation for the chain Cu(1) down to 4 K with no change below T_c ; a condensation of the free carriers in the chain is not seen at Cu(1). They found a magnetic relaxation mechanism for Cu(2) of the plane and claimed an opening of a spin gap at 130 K in the underdoped

$YBa_2Cu_4O_8$ (see I. Mangelshot, this conference). On the other hand, the saturation of the plane resistivity starts around 150 K and the question arises if there is a connection between spin gap and conductivity.

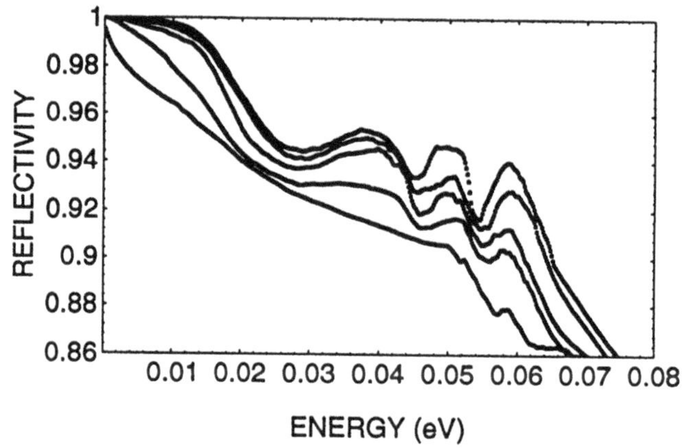

Figure 3: Near normal reflectivity of $YBa_2Cu_4O_8$ (T_c = 72 K) for $\vec{E}_{light} \parallel$ **a** - axes. The temperatures are from top to bottom : 6 K, 30 K, 50 K, 65 K, 95 K.

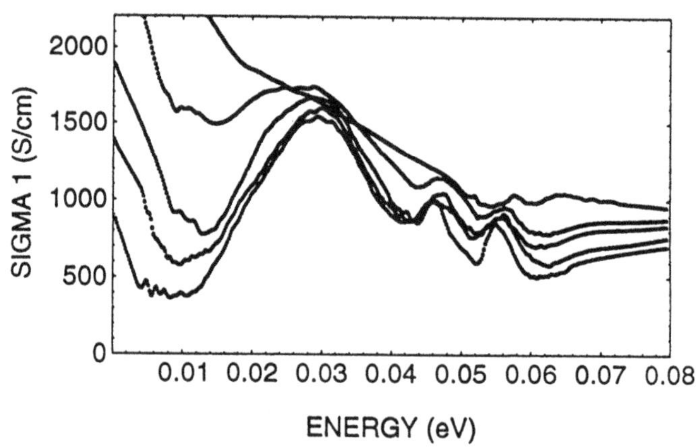

Figure 4: Optical conductivity $\sigma_1^a(\omega)$ for the **a** - direction. The temperatures are from top to bottom : 95, 65, 50, 30, 6 K.

In fig. 3, we show the reflectivity of 124 at different temperatures with the light polarized along the a axis. Below Tc an additional plasma edge develops around 20 meV. At the energies of 45 meV and 54 meV, an absorption takes place with decreasing temperature. Such a signature is expected for a London model as discussed below.

On the reflectivities (4 meV - 12 eV), we performed the Kramers-Kronig transformation to evaluate the optical constants. The dynamical (optical) conductivity σ_1 is depicted in fig. 4. With lower temperature, the spectral weight becomes smaller, i.e. there is a condensation to the superconducting state represented by a delta function.

3 Discussion

3.1 Frequency Dependent Scattering

The conductivity in the mid and near IR is not Drude like ($\sigma \propto \omega^{-2}$). Especially the CuO_2 plane reveals a strong spectral weight at 1 eV, compared to zero frequency. An approach to the IR conductivity, in contrast to interband transitions, is to assume an inelastic scattering of the free carriers in the mid IR by a non-phonon excitation spectrum (electron-spin, electron-electron etc.). The underlying excitations need not to be infrared active themselves; but, the interactions can be seen by an unusual conductivity of the carriers involved. Therefore, we postulate a frequency dependent scattering rate $\Gamma^*(\omega)$ for the optical conductivity:

$$\sigma(\omega) = \frac{ne^2}{m^*} \cdot \frac{1}{\Gamma^*(\omega) - i\omega}$$

with $m^*(\omega) = m(1+\lambda)$ a renormalized mass and $\Gamma^*(\omega) = \Gamma/(1+\lambda)$ a renormalized scattering rate (fig.5).

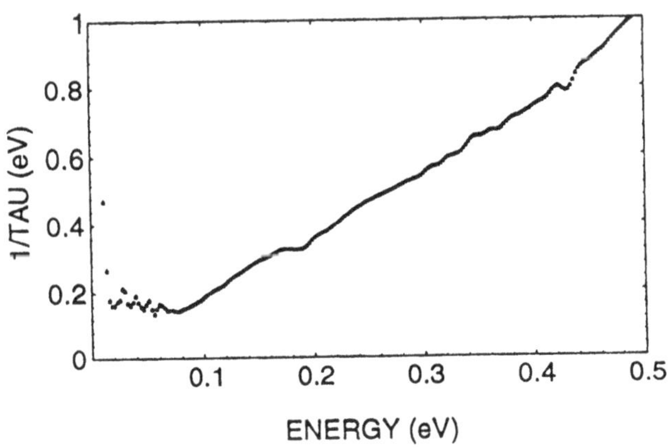

Figure 5: Frequency dependent scattering rate $\Gamma^*(\omega)$ for the a- direction of $YBa_2Cu_4O_8$.

An astonishing linearity comes out between 0.05 and 0.5 eV for a 72 K sample of 124. For $YBa_2Cu_{3.5}O_\delta$ we have even found [5] a linearity up to 0.8 eV. The linearity of $\Gamma^*(\omega)$ has been thought of as evidence of a nearly localized Fermi liquid, a phenomenological model, proposed by Varma *et al.* [6]. They predicted the universal relation $\Gamma^*(\omega) \propto \hbar\omega + k_B T$, i.e. a dc conductivity reciprocal in temperature and linear in the frequency ω at constant temperature. However, there is no correlation of the optical scattering rate and T_c as listed in table 1. Therefore, the interaction responsible for the scattering seems not being related to the interaction resulting in superconductivity. The same is true for the static conductivity (see H. Takagi, this conference).

Table 1: Strength of the frequency dependent scattering rate : $\Gamma^*(\omega) = A \cdot \hbar\omega$.

	T_c (K)	A
$YBa_2Cu_3O_7$ [1]	91	0.6
$YBa_2Cu_{3.5}O_{7.5-x}$	72	1.1
$YBa_2Cu_{3.5}O_{7.5-x}$	30	1.1
$YBa_2Cu_4O_8$	81	1.8

3.2 Spinflips, Magnetic Bipolarons

The HTSC superconductors differ in two aspects from a common superconducting metal. First, the bonding is at least partly of an ionic nature. Second, the CuO_2 planes show short range ordered antiferromagnetic correlated regions of the magnetic moments of the Cu^{2+} ions. As regards the ionic binding of oxides, one should expect a polaronic nature of the charge carriers. On the other hand, Mott [7] has pointed out that holes in an antiferromagnetic matrix polarize the surroundings, resulting in an ferromagnetic region as sketched in fig. 6.

Hence, as a first heuristic approach to the HTSC, one may anticipate magnetic polarons in the CuO_2 planes. Indeed, Wachter and Degiorgi [8] have assumed, that superconductivity in the HTSC is based on a Bose -like condensation of bipolarons. In the antiferromagnetic matrix, the magnetic polarons can combine to form bipolarons as exchange energy is gained; only seven broken magnetic bonds for a bipolaron instead of eight ones for two single polarons (see fig. 6). As argued by Wachter and Degiorgi, the binding mechanism of the bipolarons is a combined effect of gained magnetic exchange energy, lattice energy (due to a Jahn-Teller effect) and Coulomb repulsion.

The transport behavior of the polarons should be a diffusive motion in contrast

[1] For the a direction of untwinned 123; Z. Schlesinger *et al.* Phys. Rev. Lett. 65 (1990),801.

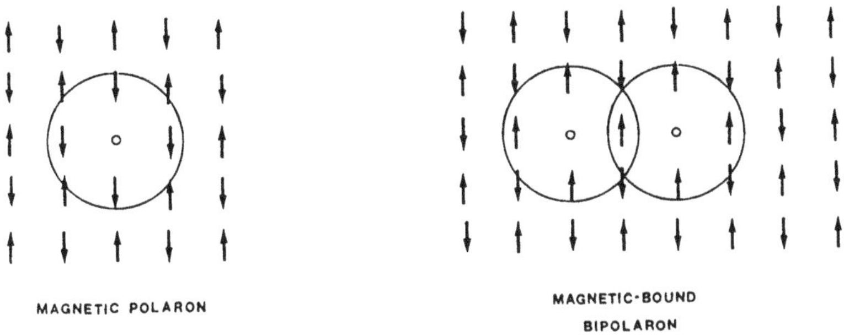

Figure 6: Model of ferromagnetic polarons and bipolarons in a antiferromagnetic matrix. The binding energy results from competing effects such as Jahn-Teller splitting, Coulomb repulsion, gained exchange energy etc.

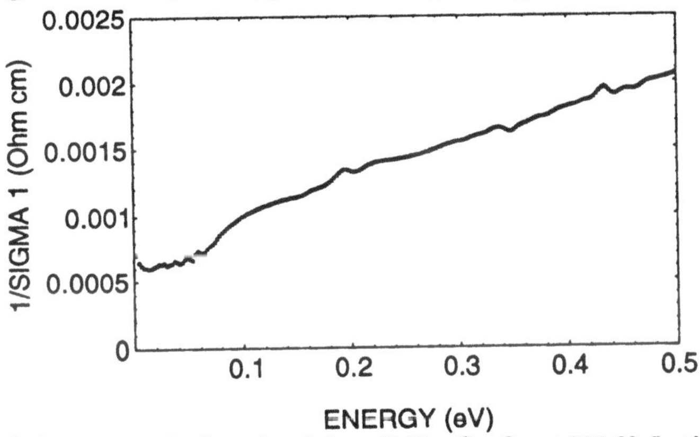

Figure 7: Inverse optical conductivity of $YBa_2Cu_4O_8$ at 300 K. In the energy region $\omega > J \sim 0.1$ eV, we have found $\sigma \propto \omega^{-1}$, as expected for a diffusive motion of holes.

to a propagating one. A theoretical prediction of this spin-flip diffusive motion has been made by Rice and Zhang [9]. The optical conductivity for the energy range above the magnetic exchange energy J (about 0.1 eV) has been calculated to be inverse to the frequency : $\sigma \propto \omega^{-1}$. There is a good agreement with our experiment on 124 (fig. 7). Thus, the unconventional optical behavior in the IR energy range above 0.1 eV could be understood by a diffusive motion of magnetic polarons throughout the CuO_2 plane.

3.3 Superconductivity

With the f sum rule, it is possible to calculate the amount of condensed carriers in the superfluid phase. For the superconducting state, we have proposed [5] the superconducting carriers to be described by the London model. Thus, the optical conductivity of the condensed carriers should be given by $\sigma_1(\omega) = A \cdot \delta(\omega)$, where $A = \pi n_s e^2/2m^* = \frac{1}{8} \cdot \omega_{ps}^2$ with n_s the density of superconducting carriers and m^* their effective mass; ω_{ps}^2 is called the plasma frequency of the condensed carriers. In the superconducting state, an absorption is manifested itself by a dip in the reflectivity as seen in our measurements (fig. 3). A comprehensive presentation is given elsewhere [10].

Here, we show the analysis of the thermal occupation of the superfluid ground state. For fig. 8, the f sum rule has been used to determine the superconducting fraction of the free carriers. The thermal occupation has been normalized to the value at 6 K. If we postulate a Bose condensation of e.g. the magnetically bound bipolarons, the thermal occupation should follow the two-fluid model : $n_s \propto 1 - (\frac{T}{T_C})^4$ [11]. The results show a good consistency with the concept of a Bose condensation.

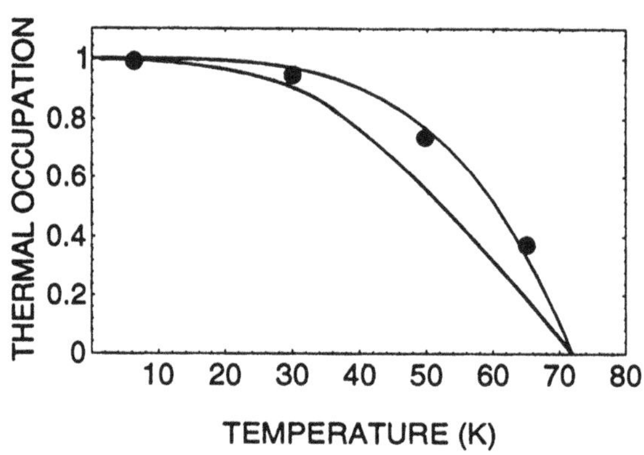

Figure 8: Thermal occupation of the superfluid ground state as deduced from f sum calculations. The two fluid picture (upper plot) gives a good concordance. The lower curve represents BCS.

As seen in fig. 3, the reflectivity does not reach 100 %. We expect that there are left some not condensed carriers at 6 K responsible for absorption; in the plane, two kinds of carriers might exist . Further, any excitation with an energy $\omega < \omega_{ps}$ results in an absorption. So, an interpretation of the structure around 20 meV has to address the question, if the structure is due either to an electronic gap (in energy space) or the first absorptive excitation.

4 Conclusion

The stoichiometric system $YBa_2Cu_4O_8$ shows probably no chemical phase separation. The intrinsic properties of the CuO_2 have been probed by the static and dynamic conductivity. Within the frame of the hypothesis of magnetic polarons, normal as well as superconducting features can be understood. The unusual optical conductivity manifests the diffusion of the holes through the CuO_2 plane. The temperature dependence of $\sigma(\omega)$ may be regarded as a Bose condensation of bipolarons.

Acknowlegements

The authors would like to thank Prof.Dr. E.Kaldis for sample preparation as well as continuous stimulating discussions. They are also indebted to Dr. Karpinski, Dr. Schwer and J.Müller.

References

[1] J.G. Bednorz and K.A. Muller.
Z. Phys. B *64*, (1986). 189.

[2] J.Karpinski, S.Rusiecki, E.Kaldis, B.Bucher, and E.Jilek.
Physica C *160*, (1989). 449.

[3] B.Bucher, J.Karpinski, E.Kaldis, and P.Wachter.
Physica C *167*, (1990). 324.

[4] H. Zimmermann, M. Mali, D. Brinkmann, J. Karpinski, E. Kaldis, and S. Ruciecki.
Physica C *159*, (1989). 681.

[5] B.Bucher, J.Karpinski, E.Kaldis, and P.Wachter.
Phys. Rev. B *45*, (1992). 3026.

[6] C.M. Varma, P.B. Littlewood, S. Schmitt-Rink, E. Abrahams, and A.E. Ruckenstein.
Phys. Rev. Lett. *63*, (1989). 1996.

[7] N.F. Mott.
Metal - Insulator Transition, Taylor & Francis Ltd, London, 1974.

[8] P. Wachter and L. Degiorgi.
Solid State Commun. *66*.

[9] T.M. Rice and F.C. Zhang.
Phys. Rev. B *39*, (1989). 815.

[10] B.Bucher and P.Wachter. *in preparation.*

[11] D.R. Tilley and J. Tilley.
Superfluidity and Superconductivity, Adam Hilger Ltd, Bristol, 1986.

PHASE SEPARATION IN THE SUPERCONDUCTING $La_2CuO_{4+\delta}$ PHASES ($0 < \delta < 0.09$) PREPARED BY ELECTROCHEMICAL OXIDATION

J-C. GRENIER, A. WATTIAUX and M. POUCHARD
Laboratoire de Chimie du Solide du C.N.R.S., Université de BORDEAUX I,
33405 Talence Cédex, France

ABSTRACT

$La_2CuO_{4+\delta}$ materials ($0 \leq \delta \leq 0.09$) have been prepared using a new way, "the electrochemical oxidation". The amount (δ) of additional oxygen depends upon the experimental conditions of polarization and can be controlled. The structural characterization (X-ray and neutron diffraction and T.E.M.) as well as the transport and magnetic properties *vs.* δ show the existence of two phases and of an insulating-metal transition for $\delta \simeq 0.055$. Due to the high mobility of the interstitial oxygen a phase separation was evidenced in the insulating as well as in the metallic phases; it is ascribed to the ordering of these oxygen atoms at low temperature. Moreover in the latter materials ($\delta \geq 0.06$), two superconducting transitions were observed ($T_c = 32$ K and $T_c = 44$ K) and assigned to definite compositions of oxygen excess.

1. Introduction

Among the superconducting cuprates, La_2CuO_4 ($T_c \simeq 40$ K)[1] is especially interesting as its structure is relatively simple and relevant to achieve basic researches. Occurrence of superconductivity in this material requires the formation of holes in the "CuO_2" planes, which can be obtained using various ways:

- substitution of alkaline-earth or lead for lanthanum[2-4] ,
- oxygen (or anion) overstoichiometry obtained under high oxygen pressure synthesis[1] or by fluorination or chlorination[5-7].

Recently we developed in Bordeaux the electrochemical oxidation as a new method for preparing oxides containing cations with high oxidation state such as the fully oxidized cubic perovskites $SrFeO_3$ or $SrCoO_3$ [8,9] and the oxygen overstoichiometric $La_{2-x}A_xMO_{4+\delta}$ phases (A = Nd; M = Ni or Cu) with high values of δ [10-13].

Using this method we first prepared the superconducting $La_2CuO_{4.09}$ phase, which was confirmed by various authors [10,12,14,15]. After some controversy, the insertion of oxygenated species into the network of these materials was finally demonstrated on the basis of their physical characterization [16,17]. Nevertheless these materials raise indeed many questions such as:

- range and control of the oxygen stoichiometry (δ) or in other words the hole concentration.

- the nature and localization of the interstitial oxygen atoms - existence or not of structural ordering - in correlation with phase segregation phenomena and the implication in superconductivity.

Concerning the preparation of these materials, further remarks can be made. High oxygen pressure synthesis is carried out at high temperatures for which the thermodynamical equilibrium is usually reached. The parameters governing this solid-gas reaction are p_{O2} and T (p_{O2} < 20 kbars, La_2CuO_4 being decomposed at higher pressures). However one should point out the contradictory features of this method, i.e. the insertion of oxygen in the network while a mechanical strain is applied. Fluorination and chlorination using F_2 or Cl_2-gas treatment lead also to superconducting La_2CuO_4 materials; the reaction is achieved under low pressure (1-6 bars) at relatively low temperature 200-300 °C but these strong oxidizing agents may cause a partial decomposition of the starting material. Therefore "soft reactions" are worthier and electrochemical oxidation appeared as a relevant technique for this purpose. Using the redox OH^-/O_2 couple at pH ≃ 14, at room temperature (T = 20 °C), the oxygen activity obtained under anodic polarization can be very high (more than 10^{10} equivalent O_2 atm. for an overpotential of 300 mV).

However this electrochemical reaction achieved at room temperature belonging to the "chimie douce" processes, is not expected to lead systematically to the thermodynamically most stable phase as assumed for heterogeneous equilibria which are generally investigated at high temperatures. In other words, kinetics plays a very important role in the formation of metastable phases provided that their free energy is lower than that of the starting materials and, as often encountered in intercalation reactions, ordering of intercalated species may occur [18]. Such phenomena were studied in $La_2CuO_{4+\delta}$.

2. Experimental and characterization of the samples.

The electrochemical measurement procedure has been described elsewhere in detail[10,11]. Experiments were carried out at 20°C in air in a two compartment cell filled with a 1N KOH solution, using a three electrode device (Fig.1):

- the reference electrode was a HgO/Hg electrode filled with 1N KOH (E_{th} = +0.098 V/SHE).

- the counter electrode was a gold foil of large area.

- a rotating disk electrode was used as working electrode; the disk was a polished ceramic (ø 8-32 mm; ≈ 2 mm thick; 0.4-10 g) whose density should be higher than 85-90 % .

The starting material La_2CuO_4 was prepared by the solid-state reaction of a stoichiometric mixture of calcinated La_2O_3 and CuO , fired in air at 1050 °C, for 18 h. Then the sample was ground, pelleted as a disc and annealed in air during 16 h and finally quenched.

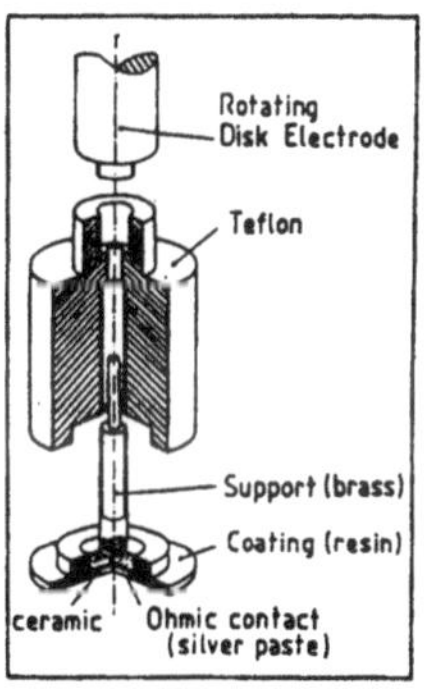

Fig. 1

Schematic representation of the rotating working electrode

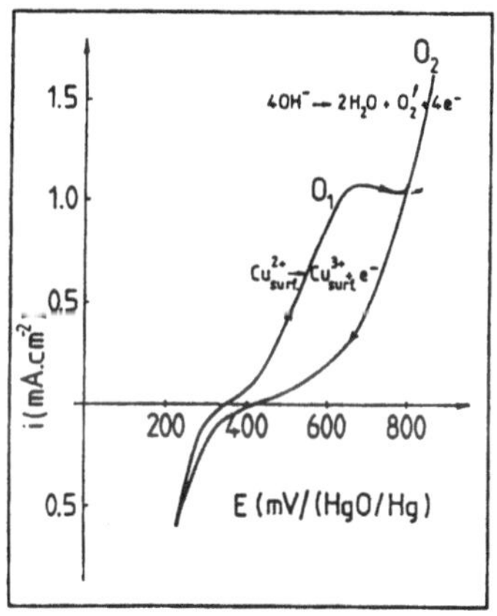

Fig. 2

Cyclic voltammogram for La_2CuO_4 in 1N KOH (6 $mV.s^{-1}$)

The cyclic voltammograms (I,E) were obtained by means of potentiodynamical device. A typical voltammogram is represented in Fig. 2. One can observe:

- an oxidation wave (O_1) arising between 400 and 800 mV and corresponding to the following overall reaction:

$$A_2MO_4 + 2\delta\ OH^- \longrightarrow A_2MO_{4+\delta} + \delta\ H_2O + 2\delta\ e^-$$

- above 800 mV the oxygen evolution (O_2) occurs according to the reaction :

$$4\ OH^- \longrightarrow 2\ H_2O + O_2^{\nearrow} + 4\ e^-$$

The applied potential for the electrochemical oxidation was chosen as that of the beginning of the plateau.

- a reduction peak (R_1) in the range 200-400 mV

For preparing $La_2CuO_{4+\delta}$ materials, experiments were performed at 25 °C, in air, under an anodic potential of 450 mV for various polarization times. Afterwards the pellets were washed in water, then in alcohol and finally dried. The oxygen content was determined the iodometric titration of the oxidation state of copper. In addition the absence of protons was checked using the proton N.M.R. and I.R. spectroscopy [17]. Before oxidation, the composition of the starting material was $La_2CuO_{4.01}$ and the amount of oxygen excess (δ) monotonously increases with polarization time (Fig. 3) up to the value $\delta \simeq 0.09$. In the same time the O.C.V. measurement whose value increases with δ (Fig. 3 insert), can be an easy way for controlling the oxygen excess.

Detailed electrochemical studies have emphasized the following points: - this reaction can be considered as an intercalation process; indeed the electrochemical reduction of $La_2CuO_{4.09}$ at 200 mV for 15 hours leads to a complete deintercalation of oxygen, i.e. to $La_2CuO_{4.00}$.

- the value of the diffusion coefficient of electroactive species was determined at 300 K close to 10^{-12} $cm^2.s^{-1}$ [19]. This value appears quite different from the usual values of $D_{O^{2-}}$ ($\simeq 10^{-17}$ $cm^2.s^{-1}$) or of D_{OH^-} (5.10^{-5} $cm^2.s^{-1}$), demonstrating the presence of very mobile "electroactive species".

X-ray diffractograms of the ground and washed samples showed well crystallized single phase materials. All of them were indexed with an orthorhombic unit cell (Bmab space group). Fig. 4 displays the evolution of the cell parameters as a function of the oxygen excess (δ). As expected, the introduction of interstitial oxygen atoms in the La_2O_2 layers results, in a first stage ($\delta \leq 0.05$) in relaxing the equatorial constraints in the *a,b* plane leading to a pseudo-tetragonal symmetry (a $\simeq$ b; b/a decreasing from 1.011 down to 1.002). Simultaneously the c parameter increases due to a steric effect in the La_2O_2 layers. For $\delta \geq$ 0.05 the insertion of oxygen causes again a structural distortion, the b/a ratio increasing up to 1.015. Reversely the c parameter increases only very slightly, which suggests that the structure has reached a stable relaxation state along the $\vec{c}$ direction. The increase of the unit cell volume ($\Delta V/V \simeq + 0.6$ %) clearly supports the introduction of oxygen in the network.

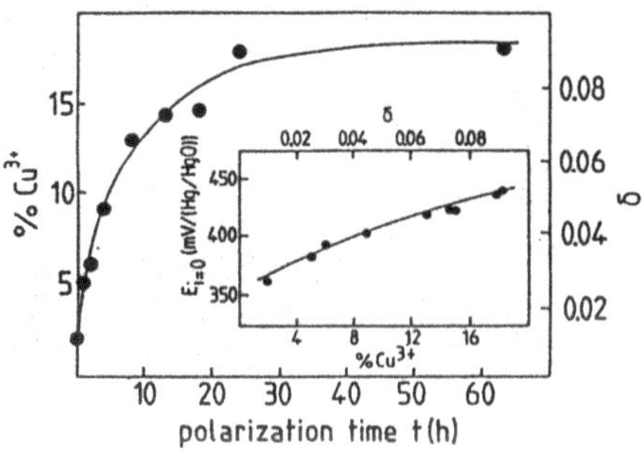

Fig. 3 Polarization time dependence of of the content of Cu^{3+} and insert evolution of the O.C.V. vs Cu^{3+} concentration.

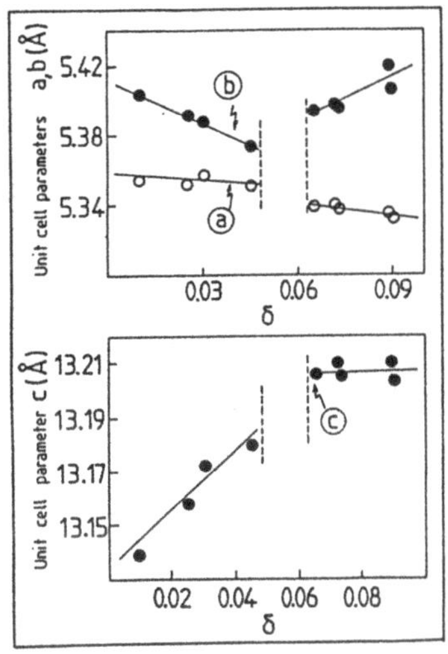

Fig. 4 Evolution of the unit cell parameters as a function of oxygen excess at 300 K.

Results of Rietveld refinement of neutron powder diffraction data on $La_2CuO_{4.09}$ at 300 and 4.2 K have corroborated previous works of C. Chaillout et al.[20] and those of Jorgensen et al.[21] on $La_2NiO_{4+\delta}$ concerning the location of the interstitial oxygen in a site surrounded by four lanthanum and four oxygen atoms (Fig. 5). However this refinement in the Bmab space group that supposes a random occupancy of the interstitial site, does not take into account small diffraction peaks. This calculation leads to average atomic positions, the real microstructure being much more complex as evidenced by electron microscopy.

Electron microscopy experiments of $La_2CuO_{4.09}$ were performed at low temperature (100 K) in order to prevent degradations of the sample

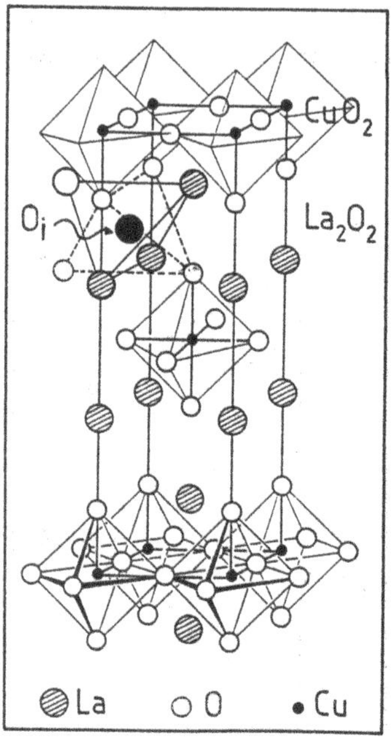

Fig. 5 Idealized structure of La_2CuO_4 showing the interstitial site of the additional oxygen O_i

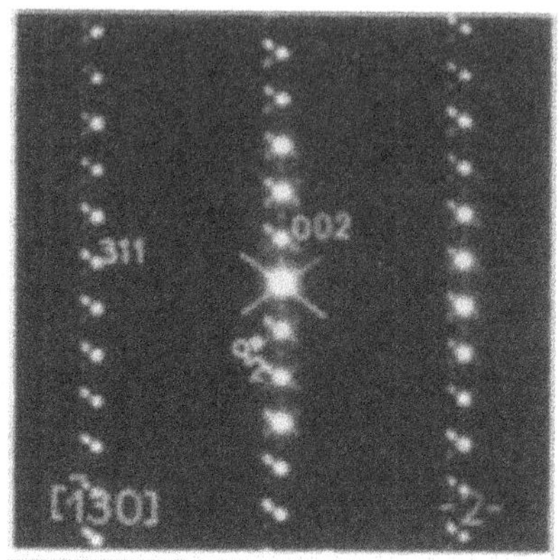

Fig. 6 E.D. pattern of $La_2CuO_{4.09}$ exhibiting incommensurability.

caused by the electron beam under high vacuum. In addition to the main spots corresponding to the basic structure of La_2CuO_4, additional peaks were observed characterizing B or C-type Bravais lattices. Moreover weak satellite spots were also seen in the latter crystals indicating the presence of microdomains showing incommensurability (Fig.6). These superstructure spots were assigned to the local ordering of the additional oxygen atoms involving an inhomogeneous distribution of oxygen.

In a recent study of electrochemically prepared $La_2NiO_{4+\delta}$ phases ($0.17 \leq \delta \leq 0.25$), such ordering was also evidenced and analogies between both systems should be done [12]. Indeed, various superstructures either commensurate or incommensurate have been found in these phases. On the basis of the obtained E.D. patterns a structural model has been proposed to account for the commensurate superstructures of a series of phases, $La_{2n}Ni_nO_{4n+1}$ (δ = 1/n, n ≥ 4). In this structural model the interstitial oxygen atoms are located along the [111] directions of the I4/mmm-type unit cell. Idealized structures have been depicted for the δ = 0.17 (n = 6) and δ = 0.25 (n = 4) compositions.

On the other hand, E.D. patterns of incommensurate structures have also been observed and described using a q^* modulation vector in addition to the three fundamental vectors a^*_t, b^*_t and c^*_t of the I4/mmm structure. Moreover it was shown that the modulation vector of the incommensurate structures lies between the two "modulation vectors" that can be defined for describing the superstructures of the commensurate phases (δ = 0.17 and δ = 0.25 respectively) in the reciprocal plane corresponding to the $[\bar{3}11]_t$ zone axis so that we assumed that a direct relation exists between q^* and δ.

Then a neutron diffraction study of $La_2NiO_{4.25}$[22] showed that the refinements with the Fmmm space group give only an average position of the atoms and that, in order to take into account the weak additionnal diffraction peaks, a larger unit cell (C2 monoclinic) should be considered (Fig. 7).

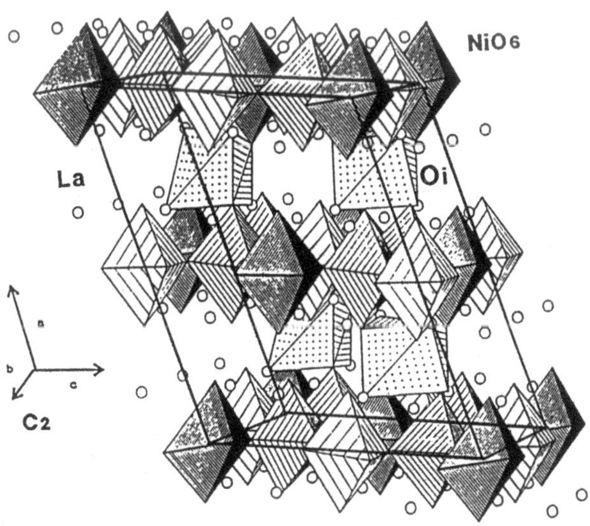

Fig. 7 Structure of $La_2NiO_{4.25}$ (n = 4; $La_8Ni_4O_{17}$; C2 monoclinic)

The excess oxygen ordering was confirmed and therefore we can expect that similar ordering occur in $La_2CuO_{4+\delta}$ phases but corresponding to higher values of n (n ≥ 8 ?) as the oxygen excess is smaller.

In addition on the basis of the structure refinement data, the existence of $[O_3^{5-}]$ polyoxides suggesting the delocalization of one hole over three oxygen atoms has been proposed and correlated to the formation of a C.D.W. in the NiO_2 plane. Such a C.D.W. implies electron transfers

as: $Ni^{2+} + 3O^{2-} \rightleftarrows Ni^{+} + O_3^{5-}$ [23].

or $2\ Ni^{2+} \rightleftarrows Ni^{+} + Ni^{3+}$

3. Transport properties.

Typical thermal evolutions of the electrical resistivity for the $La_2CuO_{4+\delta}$ phases ($0 \le \delta \le 0.09$) are given in Fig.8. The behavior of the resistivity of these materials, at least in the temperature range 200 ≤ T ≤ 300 K shows a drastic change with the composition. For $\delta \le 0.04$ the materials behave as semiconductors whereas for $\delta \ge 0.06$ they are metallic. One should point out that, at 298 K, a transition in the evolution of the unit cell parameters is also noticed close to this

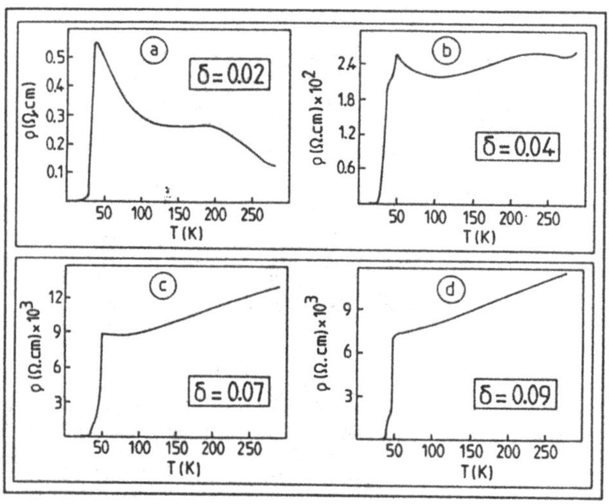

Fig. 8 Temperature dependence of the resistivity for various compositions of the $La_2CuO_{4+\delta}$ system.

composition. Therefore, one can define a critical compositon $\delta_c \simeq 0.05$ that would separate the composition range of an insulating orthorhombic phase O_I ($\delta < \delta_c$) from that of a metallic orthorhombic phase O_M ($\delta > \delta_c$).

3.1. The insulating phase O_I ($\delta < \delta_c$).

The thermal evolution of the electrical resistivity of $La_2CuO_{4.00}$ characterizes an insulator. The introduction of interstitial oxygen that involves a mixed valency of copper causes a large change of behavior with regard to the small values of the oxygen excess. For each composition, a semiconductor-type behavior is observed at room temperature (Fig. 8a-b), then, around 200 K, a hump occurs. This broad maximun was usually assigned to antiferromagnetic ordering eventhough it seems rather independent of the composition.

At decreasing temperature, a more complex evolution appears: the resistivity first decreases then increases circa 120 K before suddenly lessening in the vicinity of 50 K. For $\delta < 0.04$ the resistivity never becomes equal to zero but a superconducting behavior is observed for $\delta \simeq 0.04$ (Fig. 8 b).

As previously mentioned, J.D. Jorgensen et al. were the first to propose a phase separation at decreasing temperature (below room temperature) on the basis of neutron diffraction data [24]. More recently, J. Ryder et al.[25], using resistivity measurements, evidenced this phase segregation in $La_2CuO_{4+\delta}$ (the value of δ can be estimated close to 0.035 according to the values of the resistivity. We carried out similar experiments on $La_2CuO_{4.02}$; the resistivity was measured at increasing temperature either after "quenching" the sample down to liquid helium or after slow cooling (Fig. 9). An hysteresis can be clearly seen, which characterizes the effect of the thermal treatment: only data above 250 K are matching. Following J. Ryder et al., the existence of such an hysteresis can be interpreted in terms of a spinodal decomposition that would occur when the temperature decreases. Then, below a certain temperature, namely T_S, this decomposition leads to the formation of metallic domains (that would become superconducting at lower temperature) within an insulating magnetic phase. In agreement with this assumption, one can notice that an increasing amount of this metallic phase results in lessening the whole value of the resistivity when δ increases, and even in observing an onset of superconductivity for compositions $0.01 \le \delta < 0.04$. For $\delta \simeq 0.04$ these domains begin to percolate and the zero-resistivity appears below $T_c \simeq 30$ K. Therefore the resistivity hump observed around 200-220 K has its origin in this spinodal decomposition rather than in a magnetic ordering as firstly suggested.

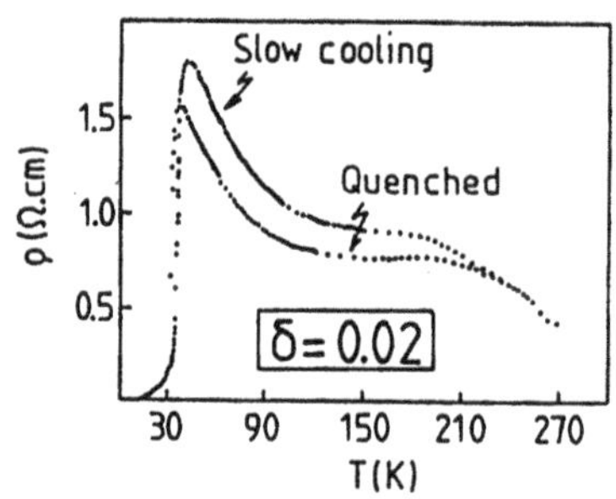

Fig. 9 Temperature dependence of the electrical resistivity of $La_2CuO_{4.02}$ (slow or fast cooling rate)

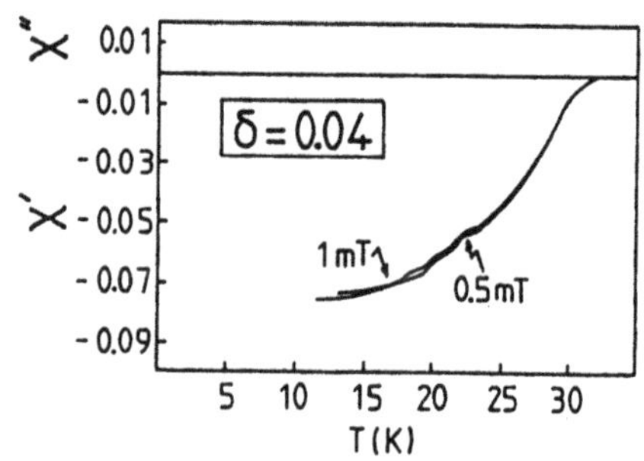

Fig. 10 Temperature dependence of the a.c. susceptibility of $La_2CuO_{4.04}$.

The a.c. susceptibility of $La_2CuO_{4.04}$ was measured under various magnetic fields. The experimental results (Fig. 10) show superconductivity to occur starting from 32 K. One can also notice that, under our experimental conditions, measuring χ'' does not reveal appreciable losses. In addition the amplitude of the real part of the susceptibility χ' remains very small and is independent upon the magnetic field. This behavior can be ascribed to the presence of superconducting domains of very small size whose response is nearly ideal.

The positive value of the thermoelectric power (α) (Fig. 11a-b) characterizes materials containing essentially p-type carriers. One can also observe a change in the thermal behavior of α in the vicinity of T_s from a temperature dependent to non-dependent behavior, which could be correlated to the spinodal decomposition.

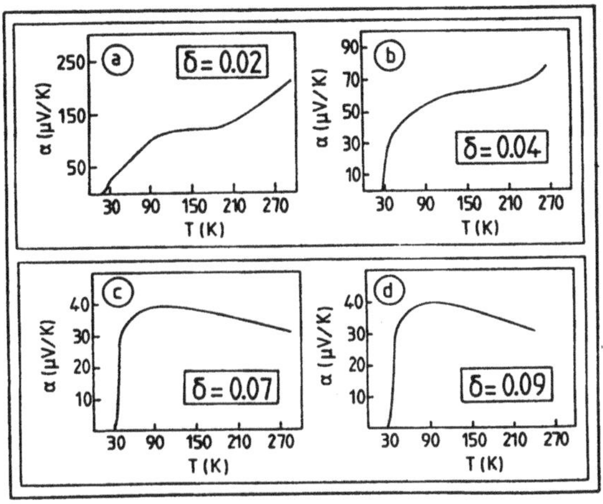

Fig. 11 Temperature dependence of the Seebeck coefficient for various compositions of the $La_2CuO_{4+\delta}$ system.

Above T_S the increase of the thermoelectric power with temperature suggests a hopping mechanism of holes from site to site whose energy lies within an energy band greater than kT.

Below T_S the evolution of the Seebeck coefficient of materials containing both insulating and metallic domains is difficult to explain. However one should point out that for the $\delta \simeq 0.04$ composition for which we assumed the metallic domains to percolate, the behavior of α between 100 and 200 K is very close to that observed for the metallic phase O_M (see further on).

From these data, one can also determine around 90-120 K another characteristic temperature separating the range where α is temperature independent from the range where α is decreasing with T. Following a recent work of J.B. Goodenough et al.[26] on the $La_{2-x}Sr_xCuO_4$ system, this temperature, namely T_ρ, would be the temperature below which dynamic charge fluctuations occur; these fluctuations would be closely related to Cu-O distance fluctuations or in other words to covalency fluctuations, resulting in the formation of domains more or less hole enriched. In the vicinity of T_ρ the resistivity stops to decrease with T and tends to a minimum value, then increases before reaching the critical temperature T_c.

3.2. The metallic phase O_M ($\delta_M \geq 0.06$).

For $\delta \geq 0.06$, the electrical resistivity (Fig. 8c-d) increases with increasing temperature (as for usual metallic compounds), at least above T_ρ whose determination is more difficult. T_ρ seems to decrease from 80 down to 50 K when δ increases from 0.06 to 0.09. In this narrow range of temperature, the increase of resistivity at decreasing temperature can be ascribed, as previously mentioned [26], to the occurrence of charge fluctuations in the normal state just before the superconducting transition.

The a.c. susceptibility of $La_2CuO_{4.09}$ was measured in the same conditions as for $La_2CuO_{4.04}$; diamagnetism appears below 44 K (Fig. 12). Between 32 and 44 K, the variation of χ" reveals non measurable magnetic losses and for magnetic fields ranging between 0.01 and 0.10 mT, the field independent real part of the susceptibility remains very small. For temperature lower than 32 K, χ' indicates a superconducting volume markedly bigger and field dependent.

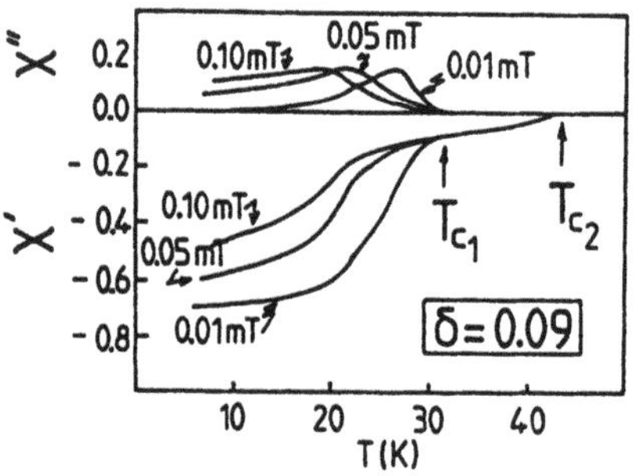

Fig. 12 Temperature dependence of of the a.c. susceptibility of $La_2CuO_{4.09}$

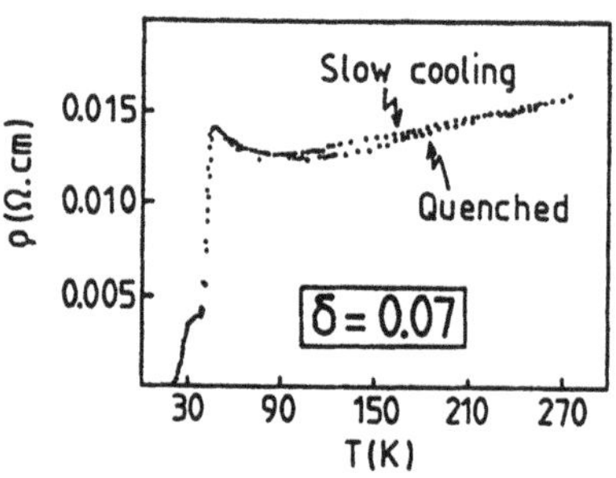

Fig. 13 Temperature dependence of the electrical conductivity of $La_2CuO_{4.07}$ (slow and fast cooling rate)

These data as well as those of the electrical resistivity (Fig. 8c-d) give evidence of two superconducting transitions that could be assigned to the presence of two phases of different compositions , a percolating one whose transition at 32 K is similar to that observed for $La_2CuO_{4.04}$, another one that is minority and non percolating, would be superconducting about 44 K. Therefore one should assume the existence of two metallic phases in the normal state. Resistivity measurements for the $La_2CuO_{4.07}$ composition performed in similar conditions as for $La_2CuO_{4.02}$ did not reveal any hysteresis phenomenon (Fig. 13), neither in the metallic state, nor during the superconducting transition. This result is not completely surprising insofar as it is hardly that the spinodal decomposition into two metallic phases could be observed using this technique as both electrical resistivity values are probably very close.

Electron microscopy observations of $La_2CuO_{4+\delta}$ were carried out as a function of temperature . B-type crystals show apparently no evolution whereas the intensity of the (0,0,2n+1) additional spots as well as the satellite spots of the C-type crystals (Fig. 6) decreases with increasing temperature. Around 250 K, all these weak spots disappear and only the basic spots of the La_2CuO_4 network remain. At decreasing temperature, these spots again appear progressively.

Therefore one may conclude that the occurrence of additional spots of weak intensity below 250 K suggests also the existence of a spinodal decomposition into domains in the metallic phase. This decomposition probably results from an easy migration of the interstitial oxygen atoms leading to stable local ordering. This was recently corroborated by Allan et al.[27] who calculated the migration energy of these oxygen atoms and, assuming O^{2-}, found a rather low value, $E \simeq 0.48$ eV.

The thermal evolution of the thermoelectric power of the metallic phase (Fig. 11c-d) does not exhibit at high temperature the usual behavior of a metal having p carriers. One can indeed observe a decrease of α with increasing temperature instead of the expected behavior given by the relation: $$\alpha = \frac{\pi^2 k^2 T}{3e}\left(\frac{d\,ln\,\sigma(E)}{dE}\right)$$

The thermal evolution of the thermoelectric power of superconducting materials in the normal state show some similarities; more especially it was very often noticed a decrease of α with temperature and even, in some cases, a change in the sign as for instance for $La_{1.86}Sr_{0.14}CuO_4$ whose value of α becomes negative about 450 K or for

$Bi_2Sr_2CaCu_2O_y$ at 180 K [27,28]. In a recent work, J.B. Goodenough et al.[26] claimed that this behavior could characterize an enhancement of the effective mass of the carriers resulting from electron-phonon interactions in a narrow band. This behavior coud be also interpreted in terms of a conduction mechanism implying the existence of a peculiar form for the anionic and cationic bands, especially in a 2D system.

4. Magnetic properties.

The thermal evolution of the magnetic susceptibility of the stoichiometric compound $La_2CuO_{4.00}$ shows a maximun around 310 K (Fig. 14a). In agreement with previous works[30], it is ascribed to the three-dimensional antiferromagnetic ordering (T_N) and characterizes strong magnetic interactions with regard to a material of essentially two-dimensional character. Above T_N the increase of the magnetic susceptibility is also fully consistent with a two-dimensional behavior and the existence of short range magnetic couplings in the "CuO_2" plane up to the vicinity of 1000 K. Inelastic neutron scattering studies have also given evidence of very strong two-dimensional magnetic interactions, the calculated value of J being close to 1000 K[31].

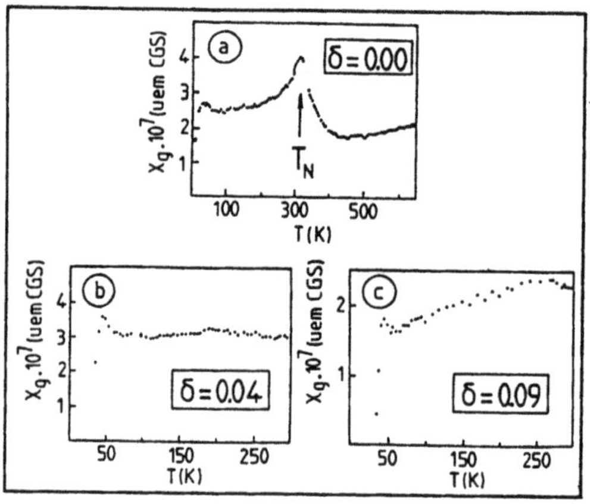

Fig. 14 Temperature dependence of the magnetic susceptibility of $La_2CuO_{4+\delta}$ (δ = 0.00, 0.04 and 0.09)

For the oxidized phases, the insertion of oxygen results in the formation of holes in the "CuO_2" planes. The magnetic ordering is then "geometrically" frustrated and the Néel temperature drastically decreases with δ. Thus, for $\delta = 0.02$, a magnetic transition is observed around 250 K: it was correlated in early works to the maximum of resistivity and considered as the magnetic ordering temperature[32]. However one should note that this transition seems to be almost independent of the composition for $0.01 \leq \delta \leq 0.03$ and has nearly vanished for $\delta \simeq 0.04$ (Fig.14b). Therefore it cannot be related to magnetic dilution but rather is consistent with the spinodal decomposition. For low values of δ - i.e. for small amounts of trivalent copper - the system is unstable and decomposes into hole poored domains in which the magnetic ordering can occur without frustrations and hole rich domains of metallic character correlated to the intercalated oxygen atoms.

For $\delta \geq 0.04$, the metallic character of the material is enhanced and the number of magnetic domains formed during the spinodal decomposition decreases. No magnetic transition occurs and a Pauli-type magnetic susceptibility, almost temperature independent, is observed.

For the metallic phase, even under a high magnetic field (1.8 T), the susceptibility becomes negative, showing that the superconducting state still exists below 44 K (Fig. 14c).

The magnetization of the materials undergoing a superconducting transition ($\delta \geq 0.04$) was measured below T_c, in the superconducting state, under a constant weak field (B = 10 mT) as a function of

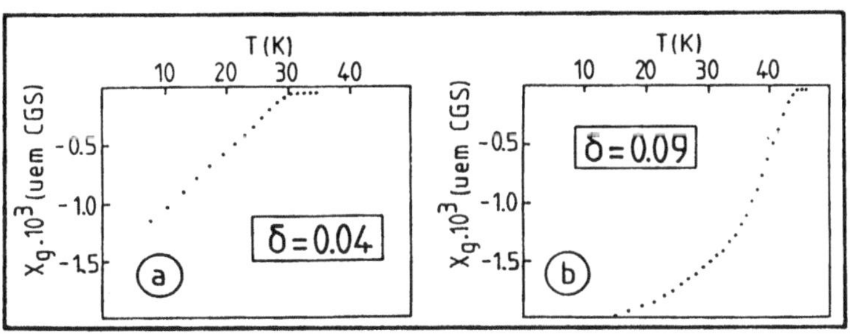

Fig. 15 Temperature dependence of the magnetic susceptibility of $La_2CuO_{4+\delta}$ (δ = 0.04 and 0.09) measured under 10 mT.

temperature. The $La_2CuO_{4.04}$ phase becomes diamagnetic below 30 K (Fig. 15a), which is in agreement with the a.c. susceptibility measurements (cf. Fig. 10). Conversely, for the $\delta \geq 0.06$ materials the superconducting behavior appears at 44 K (Fig. 15b) but one should point out that, under a static field, only one transition is observed. In addition the value of the diamagnetic susceptibility or else the superconducting volume seems to increase with the amount of oxygen overstoichiometry (δ).

The magnetization of $La_2CuO_{4.09}$ *vs.* magnetic field, at 4.2 K, (Fig. 16) shows a linear variation up to 64 mT that can be considered as the value of H_{C1}. A maximun of the magnetization is observed around 100 mT and the material remains still superconducting at 2 T.

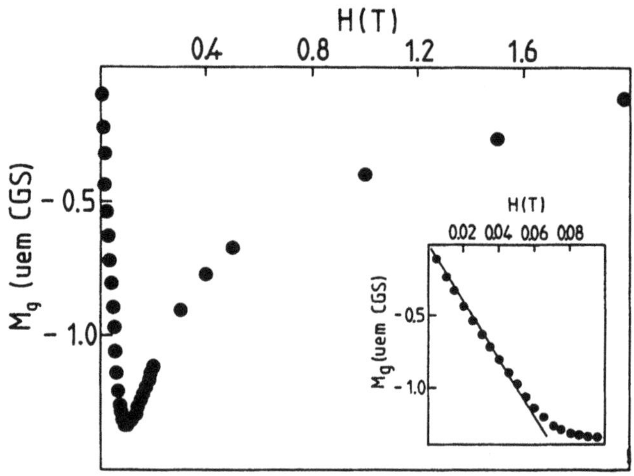

Fig.16 Evolution with the magnetic field of the magnetic moment of $La_2CuO_{4.09}$ at 4.02 K.

- **Tentative phase diagram**

A tentative phase diagram taking into account all the obtained results can be proposed (Fig. 17); its understanding requires some additional explanations.

At room temperature, X-ray diffraction results have led to propose the existence of two phases, the first one (O_I) is insulating ($\delta \leq 0.05$), the other one (O_M) is metallic ($\delta \geq 0.06$). No evidence of a two-phase domain was found, which means that the transition between both phases occurs in a narrow composition range.

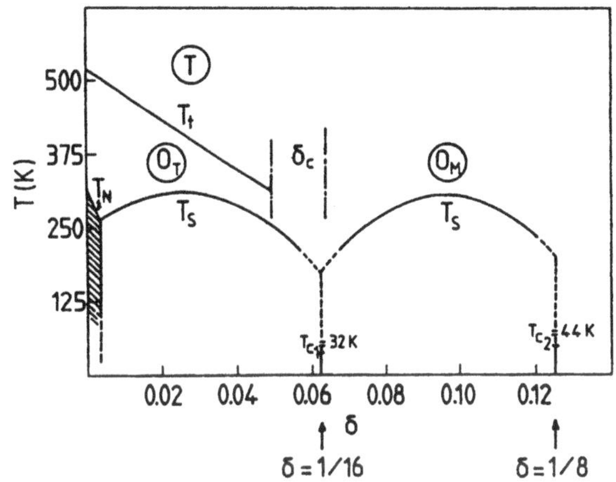

Fig. 17 Tentative phase diagram for the system $La_2CuO_{4+\delta}$ system.

As shown by magnetic measurements, the long range magnetic ordering temperature (T_N) drastically decreases as soon as oxygen is inserted or in other words as soon as formally trivalent copper is formed.

At decreasing temperature a spinodal decomposition occurs resulting in the formation of various phases. This decomposition, evidenced by resistivity measurement for the O_I phase, is likely to occur also for the O_M metallic phase as suggested by the electron microscopy study. It could involve a migration of oxygen as O^- species whose mobility woud be high enough to allow such a diffusion process. This assumption seems reasonable on account of:

i) the value of the diffusion coefficient of the electroactive species determined by means of electrochemical measurements ($D \propto 10^{-11}$-10^{-12} $cm^2.s^{-1}$), which coud result from the presence, at least temporarily of O^- species.

ii) the following equilibrium $Cu^{III} + O^{2-} \rightleftarrows Cu^{II} + O^-$ involving an electronic transfer whose existence is now doubtless in this kind of materials[33,34].

At low temperature this spinodal decomposition would result in the formation of the following phases, O_I, φ_1 and φ_2:

- the O_I phase, $La_2CuO_{4+\varepsilon}$, with an oxygen composition ε close to 0.01, which would exhibit no more long range magnetic ordering but only short range spin fluctuations.

- the φ_1 phase whose superconducting transition temperature is 32 K; its composition is unknown but, on the basis of electrical resistivity measurements, it could correspond to δ values close to 0.06.

- the φ_2 phase whose superconducting transition temperature is 44 K, is oxygen riched. On the basis of the electron microscopy study of $La_2NiO_{4+\delta}$ phases[12], one can imagine that a less oxidized phase of composition close to $1/8 \simeq 0.125$ may exist. Then, supposing this structural model is valid, the φ_2 phase could have a composition close to $\delta \simeq 0.125$. In the same way, the φ_1 phase woud correspond to $\delta = 1/16 \simeq 0.063$.

Both phases having an almost steady superconducting temperature as δ varies, exist over a very narrow range of composition and could be "line phases". Moreover on the basis of our preliminary T.E.M. studies, one can assume that these phases are formed during the spinodal decomposition of the metallic phase, thanks to oxygen migration and would correspond to a peculiar ordering of the additional oxygen atoms.

- CONCLUSION.

All these results allow to draw the following conclusions:

a) - the way for preparing $La_2CuO_{4+\delta}$ plays obviously an important role with regard to the physical properties of these materials more especially superconductivity.

b) - the electrochemical oxidation was shown to be relevant for increasing the amount of oxygen overstoichiometry. Using this process, oxygen atoms can be inserted into La_2CuO_4 in interstitial positions of the La_2O_2 layers.

c) - the reduction of oxygen atoms leads to the formation of holes $[p] = 2\delta$, which are mainly in the CuO_2 planes but which could be partially located in the La_2O_2 layers as delocalized holes in polyoxides $(O_n)^{2n-1}$.

d) - strong similarities exist between the $La_{2-x}Sr_xCuO_4$ and $La_2CuO_{4+\delta}$ system as for instance the existence of an insulator-metal transition for a hole concentration [p] ≃ 0.011, the metallic phase showing superconductivity at low temperature.

e) - due to high oxygen mobility below room temperature, phase separation phenomena occur in this system.

For the semi-conducting phase O_I, a classic spinodal decomposition into $La_2CuO_{4+\varepsilon}$ ($\varepsilon \simeq 0$) and $La_2CuO_{4+\delta}$ ($\delta \simeq 0.06$) is observed . For the metallic phase O_M, the phase separation leads to oxygen ordering into microdomains. According to a structural model proposed for $La_2NiO_{4+\delta}$ phases ($\delta \longrightarrow 0.25$), two phases having definite compositions could be formed corresponding respectively to:

- $\delta = 1/16 \simeq 0.063$, tetragonal average symmetry with $T_c = 32$ K.
- $\delta = 1/8 \simeq 0.125$, orthorhombic average symmetry with $T_c = 44$ K.

- **Acknowledgements:** The authors are very grateful to B. Darriet, A. Demourgues, P. Dordor; J-P. Doumerc, J. Etourneau, N. Lagueyte, J-C. Park and F. Weill (Laboratoire de Chimie du Solide) for their participation in this work and to J.B. Goodenough (The University of Texas) for fruitful discussions.

- REFERENCES.

1 J. Beille, R. Cabanel, C. Chaillout, B. Chevalier, C. Demazeau, F. Deslandes, J. Etourneau, P. Lejay, C. Michel, J. Provost, B. Raveau, A. Sulpice, J.L. Tholence and R. Tournier, *C. R. Acad. Sc. Paris*, **304(II)** (1987), 1097.

2- J.G. Bednorz and K.A. Müller, *Z. Phys.* **B 64**, (1986), 189.

3-J.M. Tarascon, L.H. Greene, W.R. Mackinnan, G.W. Hull and T.H. Geballe, *Science*, **235**, (1987), 1373.

4- A. Maignan, C. Michel, M. Hervieu, J. Provost, D. Groult and B. Raveau *Physica C*, **191**, (1992), 333.

5- B. Chevalier, A. Tressaud, B. Lépine, K. Amine, J.M. Dance, L. Lozano, E. Hickey and J. Etourneau, *Physica C*, **167**, (1990), 97.

6- A. Tressaud, C. Robin, B. Chevalier and B. Lépine, *J. Less-Common. Met.*, **164/165**, 832 (1990)

7- A. Tressaud, C. Robin, B. Chevalier, L. Lozano and J. Etourneau, *Physica C*, **177**, (1991), 330

8- A. Wattiaux, L. Fournés, A. Demourgues, N. Bernaben, J.C. Grenier and M. Pouchard,
Sol. State Comm., 77(7), (1991), 489

9- P. Bezdicka, J.C. Grenier, A. Wattiaux, M. Pouchard and P. Hagenmuller,
Z. anorg. allg. Chem. (submitted)

10- A. Wattiaux, J.C. Park, J.C. Grenier and M. Pouchard,
C.R.Acad.Sci. Paris,**310(II)**, (1990) 1047

11- J.C. Grenier, A. Wattiaux, N. Lagueyte, J.C. Park, E. Marquestaut, J. Etourneau and M. Pouchard,
Physica C, **173(3-5)**, (1991) 139

12- A. Demourgues, F. Weill, J.C. Grenier, A. Wattiaux and M. Pouchard,
Physica C, **192**, (1992) 425

13- F. Arrouy, A. Wattiaux, C. Cros, G. Demazeau, J-C. Grenier, M. Pouchard and J. Etourneau
Physica C, **175**, (1991) 342

14- P. Rudolf, W. Paulus and R. Schöllhorn,
Adv. Mater., **3(9)**, (1991) 438

15- R. Suryanarayanan, O. Gorochov, M.S.R. Rao, L. Ouhammou, W. Paulus and G. Heger
Physica C, **185**, (1991) 573

16- N. Lagueyte, A. Wattiaux, J.C. Park, J-c. Grenier, L. Fournes and M. Pouchard
J. Phys. III France, **1** (1991), 1755.

17- J.C. Park, P.V. Huong, M. Rey-Lafon, J-C. Grenier, A. Wattiaux and M. Pouchard
Physica C, **177**, (1991) 487

18- M.S. Whittingham, in "Intercalation Chemistry", (M.S. Whittingham and A.J. Jacobson, Eds), (1982), 1.

19- A. Wattiaux, J-C. Grenier and M. Pouchard (to be published)

20- C. Chaillout, J. Chenavas, S.W. Cheong, Z. Fisk, M. Marezio, B. Morosin and J.E. Schirber,
Physica C, **170**, (1990), 87.

21- J.D. Jorgensen, B. Dabrowski, Shiyou Pei, D.R. Richards and D.G. Hinks,
Phys. Rev. B, **40**(4), (1989) 2187

22- A. Demourgues, A. Wattiaux, J-C. Grenier, M. Pouchard, J-L. Soubeyroux, J-M. Dance and P. Hagenmuller,
J. Sol. State Chem., (submitted)

23- A. Demourgues, F. Weill, B. Darriet, A. Wattiaux, J-C. Grenier and M. Pouchard,
J. Sol. State Chem. (submitted)

24- J.D. Jorgensen, B. Dabrowski, Shiyou Pei, D.G. Hinks, L. Soderholm, B. Morosin, J.E. Schirber, E.L. Venturini and D.S. Ginley,
Phys. Rev. B, **38(16)**, (1988) 11337

25- J. Ryder, P.A. Midgley, R. Exley, R.J. Beynon, D.L. Yates, L. Afalfiz and J.A. Wilson,
Physica C, **173**, (1991) 9

26- J.B. Goodenough, J.S. Zhou, K. Allan,
J. Mater. Chem., **1(5)**, (1991) 715.

27- N.L. Allan and W.C. Mackrodt
Phil. Mag., **64(5)**, (1991) 1129.

28- T. Ohtani, K. Kobatake and T. Takehara,
Physica C, **179**, (1991) 376

29- M.F. Crommie, Amy Y.Liu, Marvin L. Cohen and A. Zettl,
Phys. Rev. B, **41**, (1990) 2526

30- M.F. Hundley, R.S. Kwok, S.-W. Cheong, J.D. Thompson and Z. Fisk,
Physica C, **172**, (1991) 455

31- G. Shirane et al.,
Phys. Rev. Lett., **59**, (1987) 1613

32- P.M Grant, S.S.P Parkin, V.Y. Lee, M. Engler, G. Lim and R.D. Dabrowski,
Phys. Rev. Lett., **58(23)**, (1987) 2482

33- A. Fujimori, E. Takayama-Muromachi, Y. Ushida and B. Okai,
Phys. Rev. B, **35**, (1987) 8814

34- A. Bianconi, A. Congiu Castellano, M. De Santis, P. Delogu, A. Gorgano and R. Giorgi,
Sol. State Comm., **63**, (1987) 1135

NEUTRON SCATTERING STUDY OF THE $YBa_2Cu_3O_{6+x}$ SYSTEM

P. Burlet[1], L.P. Regnault[1], J. Rossat-Mignod[1,2], P. Bourges[2], C. Vettier[3], J.Y. Henry[1], V.P. Plakthy[1], C. Marin[1].

1 Centre d'Etudes Nucléaires de Grenoble, DRFMC/MDN, 85X, 38041 Grenoble cedex,France

2 Laboratoire Léon Brillouin, CEA-CNRS, CE Saclay, 91191 Gif-sur-Yvette cedex, France

3 European Synchrotron Research Facility, BP 220, 38043 Grenoble cedex, France

Abstract

We report on neutron scattering experiments carried out on single-crystals of the $YBa_2Cu_3O_{6+x}$ system. The oxygen ordering in the ortho-II phase $YBa_2Cu_3O_{6.45}$ has been studied by neutron diffraction. The atomic displacements associated to this ordering has been derived. The spin dynamics has been systematically investigated in the six typical regimes : the pure and doped AF states (x = 0.15, 0.37), the weakly doped (x = 0.45, 0.51), the heavily doped (x = 0.69, 0.92) and the overdoped (x = 1) metallic states which all exhibit a superconducting ground state. The hole doping strongly affects the AF-order, dynamical AF-correlations remain in the metallic state. An energy gap has been discovered in the spin excitation spectrum of any superconducting samples, however, it becomes weaker close to the I-M transition. A quite unusual T-dependence of the spin excitation spectrum was found.

1. Introduction

Since the discovery of superconductivity in copper oxide materials by J.G. Bednorz and K.A. Muller [1] a huge amount of experimental and theoretical works has been reported [2]. However there is not yet a well accepted understanding of their unusual physical properties in the normal state and the origin of the physical mechanism involved in the Cooper pair formation is still controversial.

With the advances in single crystal growth careful and quantitative experimental works on well characterized high quality materials are becoming available [3] and are establishing on firm grounds that cuprate oxydes are not "classical" metals but they must be considered as strongly correlated electronic systems

Macroscopic measurements of transport (resistivity, Hall effect, thermoelectric power), magnetic (susceptibility, magnetization) and thermodynamical (specific heat) properties were fundamental in establishing the unusual properties of these materials. However microscopic measurements were of crucial importance to get more insight in the physics of these high T_c materials.

First of all high energy absorption spectroscopies gave unvaluable informations to clarify the electronic structure. These materials actually are charge transfer Mott insulators [4] with strong electronic correlations, the metallic state is reached by transfering, from charge reservoir layers, holes in oxygen p-orbitals of the electronically active CuO_2 layers [5].

This charge transfer is obtained by doping either by substitution of cations or by changing the oxygen content. This is generally associated to rather important structural modifications and detailled crystal structure studies are needed.

The second important piece of information is the knowledge of the charge excitation spectrum. In this respect, photoemission experiments and optical (Infrared and Raman scattering) studies clearly have established the strongly correlated nature of the electronic properties : strong reduction of the bandwidth [6], excitence of mid-infrared gap states [7].

The third piece of information which is crucial for the ground to built an appropriated theory is the knowledge of the spin excitation spectrum. Nuclear magnetic resonance (NMR) and inelastic neutron scattering (INS) experiments, by establishing the existence of spin excitations, have provided master piece informations which prove the strongly correlated nature of the CuO_2 electronic planes. They are actually complementary techniques, NMR is probing locally with high accuracy the zero-frequency spin fluctuation spectrum at different sites (Cu, O, Y,...), whereas INS gives informations on both spatial and temporal spin fluctuations over the whole crystal. Neutron scattering actually is the unique probe to get both the wave vector and the energy dependences of the spin excitation spectrum, i.e. the imaginary part of the dynamical spin susceptibility $\chi(\vec{q},\hbar\omega)$, which is needed to interpret correctly NMR results. The power of the technique is moderated by the weakness of the scattering which requires large single-crystal samples.

As far as INS measurements are concerned, an important activity started since the discovery of high T_c materials in Europe, USA and Japan but it is during these last two years that very important results were obtained in the metallic and superconducting state as the evidence of a spin gap in the $YBa_2Cu_3O_{6+x}$ system [8] and the existence of incommensurate peaks in the $La_{2-x}(Ba,Sr)_xO_4$ system [9] at low temperatures.

We have focused our interest, from the beginning, on the $YBa_2Cu_3O_{6+x}$ system because we were able to produce a large single crystal (~ 0.3 cm^3) of good quality in which the oxygen content can be changed from O_6 to O_7 with good accuracy and homogeneity. During the last five years a continuous investigation has been carried out, starting from the determination of the antiferromagnetic (AF) ordering, the magnetic phase diagram and a systematic study of the spin dynamics in the six characteristic regimes with increasing difficulty from the pure (x = 0.15) and doped (x = 0.37) AF-states, the weakly doped metallic state (x = 0.45 and 0.51), the heavily doped metallic state exhibiting superconductivity below T_c= 60 K (x = 0.69) and T_c = 90 K (x = 0.92), and the overdoped metallic state (x = 1). In this paper we report first on a structural study of the Ortho-II phase $YBa_2Cu_3O_{6.45}$ in which the oxygen ordering in Cu1 planes induces important atomic displacements. Then the results obtained by I.N.S. will be presented , a special emphasis will be payed on the recent and more accurate results obtained on the superconducting samples with x = 0.51, 0.92 and 1 using the high flux triple-axis spectrometer 2T at the Laboratoire Léon Brillouin. We have to mention that a detailed study as been carried out recently by Tranquada et al [10] on a sample with x = 0.6.

2. Crystal structure of the ortho-II phase $YBa_2Cu_3O_{6.45}$.[11]

The crystal structure of the ortho-I phase at x=1 has been intensively studied, mainly by neutron diffraction, and determined very accurately as well as that of the non superconducting tetragonal phase at x=0. The structure of the ortho-II phase which exits in the vincinity of x=0.5 is not yet well known . In a number of electron diffraction experiments [12] superstructure reflexions (h/2,k,l) with h=2n+1 have been observed that give evidence of a unit cell doubling. This superstructure was explained by an ordering of full and empty chains of oxygens in the Cu(1) planes. No additional information was obtained in a few X-Ray[13] and neutron diffraction[14] experiments and the atomic displacements which should accompagny the oxygen ordering are still unknown.

In a neutron diffraction experiment carried out at the Siloë reactor of C.E.N. Grenoble on a single crystal of $YBa_2Cu_3O_{6.45}$ with Tc=39 K we can measure these parameters [11].

Q-scans along the (h,0,0) and (0,k,0) directions around the superlattice (0.5,0,1) position are given in Figure.1. The fit of these experimental profils, taking into account the instrumental resolution, gives a short correlation length ξa=94(8) Å across CuO chain in the basal plane while along the chain direction no broadening is observed and only a lower limit of the correletion length ξb=350 Å can be deduced.

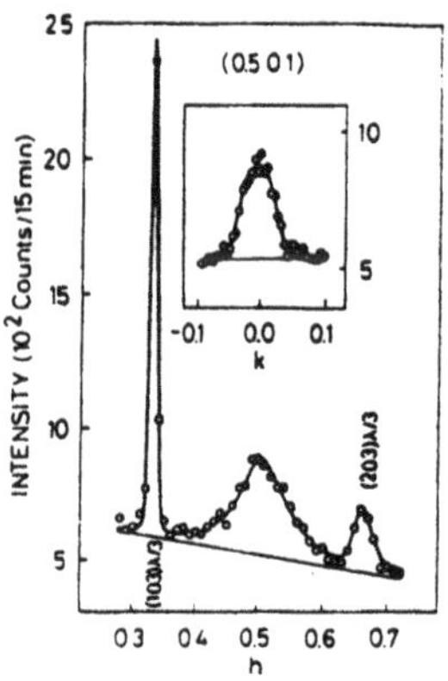

Fig. 1. Scan through the (0.5,0,1) superstructure reflexion along the (h,0,0) and 0,k,0) directions.

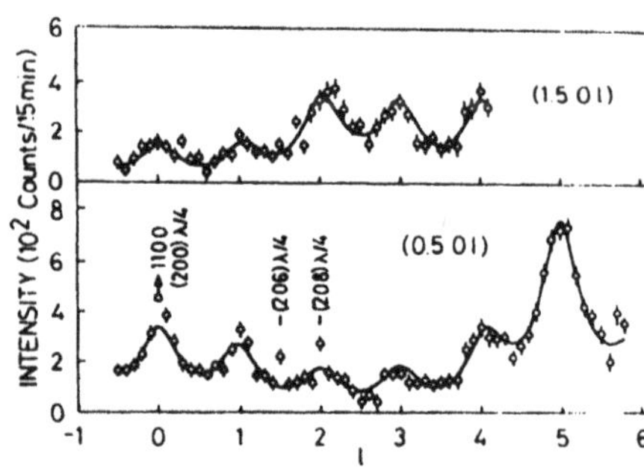

Fig. 2. (0.5 0 l) and (1.5 0 l) scans

Moreover Q-scans along (0.5,0,l) and (1.5,0,l) direction (Fig. 2) shows that the surperstructure peaks have very different intensities which proves that other atoms than the oxygens in the chains are involved because, if not, all the superstructure reflexions would have the same structure factor and then equivalent intensities. An other important fact in figure 2 is that the superstructure peaks are rather broad which implies short correlation lengths along the c axis.The (0.5,0,l) and (1.5,0,l) scans have been fitted using a structure factor taking into account the ordering of the oxygens and atomic displacements compatible with the symmetry and the wave-vector k=[1/2 0 0] of the superstructure. A correlation length ξc=43(2) Å and atomic displacements along the x direction δx = 0.038(2) and 0.010(3) Å of the Ba and Y atoms respectively and along the z direction δz =0.012(2) Å for the Cu(1) , δz = 0.038(2) Å for the apex oxygen O(1) have been determined.

A schematic representation of this superstruture is given in Figure 3 which shows the double unit cell, the displacements of the atoms from there average positions are indicated by arrows. This structure seems to be obvious. Due to the electrostatic interaction, the Ba ions are shifted toward the oxygen-filled chains which carry an excess of negative charges. The atomic positions in the vincinity of complete and vacant chains are similar to that of the x=1 and x=0 materials, respectively. In particular the O(1)-Cu(2) distances are short (2.32 A) and long (2.42 A) if they are associated with filled and empty chain respectively

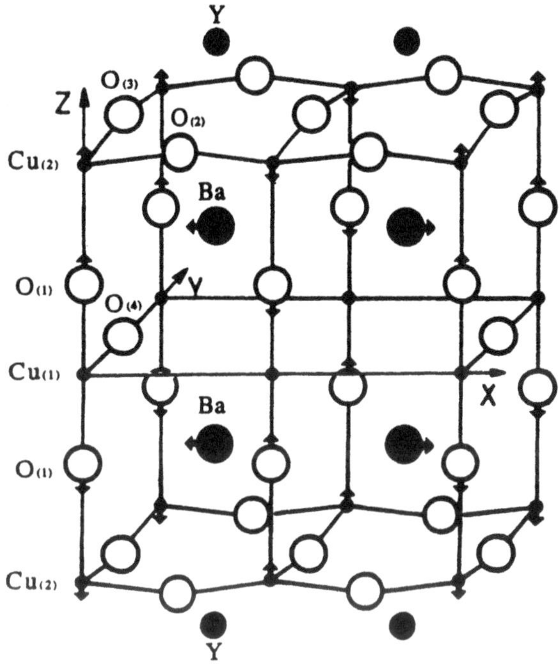

Fig. 3 The double unit cell of the ortho-II phase $YBa_2Cu_3O_{6.45}$

3. Spin dynamics in the $YBa_2Cu_3O_{6+x}$ system

3.1. The pure and doped AF-states [15, 16, 17]

The AF-structure of $YBa_2Cu_3O_6$ (T_N = 415 ± 5 K, **k** = (1/2 1/2 0), m_0 = 0.64 μ_B), shown in fig. 4, is not affected by oxygen doping up to x = 0.20. Only copper atoms in CuO_2 layers (Cu(2) planes) carry a magnetic moment and the coupling between two layers is antiferromagnetic . Cu atoms in Cu(1) planes are indeed in a non magnetic Cu^+ state.

For x > 0.20 magnetic Bragg peak intensities, reported in fig. 5, show that both T_N and m_0 decrease, rather smoothly up to x = 0.35 and more abruptly above resulting in a sharp destruction of the long range AF-ordering at x_c = 0.41 ± 0.02. From these data the AF-part of the phase diagram was deduced (see fig. 6), the metallic part was obtained by measuring the superconducting phase transition.

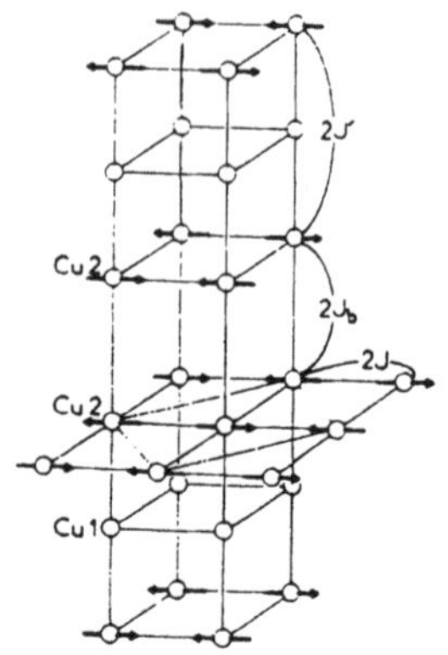

Fig. 4. Antiferromagnetic structure of $YBa_2Cu_3O_6$.

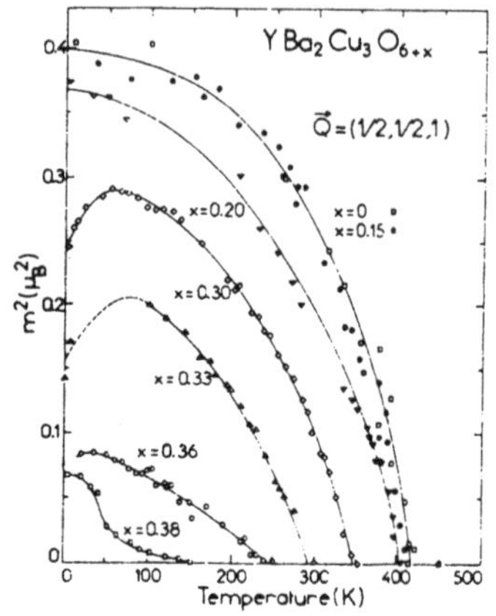

Fig. 5. Intensity of the magnetic Bragg peak (1/2, 1/2, 1) as a function of temperature for various oxygen contents.

A detailed investigation of a sample with x = 0.37 has shown that magnetic Bragg peaks coexist together with rods of magnetic scattering. So AF-polarons, created by the oxygen 2p-holes, induce some disorder which mainly consists in a rotation of the local moment direction within the basal plane. From the in-plane correlation length we can deduce that a hole concentration of about 2 % is large enough to prevent any 3d long range AF-order.

At low temperature a reentrant behaviour of the AF-order parameter takes place which could be explained by the concept of magnetic two level systems [18] produced by the AF- polarons. At large doping and at high temperatures the AF-order parameter exhibits a quite unusual T-dependence which could be attributed to the hole motion by inducing additional quantum spin fluctuations.

In the AF-state the spin excitation spectrum, which is well described by the spin wave theory, allow us to determine the interaction parameters. A large in-plane spin wave velocity c_0 = 1 eVÅ has been deduced, yielding a Cu-Cu superexchange interaction $J_{Cu.Cu}$ = 0.15 eV. The coupling between the two Cu(2) layers is rather large ($J_b \simeq 10^{-2}$ - $10^{-1}J$) but the coupling between bilayer is very weak ($J' = 10^{-5}J$), as well as the planar anisotropy ($\Delta J/J = 10^{-4}$).

In the doped AF-state, the small amount of p-holes strongly modifies the spin dynamics. The disorder in the spin direction, induced by AF-polarons, gives rise to a strong damping of in-plane spin excitations and a renormalization of the spin wave velocity and the anisotropy gap, as shown in fig. 7. The renormalization is found to be q-dependent and decreases as q becomes larger than the reciprocal of the correlation length. Therefore at the critical hole concentration n_h^c = 2 %, the spin wave velocity is expected to vanish together with the spin-stiffness constant.

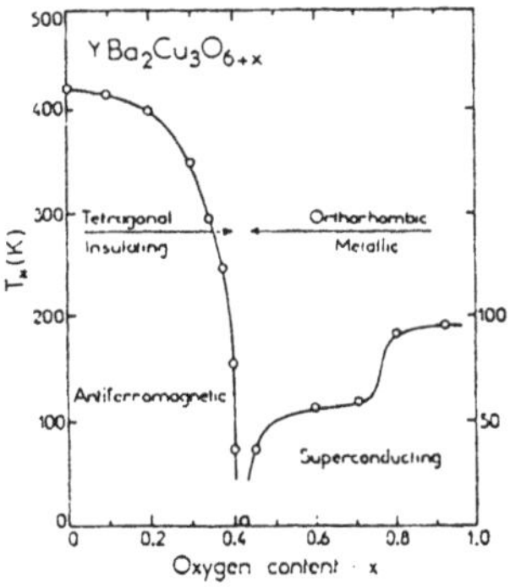

Fig. 6. Phase diagram of the $YBa_2Cu_3O_{6+x}$ system as a function of the oxygen content.

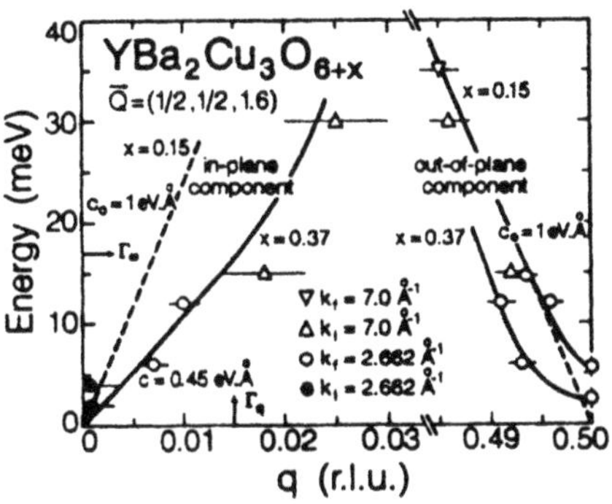

Fig. 7. Excitation energies of in-plane and out-of-plane spin components in the pure (x = 0.15) and the doped (x = 0.37) AF-states.

3.2.. The weakly doped metallic state [16, 19, 20, 21, 22]

For an oxygen content larger that $x_C = 0.41$, an insulating metal (I-M) transition takes place because of a sudden transfer of a large amount of p-holes (~ 10 %) from Cu(1) into Cu(2) planes. Their sudden transfer occurs because of some oxygen ordering to form long Cu-O chains inducing then a tetratogonal-orthorhombic phase transition. Experiments carried out on samples with oxygen contents x = 0.45 and 0.51 and exhibiting superconducting transitions at $T_C \simeq 40$ K indicate that the 3d-AF ordering does not survive to such a p-hole doping. However the magnetic scattering is still observed, it consists in a broad spectrum in energy (see fig.8) weakly T-dependent which is concentrated in q-space along the 2d-AF rods, as can be seen in fig. 9 and 10 which show q-scans across and along the rod, respectively.

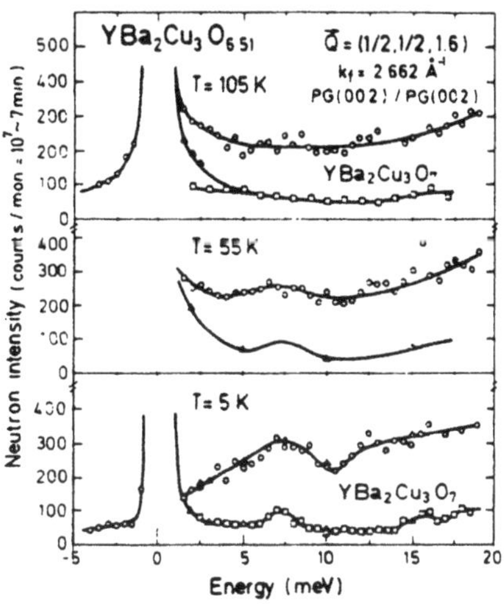

Fig. 8. Low energy scans performed at **Q** = (1/2, 1/2, 1.6) for $YBa_2Cu_3O_{6.51}$ as a function of temperature. The background (Δ) is well represented by the scattering.measured`on the fully oxygenated sample $YBa_2Cu_3O_7$.

These results prove that dynamical AF-correlations persist in the metallic state. From the q-width, $\Delta q = 0.10 \pm 0.01$ r.l.u. a correlation length $\xi_{AF}/a = 2.2$ can be deduced. It must be emphasized that this q-width is energy-independent and T-independent up to room temperature. Moreover the line shape is not of Lorentzian type but rather of Gaussian type.

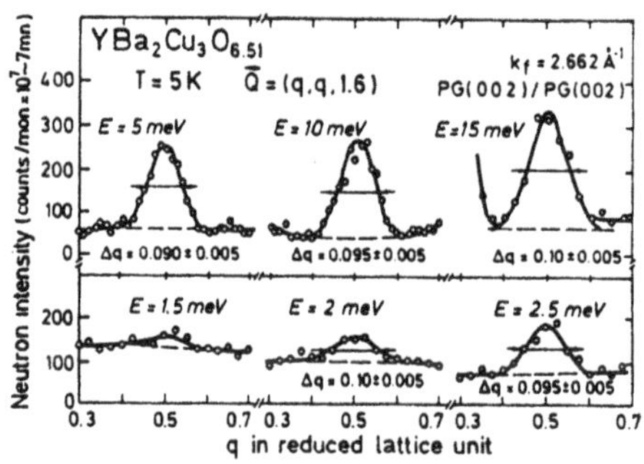

Fig. 9. Q-scans across the 2d-magnetic rod at **Q** = (1/2, 1/2, 1.6) and different energies for $YBa_2Cu_3O_{6.51}$ in the super conducting state (T = 5 K).

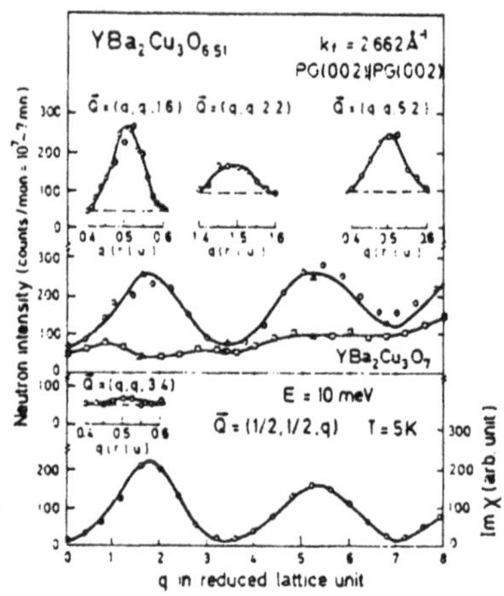

Fig. 10. Q-scans along the 2d-magnetic rod **Q** = (1/2, 1/2, q), at $\hbar\omega$ = 10 meV for $YBa_2Cu_3O_{6.51}$ at T = 5 K. The background (Δ) value, together with measurements on $YBa_2Cu_3O_7$ yield on almost complete AF-coupling between the two Cu(2) layers.

Another very important result is the strong intensity modulation observed in q-scans along the rod (see fig. 10) which demonstrate that the AF-coupling between copper spins in the two adjacent Cu(2) layers is not affected in the metallic state even up to room temperature. This result could mean a certain coherence between the two Cu(2) layers in the hole motion within CuO_2 planes.

Actually the physical quantity of interest we can extract from the measured magnetic cross section is the imaginary part of the dynamical susceptibility $\chi(\vec{Q}, \hbar\omega)$ according to the following expressions :

$$\frac{d^2\sigma(\vec{Q},\hbar\omega)}{d\Omega\, d\hbar\omega} \propto \sum_{\alpha=x,y,z} (1 - \hat{Q}_\alpha^2)\, S^{\alpha\alpha}(\vec{Q},\hbar\,\omega)$$

where the dynamical structure factor $S^{\alpha\alpha}(\vec{Q},\hbar\omega)$ is the Fourier transform of the correlation function of the α-spin components with

$$S^{\alpha\alpha}(\vec{Q},h\,\omega) = \frac{1}{\pi}\,\frac{1}{1-\exp(-h\omega/kT)}\,\mathrm{Im}\,\{\chi(\vec{Q},h\,\omega)\}$$

By combining low and high energy scans the T-dependence of Imχ has been determined (see fig. 11). At low - T, Imχ exhibits a broad maximum, from which a characteristic energy $\Gamma_\omega \approx 18$ meV can be estimated, and falls off at high energies much faster than a simple Lorentzian function, the scattering above 50 meV being quite small. By increasing the temperature the low energy part of the spectrum is strongly depressed in a non trivial manner, a simple hω/kT law is not obeyed. From 200 K to 287 K, the maximum of Imχ is T-independent, yielding a quasi-particule characteristic energy $\Gamma_\omega \simeq 28$ meV, at room temperature.

The low energy part of the spectrum (see fig. 9 and 11) clearly shows that an energy gap is opening in the spin excitation spectrum as the temperature is decreased. It must be noted that the spin gap has a quite small value only of the order of kT_C which may indicate a gapless superconductivity on approaching the I-M transition.

In conclusion, in the weakly doped regime, the spin excitation spectrum is not that expected for a disordered AF-system but it reflects the behaviour of the quasi-particles which undergo a superconducting transition at low-T with a quite small spin gap value. The small value of the characteristic energy ($\simeq$ 28 meV) results certainly in many body effects due to strong correlations.

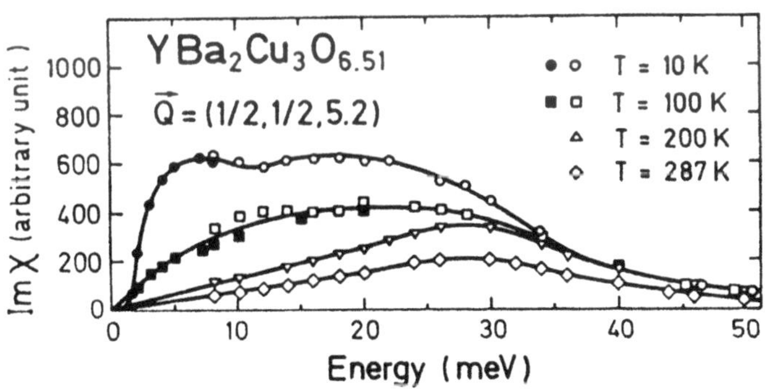

Fig. 11. Im$\{\chi(\vec{Q},h\omega)\}$ as a function of the energy for $YBa_2Cu_3O_{6.51}$ for increasing temperature.

3.3 The heavily doped metallic regimes

3.3.1 The 60 K-superconducting phase [20, 21, 22]

A sample with an oxygen content x = 0.69 (T_C = 59 K), at the end of the T_C = 60 K plateau, has been investigated and a magnetic scattering has been clearly identified. The obtained low energy spectra as a function of temperature are reported in fig. 12 in the form of Imχ. The broad energy spectrum yields as characteristic energy $\Gamma_\omega \simeq 28$ meV quite similar to the previous case. However at low-T the spin excitation spectrum is not enhanced, as previously but strongly suppressed by the appearance of an energy gap with a much larger value, $E_G = 16 \pm 1$ meV, in comparison with kT_C ($E_G/kT_C \simeq 3.2$).

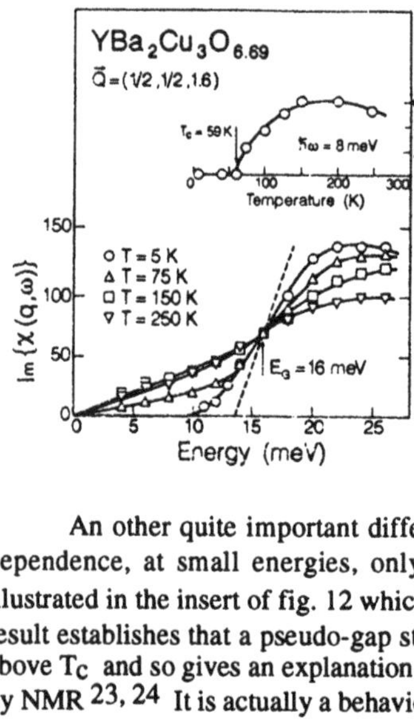

Fig. 12. Low energy part of the spin excitation spectrum of Im{$\chi(\vec{Q},\hbar\omega)$} for $YBa_2Cu_3O_{6.69}$ below (T = 5 K) and above (T = 75, 150, 250 K) the superconducting transition (T_C = 59 K). An energy gap (E_G= 16 meV) clearly persists above T_C.

An other quite important difference is the fact that Imχ recovers a linear energy dependence, at small energies, only above about 160 K. This unusual behaviour is illustrated in the insert of fig. 12 which shows the T-dependence of the slope of Imχ. This result establishes that a pseudo-gap starts to develop in the spin excitation spectrum well above T_C and so gives an explanation for the unusual T-dependence of $1/T_1T$ as measured by NMR [23, 24] It is actually a behaviour typical for the heavily doped regime.

3.3.2 The 90 K-superconducting phase [21,22]

A sample with an oxygen content x = 0.92 has been successfully investigated a year ago and the results were presented for the first time at the high T_C Conference in Kanazawa [21]. Energy scans exhibit a rather complexe lineshape (see fig. 13). However q-scans across (fig. 14) and along (fig. 15) the (1/2 1/2 l) rod, clearly show evidence for a magnetic scattering contribution at high energies and allow us to define the nuclear contribution. The T-dependence of the magnetic scattering has been determined and is plotted in fig. 16 in the form of Imχ.

Several unusual features have to be underligned. At T = 5 K, there is no measurable scattering below 25 meV giving evidence for a sharp energy gap in the spin excitation spectrum at $E_G = 28 \pm 1$ meV. More surprising is the sharp drop of the spectrum above 45 meV and the existence of a strong enhancement of the spectrum around 41 meV resulting from a narrowing of the q-width : 0.18 r.l.u. instead of 0.25 r.l.u. (see fig. 11).

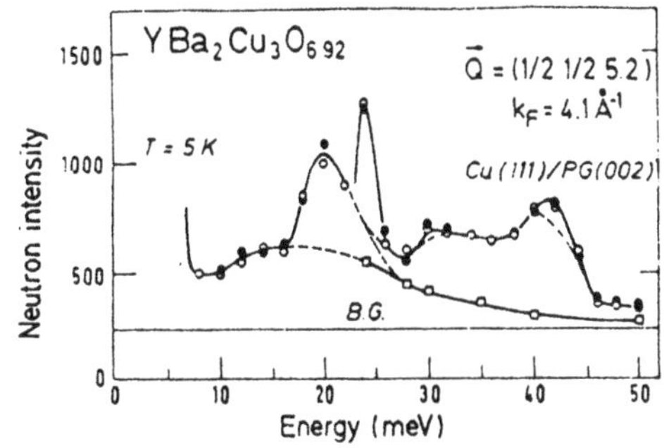

Fig. 13. Energy scans performed at **Q** = (1/2, 1/2, 5.2) for $YBa_2Cu_3O_{6.92}$ in the superconducting state (T_C = 91 K). Above the background (BG) the nuclear contribution is indicated (□) together with a spurious contribution arising from the (006) Bragg peak.

This enhancement already desappears at T = 81 K, i.e. below T_C = 91 K, and correlatively some excitations appear in the gap. It must be emphasized that the spin excitation gap is not decreasing with temperature as expected in the classical BCS theory.

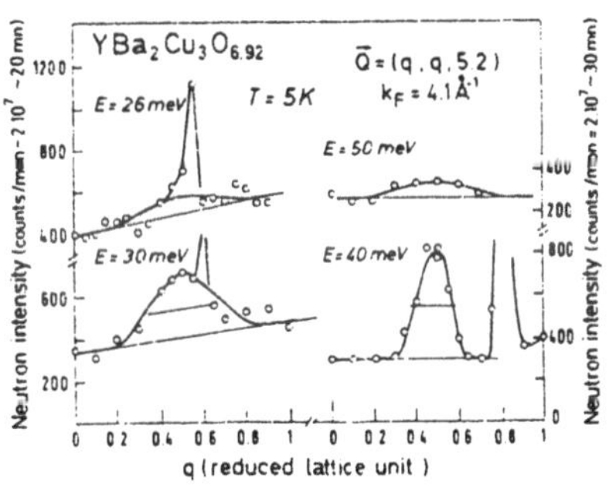

Fig. 14. Q-scans across the 2d-magnetic rod at **Q** = (1/2, 1/2, 5.2) and different energies for $YBa_2Cu_3O_{6.92}$ in the superconducting state (T_C = 91 K).

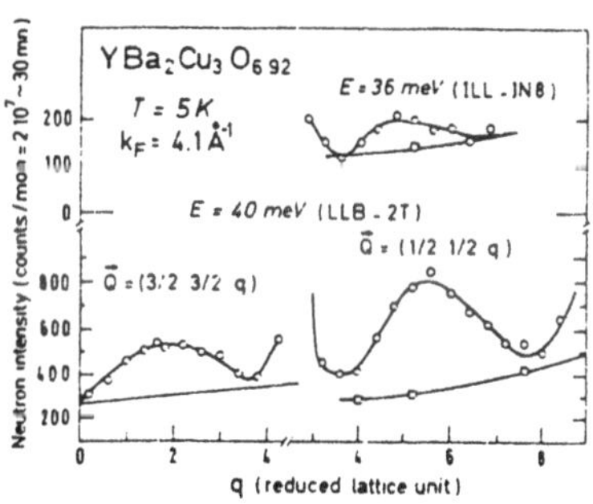

Fig. 15. Q-scans along the 2d-magnetic rod for $YBa_2Cu_3O_{6.92}$ (T = 5 K) showing the almost complete AF-coupling between the two Cu(2) layers.

At higher temperatures, the magnetic scattering is strongly depressed. However it must be noticed that the characteristic energy scale Γ_ω = 30 ± 2 meV does not change with temperature, nor does the AF-correlation length which is now quite short (Δq = 0.27 ± 0.02 r.l.u., ξ/a = 0.84 ± 0.04). It is very astonishing to find an almost complete AF-coupling between the two Cu(2) layers with a so short in-plane correlation length. This may be a signature of a singlet spin liquid state.

As in the 60 K-superconducting regime, Imχ recovers a linear energy dependence only well above T_C. Actually the slope Im$\chi/\hbar\omega$ (see fig. 17) starts to decrease below 120

130 K in excellent agreement with $1/T_1T$ NMR measurements performed on the same sample [24]. Therefore a pseudo-gap is also opening in the 90 K-phase with x = 0.92 but at a smaller temperature than for the 60 K-phase.

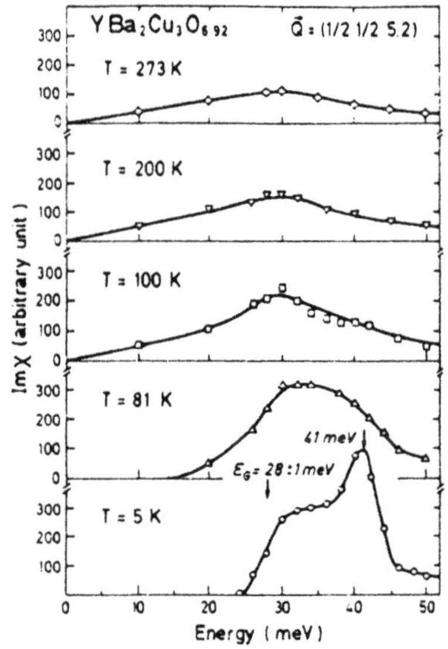

Fig. 16. $\mathrm{Im}\{\chi(\vec{Q},\hbar\omega)\}$ as a function of the energy for $YBa_2Cu_3O_{6.92}$ and increasing temperatures below and above T_c = 91 K. An energy gap (E_G = 28 meV) is clearly seen together with a strong enhancement at $\hbar\omega$ = 41 meV at lower temperatures.

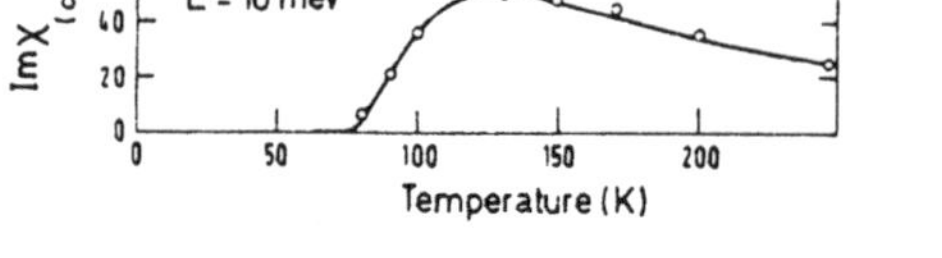

Fig. 17. Imχ at $\hbar\omega$ = 10 meV and **Q** = (1/2, 1/2, 1.6) as a function of temperature for $YBa_2Cu_3O_{6.92}$. In this 90 K-superconductor a pseudo-gap opens for T ≃ 120-130 K, i.e. well above T_C.

3.4 The overdoped metallic regime

In order to understand why $1/T_1T$ NMR measurements show a pseudo-gap effect for a sample with x = 0.92 and no effect with the fully oxygenated sample (x = 1), a INS experiment has been performed quite recently on such a sample and results are reported here. Actually a careful investigation of the phase diagram between x =0.9 and 1 has shown [25] that $YBa_2Cu_3O_7$ is already an overdoped sample because T_c is smaller by a couple of degrees than the optimum value found around x = 0.94. Moreover in O_7 the conductivity along the c-axis has been found of metallic type ($\rho_c \propto T$) [26].

Experiments were more difficult because the magnetic scattering continues to decrease. However at high energies a magnetic scattering has clearly been identified (see fig. 18). The q-width along (110) has increased by about 10 % in comparison with x = 0.92 ($\Delta q = 0.30 \pm 0.02$ r.l.u., $\xi/a \simeq 0.76 \pm 0.04$) . Moreover an important difference exists in the AF-coupling between the two Cu(2) layers which is now much weaker. The T-dependence of the spin excitation spectrum is reported in fig. 19 in the form of Imχ. At low-T the shape in energy of Imχ has changed. A sharp gap is also observed but with a slighthly smaller value : $E_G \simeq 26 \pm 1$ meV. More important at high energies the spectrum drops sharply, no measurable scattering is found at 50 meV (see fig. 18). This sharp decrease is even more pronounced at higher temperatures than at low-T because it is compensated by a strong enhancement around 40 meV due to a narrowing in q-width (Δq = 0.20 instead of 0.30 r.l.u.). This result is quite similar to that observed in the x = 0.92 sample

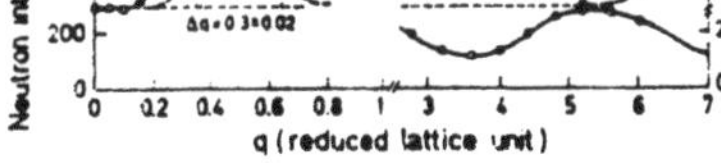

Fig. 18. Q-scans across and along the 2d-magnetic rod at **Q** = (1/2, 1/2, 5.2) and different energies for $YBa_2Cu_3O_7$ in the superconducting state (T_c = 89 K).

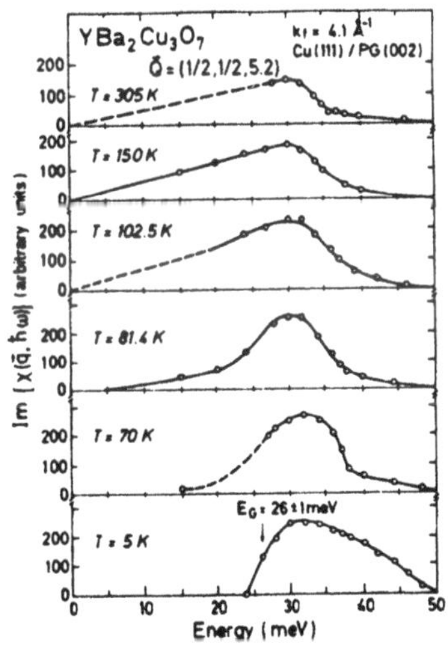

Fig. 19. Im{$\chi(\vec{Q}$,hω)} as a function of the energy for $YBa_2Cu_3O_7$ above and below T_c = 89 K. An energy gap (E_G = 26 ± 1 meV) is clearly seen together with a strong enhancement at h$\omega \simeq$ 40 meV at low temperatures.

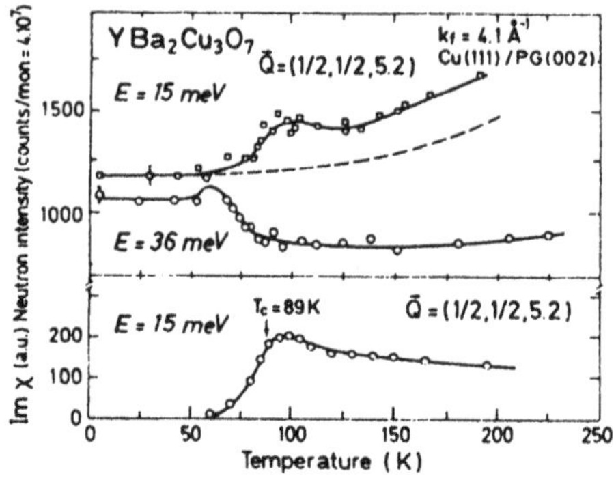

Fig. 20. Imχ at **Q** = (1/2, 1/2, 5.2) for two typical energy transfers : hω = 15 meV showing the filling of the spin gap and the absence of pseudo-gap opening aboveT_C, and hω = 36 meV showing the corallated disappearence of the strong enhancement at hω = 40 meV.

In fig. 20 we have reported the T-dependence of this enhanced scattering at E = 36 meV which actually starts to decrease around 60 K and vanishes completly around T_C = 89 K. Correlated to the suppression of this enhancement, excitations appear in the gap as shown by the T-dependence of Imχ at E = 15 meV (see fig. 20). More interesting is the fact that Imχ at 15 meV starts to decrease only around T_C in excellent agreement with NMR measurements. These results lead to the important conclusion that, in overdoped samples, the gap in the spin excitation spectrum opens around T_C which is actually the main difference with 90 K-underdoped samples together with a sharper decrease of the spectrum at high energies.

4. Conclusion

The six different regimes of the $YBa_2Cu_3O_{6+x}$ system from x = 0 to x = 1 has now been investigated. Clearly INS experiments have brought not only many valuable informations but also many questions by their unusual character. These results have to be considered as first generation experiments and certainly in many cases more detailed second generation experiments have to be carried out to clarify some open questions.
The main results which have been obtained are :

. the determination of exchange interaction parameters in the AF-state ;

. the doping by p-holes creates AF-polarons which give rise to an in-plane spin disorder and a strong renormalization of the spin wave velocity ;

. in the metallic state Im$\{\chi(\vec{Q},h\omega)\}$ has been measured in the whole T-and doping ranges and dynamical AF-correlations have always been observed, the q-width is increasing continuously with doping ;

. the characteristic energy of spin fluctuations $\Gamma_\omega \simeq$ 30 meV is quite small and weakly dependent on doping ;

. the temperature dependence of the spectrum is quite unusual especially in the weakly-doped regime : the correlation length is T-independent but the amplitude of the AF-correlated spectrum decreases as T increases ;

. the spectrum has a gaussian shape in q-space and drops rapidly at high energies, especially for O_7 ;

. a gap in the spin excitation spectrum has been found in metallic samples for any doping. However at low doping the gap is much weaker compared with kT_C. The dependence of the energy gap value as a function of T_C is given in fig. 21. At large doping a ratio $E_G/kT_C \simeq 3.5$ is reached, however the spin gap value remains much smaller than the charge gap value $2\Delta/kT_C$ which ranges from 5, in Raman spectroscopy to 7-8 in point contact spectroscopy or photo-emission measurements ;

. the spin gap energy does not decrease as function of T and then the BCS prediction is not followed ;

. a strong enhancement of the energy spectrum has been found at low T in 90 K-superconductors ;

. in the heavily doped regime, a pseudo-gap develops already above T_C, but not in the overdoped regime (x = 1) for which it is the main characteristic.

These results could be discussed in relation with the important theoretical activity in this field, unfortunatly there is not enough space here to undergo such a discussion. The two main points are actually :

i) the understanding of the origine of the energy gap : some singlet spin pairing or nesting effects

ii) the shape of the energy spectrum : the energy scale and the sharp drop at high energies.

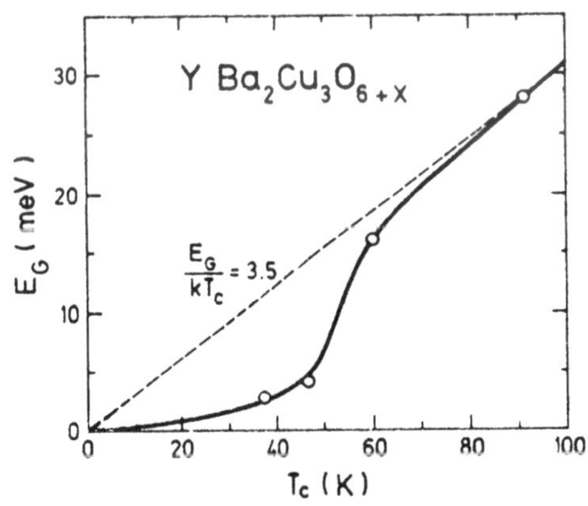

Fig. 21. Value of the spin excitation energy gap as a function of the superconducting transition temperature for the system $YBa_2Cu_3O_{6+x}$.

Acknowledgments

We acknowledge stimulating discussions with D. Grempel, M. Lavagna and S.V. Maleyev. The collaboration with the NMR group of C. Berthier throughout the course of this work has been quite fruitful.

References

1. J.G. Bednorz and K.A. Muller, Z. Phys. B **64** (1986) 189

2. See Proceedings of International Conferences on High Temperature Superconductors : Interlaken, Standford, Kanazawa.

3. B. Batlogg, Summary of Kanazawa Conference, Physica C **185-189** (1991) and Proceeding of Toshiba International School of Superconductivity, Kyoto, Eds. S. Maekawa and M. Sato,Springer Verlag (1991).

4. J. Zaanen, G.A. Sawatsky, J.W. Allen, Phys. Rev. Lett. **55** (1985) 418.

5. J. Fink et al., Physica C **185-189** (1991) 45.

6. J.C. Campuzano, G. Jennings, M. Faiz, L. Beaulaigue, B.W. Veal, J.Z. Lin, A.P. Paulikas, K. Vandervoort, H. Claus, R.S. List, A.J. Arko, R.J. Barlett, Phys. Rev. Lett. **64** (1990) 2308.

7. S. Uchida, Physica C **185-189** (1991) 28.

8. J. Rossat-Mignod, L.P. Regnault, C. Vettier, P. Burlet, J.Y. Henry, G. Lapertot, Physica B **169** (1991) 58.

9. S-W. Cheong, G. Aeppli, T.E. Mason, H. Mook, S.M. Hayden, P.C. Canfield, Z. Fisk, K.N. Clausen, J.L. Martinez, Phys. Rev. Lett. **67** (1991) 1791.

10. J.M. Tranquada, P.M. Gehring, G. Shirane, S. Shamoto, M. Sato, Preprint.

11. P. Burlet, V.P. Plakthy, C. Marin, J.Y. Henry, Physics Letters A, In press.

12. C. Chaillout, M.A. Allario-Franco, J.J. Capponi, J. Chenavas, P. Strobel, M. Marezio, Solid State Comm. **65** (1988) 283

13. R.M. Fleming, L.F. Schneemeyer, P.K. Gallagher, B. Batlogg, L.W. Rupp, J.V.Waszezak, Phys. Rev. **B37** (1988) 7030

14. T. Zeiske, R. Sonntag, D. Hohlwein, N.H. Andersen, T. Wolf, Nature, **353** (1991) 542

15. J. Rossat-Mignod, et al, J. Phys. France **49** (1988) C8-2119.

16. J. Rossat-Mignod, L.P. Regnault, M.J. Jurgens, P. Burlet, J.Y. Henry, G. Lapertot, in "Dynamics of Magnetic Fluctuations in High T_C Materials", Eds. G. Rieter, P. Horsh and G. Psaltakis, Plenum Press (1990).

17. J.M. Jurgens, PhD Thesis, University of Leiden (1989).

18. D.N. Aristov and S.V. Maleev, Z. Phys. B **81** (1990) 433.

19. J. Rossat-Mignod, L.P. Regnault, M.J. Jurgens, C. Vettier, P. Burlet, J.Y. Henry, G. Lapertot, Physica B **163**, (1990) 4.

20. J. Rossat-Mignod, L.P. Regnault, M.J. Jurgens, C. Vettier, P. Burlet, J.Y. Henrv, G. Lapertot, P. Physica B **169** (1991) 58.

21. J. Rossat-Mignod, L.P. Regnault, C. Vettier, P. Bourges, P. Burlet, J. Bossy, J.Y. Henry, G. Lapertot, Physica C **185-189** (1991) 86.

22. J. Rossat-Mignod, L.P. Regnault, C. Vettier, P. Bourges, P. Burlet, J. Bossy, J.Y. Henry, G. Lapertot, Proceeding of ICNS'91, Oxford, Physica B (1992).

23. Y. Kitaoka, K. Ishida, S. Ohsugi, K. Fujiwara, K. Asayama, Physica C **185-189** (1991) 98.

24. C. Berthier, Y. Berthier, P. Butaud, W.G. Clark, J.A. Gillet, M. Horvatic, P. Segransan, J.P. Henry, Physica C **185-189** (1991) 1141.

25. T. Graf, G. Triscone and J. Muller, J. Less-Common Met., **159** (1990) 349 ; J.Y. Genoux, T. Graf, A. Junod, D. Sanchez, G. Triscone and J. Muller, Physica **177** (1991) 315.

26. C. Ayache, Private communication.

NMR INVESTIGATION OF LOW ENERGY EXCITATIONS IN $YBa_2Cu_3O_{6+x}$ SINGLE CRYSTALS

T. AULER, C. BERTHIER, Y. BERTHIER, P. BUTAUD*, W. G. CLARK**, J. A. GILLET, M. HORVATIĆ***, P. SEGRANSAN
Laboratoire de Spectrométrie Physique (UA 08 CNRS) Univ. J. Fourier Grenoble I, B.P. 87, 38402 Saint-Martin d'Hères, France

J. Y. HENRY
C.E.N.Grenoble, DRF/SPh-MDN, 85X, 38041 Grenoble Cedex, France.

ABSTRACT.

The difference between overdoped ($dT_c/dn_h < 0$, n_h number of holes in the CuO_2 plane) and underdoped ($dT_c/dn_h > 0$) superconducting oxides is highlighted through an ^{17}O and ^{63}Cu NMR study in $YBa_2Cu_3O_{6+x}$ single crystals. The overall properties support the phase diagram based on the spin-charge separation.

1. Introduction.

The possibility of describing the whole phase diagram, from antiferromagnetism to superconductivity, by simply changing the oxygen concentration in the CuO chains[1] has promoted $YBa_2Cu_3O_{6+x}$ to a model system for high T_c superconducting oxides. Recently, a careful analysis of the phase diagram of $YBa_2Cu_3O_{6+x}$ by Graf et al.[2] has revealed that close to the maximum concentration of oxygen, one moves from an underdoped regime ($dT_c/dn_h > 0$, where n_h is the number of holes in the CuO_2 plane) with a maximum value of $T_c = 94$ K for $x = 0.94$, to an overdoped regime ($dT_c/dn_h < 0$) for $0.94 < x < 1$ with T_c decreasing down to 92 K. Indeed, a drastic difference in the temperature (T) dependence of the Cu(2) Nuclear Spin-Lattice Relaxation Rate (NSLRR) between $YBa_2Cu_3O_7$ and $YBa_2Cu_3O_{6.9}$ has already been pointed out by Horvatić et al.[3] : in the stoichiometric compound, the NSLRR divided by temperature $^{63}(T_1T)^{-1}$ continuously increases when T decreases from 300 K to T_c, and then drops abruptly, whereas in $YBa_2Cu_3O_{6.9}$ $^{63}(T_1T)^{-1}$ passes through a maximum around $T^* = 130$ K, well above T_c. This difference was first, attributed to the disorder present in the substoichiometric phase, and to its consequence on the conductivity along the c-axis. Later, $^{63}(T_1T)^{-1}$ data in $YBa_2Cu_4O_8$[4] revealed the same behaviour as for $YBa_2Cu_3O_{6.9}$, i.e., a maximum well above T_c, although the former is an ordered and stoichiometric compound, thus indicating that this behaviour is related rather to the hole concentration in the CuO_2 planes. Theoretically, it has been predicted by Suzumura et al.[5] and Nagaosa and Lee[6] in the framework of the spin-charge separation[7] that the normal state in the underdoped and overdoped regime should be quite different. For hole concentrations δ smaller than δ_c (which corresponds to the maximum of T_c), when lowering T, the system should first cross-over from a "strange metal", in which holons and spinons are decoupled, into an RVB state, and then undergo a Bose condensation at $T = T_c$. For overdoped compounds ($\delta > \delta_c$), there is a transition line below which holons condense to give a Fermi liquid, before undergoing a BCS type transition at T_c.

In this paper, we compare the NMR properties - T dependence of the spin susceptibility and of the NSLRR for Cu(2) and O(2,3) sites - in $YBa_2Cu_3O_{6+x}$ single crystals of compositions belonging either to the overdoped or to the underdoped regime. The general features are in fair agreement with the phase diagram mentioned above; we also show that they are not restricted to the 123 system, but extend to other high T_c materials, and thus put severe constraints to any comprehensive microscopic theory of these compounds.

2. Results and discussion

NMR measurements were carried out on three "porous" single crystals[8] corresponding to the following compositions: $YBa_{1.92}Sr_{0.08}Cu_3O_7$ (Sample 1, $T_c = 89$ K), $YBa_{1.93}Sr_{0.07}Cu_3O_{6.92}$ (Sample 2, $T_c = 91$ K) and $YBa_2Cu_3O_{6.52}$ (Sample 3, $T_c = 59$ K). Sample 1 and 3 were enriched in ^{17}O, whereas Sample 2 is a piece of the single crystal used for neutron experiments by Rossat-Mignod et al.[9,10]. As will be discussed below, Sample 1 and 2 belong respectively to the overdoped and underdoped regime of the so called "90 K" phase. Details of the sample characterisation and experimental set-up can be found elsewhere.[11]

We first compare the Cu(2) $(T_1T)^{-1}$ in the three samples, which are plotted in Fig. 1. (We also include the data from a $YBa_2Cu_3O_7$ oriented powder sample, obtained by grinding porous single crystals, with a T_c of 92 K [3]). There are two clearly distinct behaviours:

– for the two O7 samples ($T_c = 92$ and 89 K), $^{63}(T_1T)^{-1}$ continuously increases when T decreases from 300 K to T_c, and then drops abruptly,

– for Samples 2 ($T_c = 91$ K) and 3 ($T_c = 59$ K), $^{63}(T_1T)^{-1}$ passes through a maximum around $T^* = 130$ K, irrespective of T_c. Such behaviour was first reported by Horvatić et al.[12], and Warren et al. [13] for the 60 K phase.

These results are now fully corroborated by neutron inelastic scattering; in the case of Sample 2, the T dependence of $^{63}(T_1T)^{-1}$ and that of $\chi''(Q_{AF},\hbar\omega = 10\text{ meV})$[10] as given by neutron inelastic scattering are quite similar. (Fig.2)

Remembering that $(T_1T)^{-1} \propto \sum_q |A_q|^2\chi''(q,\omega_n)/\omega_n$ this means that the Cu(2) NSLRR is dominated by the contribution of the enhanced antiferromagnetic fluctuations (AFF) to $\chi''(q,\omega)$. Neutron data [9,10] indicate that the decrease of $^{63}(T_1T)^{-1}$ below $T^* \approx 130$ K is due to the opening of a (pseudo-)gap in the magnetic excitations, or at least to a transfer from low energy to high energy in the spectral weight of the magnetic excitations.

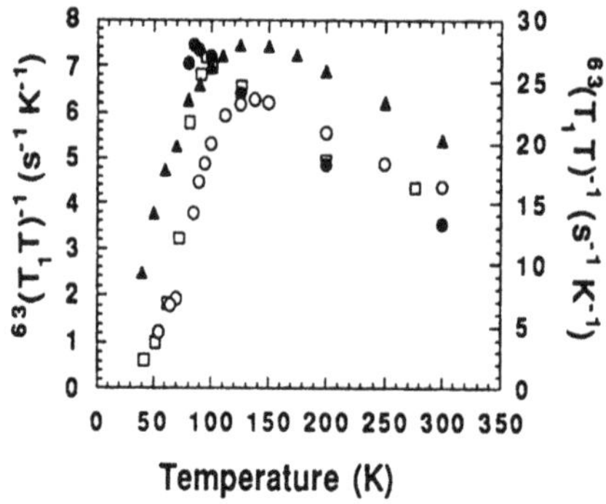

FIG. 1 Temperature dependence of $^{63}(T_1T)^{-1}$ in:

a) overdoped regime: □ $YBa_2Cu_3O_7$ (oriented powder, ref. 3); ● $YBa_{1.92}Sr_{0.08}Cu_3O_7$ (single crystal, H_0 // a,b plane)

b) underdoped regime: ○ $YBa_{1.93}Sr_{0.07}Cu_3O_{6.92}$ (single crystal, H_0 // c-axis); ▲ $YBa_2Cu_3O_{6.52}$ (single crystal, H_0 // c-axis).

The maximum of $(T_1T)^{-1}$ above T_c is a common feature in all underdoped compounds: $YBa_2Cu_4O_8$[4], as grown $Bi_2Sr_2CaCu_2O_8$ [14], and also $La_{2-x}Sr_xCuO_4$,[15] where $^{63}(T_1T)^{-1}$ seems to have a maximum around 50 K. On the contrary, in overdoped compounds: $YBa_2Cu_3O_7$, annealed $Bi_2Sr_2CaCu_2O_8$[14] and $Tl_2Sr_2CaCu_2O_8$[16] there is no maximum above T_c, i.e., the gap for the spin and charge excitations opens at the same temperature. This has recently been confirmed by neutron inelastic scattering in a $YBa_{1.93}Sr_{0.07}Cu_3O_7$ single crystal, a composition very close to our Sample 1.[17]

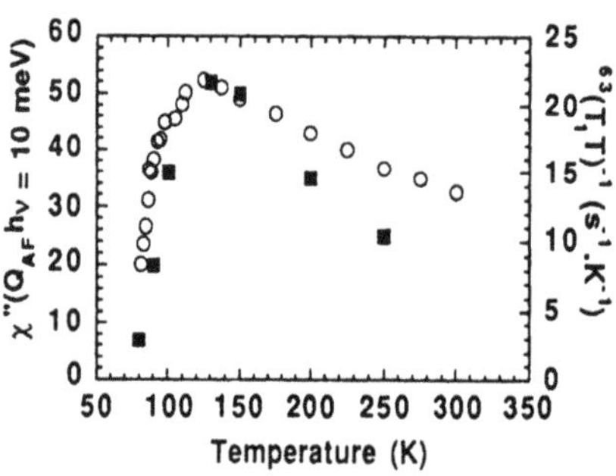

FIG. 2 Comparison of $^{63}(T_1T)^{-1}$ (○) and $\chi''(Q_{AF}, \hbar\omega = 10\ meV)$ (■) measured by NMR [11] and inelastic neutron scattering [10] in the same $YBa_{1.93}Sr_{0.07}Cu_3O_{6.92}$ single crystal.

In regard to the NMR of O(2,3) nuclei, we first consider the T dependence of the spin susceptibility. It is now fairly well established in $YBa_2Cu_3O_{6+x}$ that all nuclei of the CuO_2-Y- CuO_2 sandwich probe a unique static spin susceptibility $\chi_S(T)$, which has the same T dependence as the macroscopic susceptibility.[18-22] The $\chi_S(T)$ is directly related to the spin part of the Magnetic Hyperfine Shift (MHS) tensor ^{17}K experienced by the O(2,3) nuclei : $^{17}K_{\alpha\alpha} \propto {}^{17}C_{\alpha\alpha}\,\chi_S$, where α refers to the principal axes of the MHS and Electric Field Gradient tensors (i.e., X parallel to the c-axis, Y perpendicular and Z parallel to the Cu-O-Cu bond in the ab plane,) and $^{17}C_{\alpha\alpha}$ is the hyperfine field. Fig. 3a shows the $^{17}O(2,3)$ MHS results in Sample 1. Note that $K_{\alpha\alpha}$ increases slightly with decreasing T, which means that the spin susceptibility is not flat, but slightly increases as already noticed in Cu(2)[23] and Y[24] MHS. This appears to be the signature of slightly overdoped samples, which is the case of pure $YBa_2Cu_3O_7$.[2] In contrast, the MHS of Sample 3 shown in Fig. 3b exhibits a strong T dependence as already reported in "60 K phase" samples.[19,21,22,25] The relationship between K_{ZZ} and K_{YY} is linear [22] as expected for one spin degree of freedom.

In the overdoped phase (Sample 1) the slight increase of χ_S when T decreases from 300 K to T_c is compatible with the Pauli susceptibility of a Fermi liquid in a narrow band metal. In the underdoped regime (Sample 3) $\chi_S(T)$ decreases continuously with T; no definite theory is available yet to explain this behavior, except recent quantum dynamical Monte-Carlo calculations.[26] However, in the scenario proposed by Nagaosa and Lee ,[6] this should be due to pairing of spinons. Experimentally, it is obviously related to the opening of a (pseudo-)gap.

Fig.4a shows the T dependence of the NSLRR for O(2,3) nuclei in Sample 1 for the three orientations of the applied external magnetic field H_0.[27] The most striking feature of these data is that $^{17}(T_1T)^{-1}$ also continuously increases on decreasing T from 350 K down to T_c,[22,27] although this increase is much less pronounced than that for the Cu(2) $^{63}(T_1T)^{-1}$. (As for the MHS, the increase is more pronounced when H_0 is in the a,b plane.) Note that in $YBa_2Cu_3O_7$ only T independent behavior has been reported so far[21,28] corresponding to c-axis component of NSLRR measured on oriented powders. One possible explanation is that in these cases the T dependence has been masked by experimental errors, relatively limited temperature range and by the fact that in these samples one actually observes a small "precursor decrease" of NSLRR somewhat above T_c - which flattens out the $(T_1T)^{-1}$ vs. T dependence.

The behavior of $^{17}(T_1T)^{-1}$ in $YBa_2Cu_3O_{6.52}$ is quite different, as is shown in Fig. 4b. It continuously decreases with T, and at T_c only a small change in the slope can be noticed. Plotting $^{17}(T_1T)^{-1}$ versus K_{ZZ} demonstrates that for H_0 // Y the $^{17}(T_1T)^{-1}$ is accurately proportional to the spin part of the MHS.[22] Similar results have been reported in the 60 K phase for H_0 // X (c-axis).[19,21] However, if we now turn to the data measured with H_0 // Z, we see that such a relationship no longer holds, which has the important

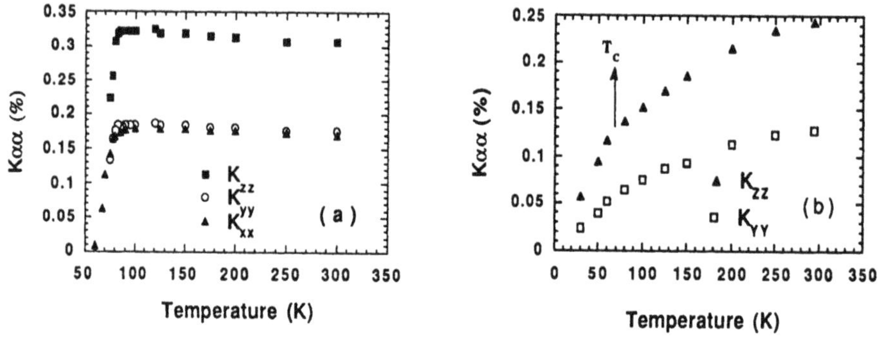

FIG. 3 Temperature dependence of the principal components of the MHS tensor for O(2,3) sites in a) $YBa_{1.92}Sr_{0.08}Cu_3O_7$, b) in $YBa_2Cu_3O_{6.52}$.

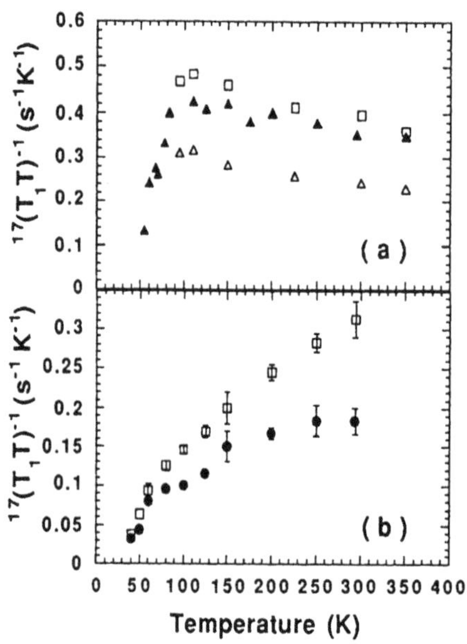

FIG. 4 Temperature dependence of $^{17}(T_1T)^{-1}$ for O(2,3) sites
a) in $YBa_{1.92}Sr_{0.08}Cu_3O_7$
(❑) H_0 // Z, (Δ) H_0 // Y, (▲) H_0 // X
b) in $YBa_2Cu_3O_{6.52}$
(●) H_0 // Z,(❑) H_0 //Y

consequence that the NSLRR anisotropy is now T dependent,[22] in contrast to the behavior observed in $YBa_{1.92}Sr_{0.08}Cu_3O_7$.[22,27] Such a T dependence of the NSLRR anisotropy for H_0 in the a,b plane was first reported by Barriquand et al.[29] in a $YBa_2Cu_3O_{6.5}$ single crystal.

It is now widely recognized that the different T dependences of the NSLRR for Cu(2) and O(2,3) result from the filtering of the AFF at the O(2,3) site, because of its symmetric position between its two Cu(2) first neighbors :[28]

$$^{17}(T_1T)^{-1} \propto \sum_q {}^{17}C^2[1 + 1/2(\cos q_a a + \cos q_b b)]\chi''(q,\omega_n)/\omega_n$$

Since in substoichiometric samples $^{17}(T_1T)^{-1}_{\alpha}$ (α=X,Y) depends linearly on $\chi_S(T)$,[21,22] a decomposition of the dynamic susceptibility has been proposed:[30]

$$\chi''(q,\omega) = \pi\chi_S(T)/\Gamma + \chi''_{AF}(q,\omega).$$[31]

Because of the filtering form factor, the contribution of $\chi''_{AF}(q,\omega)$, which is peaked around $Q = (\pi/a, \pi/a)$, vanishes and $^{17}(T_1T)^{-1} \propto \chi_S(T)$.

We now turn to the results in the overdoped regime. From Figs. 3a and 4a we see that $^{17}(T_1T)^{-1}$ increases faster than $^{17}K_{\alpha\alpha}$ when T decreases, so that these values are no longer proportional as in the "60 K phase". One possible explanation is that the filtering of the AFF is no longer as perfect, due to a shortening of the AF correlation length. Indeed, the variation of $(K^{spin}T_1T)^{-1}$ due to T dependence of AFF (measured by the copper $(T_1T)^{-1}$) can be very well accounted for if from neutron data[17] we take the corresponding q-width of χ''_{AF} and the characteristic energy $\Gamma(Q_{AF})$ and assume the Gaussian (or even better multi-Gaussian) shape of $\chi''_{AF}(q)$ corresponding to minimization of the total width of $\chi''_{AF}(q)$. Note that squared Lorentzian proposed by Millis, Monien and Pines[30] is definitely too wide, and implies variations of $(K^{spin}T_1T)^{-1}$ which are much stronger than the observed ones.

It is interesting to remark that in fact the experimental data satisfy the Korringa law K^2T_1T = const. quite well.[11] However, to view this behavior as a signature of a Fermi-liquid, we have to suppose that AFF of the Cu(2) spins do not contribute to O(2,3) NSLRR, that the relaxation is due to a single process, and that for some reason the corresponding density of states at Fermi level is slightly T dependent. It is reasonable to expect that some of these conditions are not fulfilled; for the given small T dependences of ^{17}K and $^{17}(T_1T)^{-1}$ it is then impossible to make a reliable decomposition to different contributions and for the moment we are unable to distinguish the Fermi liquid behavior even if it exists. We hope that ^{89}Y data (taken on the same sample) might help reduce the ambiguity in the analysis.

3.Conclusion

In conclusion, we have shown in $YBa_2Cu_3O_{6+x}$ that depending on the hole concentration in the CuO_2 plane, the low energy excitations around $Q = (\pi/a, \pi/a)$ and $q = 0$, as well as the static susceptibility, exhibit drastically different behaviors; in the underdoped regime ($x < 0.94$), the decrease of the static susceptibility $\chi_S(T)$, the opening of a "gap" in the spin excitations at T well above T_c, the proportionality between $^{17}(T_1T)^{-1}$ and $\chi_S(T)$, and the T dependence of the O(2,3) NSLRR above T_c are signatures of a strange metal. For $x = 1$, corresponding to a slightly overdoped compound (compared to the maximum value, T_c is decreased by only 2 %), the gap for spin and charge excitations opens at the same temperature T_c and a pseudo-Korringa behavior is recovered. The cross-over between the two regimes occurs in a very narrow range of oxygen concentration just *within* the so-called "90 K plateau" in $YBa_2Cu_3O_{6+x}$. Such different behaviors for underdoped and overdoped compounds can also be recognized in other compounds, e.g., in $YBa_2Cu_4O_8$[4] and $Bi_2Sr_2CaCu_2O_8$,[14] but at the moment, only the $YBa_2Cu_3O_{6+x}$ system offers the possibility to switch continuously from one regime to the other. These features, in particular the existence in the whole underdoped regime of two different temperatures, corresponding to the onset of the gaps for spin and charge excitations, which reduce to the same one in the overdoped case, strongly support the phase diagram based on the spin-charge separation[5,6] and should be reproduced in any other theoretical microscopic description of the normal state of the cuprate oxides.

This work was supported by the Direction des Recherches et des Etudes Techniques (contract DRET N° 89/045) and by the Ministère de la Recherche et de la Technologie (contract MRT N° 90.A.0579). One of us (M. H.) would like to acknowledge the support of the EEC project N° CI1* 0568.

References

* *Present address: Institute of Physics of the University, P.O.B. 304, 41001 Zagreb, Croatia*

** *Present address: C.R.T.B.T., CNRS, 166X, 38042 Grenoble-Cedex, France*

****Present address: Department of Physics, University of California, Los Angeles, CA 90024, USA*

1 R. J. Cava et al., Physica C **156**, 523 (1988)

2 T. Graf, G. Triscone, and J. Muller, J. Less-Common Met., **159**, 349 (1990)

3 M. Horvatić et al., Physica **C 166**, 151 (1990).

4 Zimmermann et al., J. Less-Common Met. **164&165**, 138 (1990)

5 Y. Suzumura, Y. Hasegawa and H. Fukuyama, J. Phys. Soc. Jpn. **57**, 2768 (1988); H. Fukuyama, unpublished.

6 N. Nagaosa and P.A. Lee, Phys. Rev. Lett. **64**, 2450 (1990); Phys. Rev. B **43**, 1223 (1991); *ibid.* **45**, 966, (1992)

7 P. W. Anderson, Science **235**, 1196 (1987)

8 J. Y. Henry, unpublished.

9 J. Rossat-Mignod et al., Physica **B 169**, 58 (1991)

10 J. Rossat-Mignod et al., Physica **C 185&189**, 86 (1991)

11 C. Berthier et al., Physica **C 185&189**, 1141 (1991); Appl. Mag. Res., **Vol.3/4** (1992)

12 M. Horvatić et al., Phys.Rev B **39**, 7332 (1989)

13 W. W. Warren et al., Phys. Rev. Lett. **62**, 1193 (1989)

14 R. E. Walstedt, R.F. Bell and D.B. Mitzi, Phys. Rev. B **44**, 7760 (1991)

15 S. Oshugi et al., J. Phys. Soc. Jpn., **60**, 2351 (1991)

16 F. Hentsch et al., Physica C **165**, 485 (1990)

17 J. Rossat-Mignod et al., unpublished

18 H. Alloul et al., Phys. Rev. Lett. **63**, 1700 (1989)

19 M. Takigawa et al., Phys.Rev.B **43**, 247 (1991)

20 T. Shimizu et al., Bull. Mag. Res., **12**, 39 (1990)

21 Y. Yoshinari et al., J.Phys.Soc. Jpn. **59**, 3698 (1990)

22 M. Horvatić et al., Physica C **185&189**, 1139 (1991)

23 R.E.Walstedt et al., unpublished.

24 G. Balakrishnan et al., J. Phys. C **21**, L847 (1988)

25 P. Butaud et al., Physica C **166**, 301 (1990);

26 G. Dopf, A. Mutsamura, W. Hanke Phys. Rev. Lett. **68**, 353 (1992)

27 M. Horvatić , Thesis, Zagreb (1991)

28 P.C.Hammel et al., Phys.Rev. Lett. **63**, 1992 (1989).

29 F. Barriquand, P. Odier, D. Jérome, Physica C **177**, 230 (1991)

30 A.J. Millis, H. Monien, D. Pines, Phys.Rev.B **42**, 167 (1990); H. Monien, D. Pines, M. Takigawa, Phys.Rev.B **43**, 258 (1991)

31 Note that in a usual Fermi Liquid description, $\hbar\Gamma \propto N(E_F)^{-1}$ and should vary as $\chi_S(T)^{-1}$, while here Γ is constant.

Aspects of the Spin Dynamics in the Cuprate Superconductors

H. Monien
Institute for Theoretical Physics
University of California
Santa Barbara, CA 93106, USA

Abstract:

One of the interesting aspects of the CuO superconductors is that superconductivity is happening so close to the antiferromagnetic state. The nuclear magnetic resonance and the recent neutron scattering experiments clearly indicate that magnetic correlations persist in to the heavily doped regime. In this paper we will discuss some of the details of the coupling of the nuclear magnetic spin to the conduction electron spins. Furthermore we will show that a simple band structure can explain the recent neutron scattering data in the $LaSr_xCu_{2-x}O_4$ material for the optimal concentration of $x \approx 0.15$ if the lifetime effects are included.

1 Introduction

From the theoretical perspective one of the most interesting aspects of the CuO high temperature superconductors is that they belong to a class of materials which show strong electronic correlations. In the undoped materials these correlations lead to the antiferromagnetic ordering of the electronic spins. This is why Anderson [1] pointed out early on that the physics of the CuO superconductors is best described in terms of the Hubbard model. The model Hamiltonian contains the strong on site Coulomb repulsion and a hopping term describing the kinetic energy. The theoretically fascinating aspect is how to obtain superconductivity from a purely repulsive interaction. Very close to half filling where this model describes the insulating Heisenberg antiferromagnet it is extremely well understood and the neutron scattering experiments [2] and theoretical calculations, e.g. [3], are in beautyful agreement. Away from half filling the situation is much less clear. Except in one dimensions and recently infinite dimensions the physics of the Hubbard model is not understood to a degree that one knows the correlation functions quantitatively. Even today small cluster calculations and approximations which are hard to controll in the relevant regime are the only way one is able to achieve some progress in the strongly correlated regime away from half filling [4].

Under these circumstances it is useful to know what limitations one can obtain from experiments for a theory for the CuO superconductors. The constraints obtained from experiments might tell us if a theoretical model like the Hubbard model contains even qualitatively the right physics to describe the low energy excitations of the "real materials" or if some additional physics is required to understand the basic properties of the high temperature superconductors. It is quite clear that these materials are close to many instabilities which might all be important. Thus, choosing to concentrate on a particular degree of freedom of the system, one might completely miss the relevant physics. A "worst case scenario" from a theorists point of view would be that all the possible degrees of freedom of the system contribute to the most interesting of all the instabilities of the system: the superconducting state.

In order to see "a way through the jungle", I will concentrate on the magnetic properties of the copper oxide superconductors which are very interesting in themselves. The magnetic properties of the CuO materials were studied in great detail by the nuclear magnetic resonance techniques by many groups and by neutron scattering experiments which are very difficult to perform in these compounds in comparison with standard materials. In both types of experiments one would like to obtain information about the dynamical structure factor of the electronic spins. Whereas the nuclear magnetic resonance experiments probe the local environment of one particular nucleus and therefore a wavevector average of the dynamical structure factor at very low energy transfers, the neutron scattering experiments can, at least in principle, scan all frequencies and wavevectors corresponding to the whole Brillouin zone.

The interpretation of the nuclear magnetic resonance experiments is quite involved. It requires the knowledge of the so called hyperfine Hamiltonian which describes the coupling of the electronic spin to the nuclear spin. By now it established that the spin degrees of freedom in the CuO superconductors can be described in a so called "one component picture", which means that it is sufficient to consider one spin degree of freedom per unit cell. If one assumes a band structure picture one can say more precisely that only one band plays a role for the low energy physics of the spins. The nature of the band, i.e. the percentage of copper or oxygen admixture might vary, as one moves along the Fermi surface. But the statement about one component is more general in that it does not require a Fermi liquid state. The spin degree of freedom might equally well be carried by the strongly correlated quasiparticles of a tJ - model. The question is how do the spin of the quasiparticle couple to the nuclear spins. This is discussed in the next section.

2 Transferred hyperfine interactions and band structure

In this section we discuss the derivation of a hyperfine Hamiltonian similar to the so called Mila-Rice Hamiltonian [5] which was originally introduced to understand the difference in the anisotropies of the Cu nuclear magnetic relaxation rates and the Cu Knight shift measurements in terms of uncorrelated spins. The importance of a transferred hyperfine interaction in the case where the electronic spins are strongly correlated was pointed out early on by Shastry [6]. In his paper he showed that within the theoretical framework of the so called t-J model one can understand the difference between the magnitude of the Cu and O relaxation rates if the oxygen nuclear spins are only coupled to the electronic spins via a transferred hyperfine coupling.

The phenomenological hyperfine Hamiltonian which is compatible with this assumption of a "one component picture" is:

$$H_{hf} = \sum_{i\alpha\beta} {}^{63}I_{i\alpha\beta} A_{\alpha\beta} S_{i\beta} + B \sum_{<ij>\alpha} {}^{63}I_{i\alpha} S_{j\alpha} + \sum_{<ij>\alpha\beta} {}^{17}I_{i\alpha} C_{\alpha\beta} S_{j\beta} \tag{1}$$

where ${}^{63}I_{i\alpha}$ is the the αth, $\alpha \in \{x, y, z\}$, component of the copper nuclear spin at the site i, $S_{i\alpha}$ is the αth component of the electronic spin, $A_{\alpha\beta}$ is the anisotropic copper hyperfine coupling, B is the transferred Cu hyperfine coupling and $C_{\alpha\beta}$ is the transferred oxygen hyperfine coupling.

This is the so called Mila-Rice Hamiltonian. The unusual feature is the large so called transferred hyperfine coupling B. Couplings to other nuclei can be included in a similar way. Whereas the Mila-Rice Hamiltonian was derived from a quantum-chemistry calculation on a small cluster and therefore in a very ionic picture Shastry assumed an itinerant electron system. It was argued that the success of the Mila-Rice Hamiltonian indicates that the Cu spins are nearly localized. Here I would like to point out that in an itinerant "band structure picture" one basically obtains the same results for the hyperfine Hamiltonian. These results are mostly due to Takigawa [7]. In addition to the Mila Rice Hamiltonian one finds additional terms which are of similar size as the nearest neighbor transferred hyperfine interaction but we will argue that these terms are negligible because the correlations are only short ranged.

The starting point of our calculation is the canonical band structure for the plane. We include in addition to copper $3d_{x^2-y^2}$, the oxygen $2p_x$ and $2p_y$ orbitals and the copper $4s$ orbital. Here it is sufficient to include the $4s$ orbital only in perturbation theory. The Hamiltonian has the form

$$\begin{aligned} H_0 &= \sum_{k\alpha}\{\epsilon_d d^\dagger_{k\alpha} d_{k\alpha} + \epsilon_p p^\dagger_{x,k\alpha} p_{x,k\alpha} + \epsilon_p p^\dagger_{y,k\alpha} p_{y,k\alpha}\} \\ &+ \sum_{k\alpha} 2itd^\dagger_{k\alpha}(\sin(k_x/2)p_{x,k\alpha} + \sin(k_y/2)p_{y,k\alpha}) + h.c. \end{aligned} \quad (2)$$

where $d^\dagger_{k\alpha}$ ($d_{k\alpha}$) is the copper $3d_{x^2-y^2}$ electron creation (annihilation) operator, ϵ_d is the energy of that orbital, $p^\dagger_{x,y,k\alpha}$ are the annihilation operators for the oxygen (p_x, p_y) orbitals in the plane. If the $4s$ orbital is partially filled $4s$ the electron spin has a contact interaction with the nuclear spin at site i which is given by:

$$H_{hf} = B\ {}^{63}\vec{I}_i \cdot s^\dagger_{i\alpha}\vec{\sigma}_{\alpha\beta}s_{i\beta} \quad (3)$$

where $s^\dagger_{i\alpha}$ are the annihilation (creation) operators for the $4s$ orbital at the Cu site i, $\vec{\sigma}$ is the vector of Pauli matrices and ${}^{63}\vec{I}_i$ is the Cu nuclear spin. The coupling constant B depends on atomic physics and can be found for example in [5]. The contact interaction is isotropic and therefore the transferred hyperfine coupling will be isotropic. If we believe that only one band of the three bands of our model plays a role for the low energy physics we would like to reduce hyperfine Hamiltonian to the part which is connected with this one band. Therefore we project the hyperfine Hamiltonian, Eq. 3, on to the conduction electron band:

$$(H_{hf})_{proj} = \sum_{kk'} < k\alpha|H_{hf}|k'\alpha' > c^\dagger_{k\alpha}c_{k'\alpha'} \quad (4)$$

where $c^\dagger_{k\alpha}$ are the creation and annihilation operators of the conduction electrons with momentum $\vec{k}$ and spin α.

The projected hyperfine Hamiltonian has the form:

$$(H_{hf})_{proj} = B\ {}^{63}\vec{I}_0 \cdot \sum_{ij} f^*_i f_j c^\dagger_{i\alpha}\vec{\sigma}_{\alpha\beta}c_{j\beta} \quad (5)$$

where f_i is the Fourier transform of $f_k =< 4s|k >$, the overlap between the $4s$ state and the conduction electron bands at the site i. It turns out that we only need to know this overlap. This is basically what Mila and Rice did for their ionic model. The parameters for a fit of tight binding model to the "real band structure" were given by e.g. Hybertsen and Schlüter [8]. For the energy of the Cu d orbital, the oxygen

p orbital and the Cu 4s orbital respectively they found $\epsilon_d = -2.1$eV, $\epsilon_p = -3.3$eV and $\epsilon_{4s} = -8.1$eV. For the overlaps they give the following numbers $t_{pd} = 1.3$eV, $t_{pp} = 1$eV and $t_{ps} = 3.8$eV. With this parameters we can calculate the hyperfine interactions (for details consult e.g. [9]). In addition to the terms which couple the conduction electron spin on one site to the nuclear spin on a neighboring site we find terms which describe a nuclear spin assisted spin flip + hop term. The result of the calculation can be cast into the simple form:

$$(H_{hf})_{proj} = B_{10,10}{}^{63}\vec{I}_{00} \cdot \vec{S}_{11} + B_{10,01}{}^{63}\vec{I}_{00} \cdot c^{\dagger}_{10,\alpha}\vec{\sigma}_{\alpha\beta}c_{01,\beta} + \dots \quad (6)$$

where the coupling constants are of the order $B_{10,10} \sim 48\text{kOe}/\mu_B$, $B_{10,01} = B_{10,10}$. The coupling constant $B_{10,10}$ is the transferred hyperfine coupling for the Cu nuclear spin as disussed by Mila and Rice, the spin flip hop term $B_{10,01}$ is not present in the ionic picture. The next term in the expansion is about a factor of 6 smaller then the first term and can be neglected savely. Although the nuclear spin assisted spin flip hop term is of the same magnitude as the transferred term we can neglect it in our analysis of the nuclear magnetic resonance experiments because the dynamical structure factor does not have large weight at this particular area of momentum space. Therefore we are left with the standard hyperfine Hamiltonian for the CuO plane.

If we do not allow for any other relaxation channel which might for example be produced by a coupling of the nuclear spin to the orbital motion of the electron then we know exactly which moments of the dynamical structure factor contribute to the nuclear magnetic relaxation.

3 How important is orbital relaxation?

In previous theories of the nuclear magnetic relaxation in the high temperature superconductors it was assumed the only way the nuclear spin couples to the electrons is via the electronic spin. The unusual behavior of the anisotropy of the Cu nuclear magnetic relaxation rate as a function of temperature [10], [11] raises the question if there is another relaxation channel present. In a recent paper Millis and Monien [12] discuss the possibility of another relaxation channel for the Cu nuclear spin, namely the coupling of the Cu nuclear spin to the orbital motion of the electron which was already mentioned by Warren and Walstedt in an early paper. The coupling of the nuclear spin to the orbital momentum is give by:

$$H_{\text{Orb}} = {}^{63}\gamma_n\gamma_e\hbar^2\frac{\vec{I} \cdot \vec{L}}{r^3} \quad (7)$$

where $\vec{I}$ is the copper nuclear spin, $\vec{L}$ the electron angular momentum and ${}^{63}\gamma_n$ and γ_e the gyromagnetic ratio of the copper nucleus and the electron respectively. In order to estimate the size of this effect we use again the simplest tight binding picture for the CuO band of a single plane with the same parameter as used for the calculation of the hyperfine Hamiltonian. For the calculation of the relaxation rates we need to know the matrix elements of the orbital coupling, Eq. (7), with the Wannier functions. The rates can then be calculated simply by using Fermis "golden rule". At a temperature of about 100K we find for the nuclear magnetic relaxation rate produced by *orbital currents* for a field applied parallel and perpendicular to the CuO plane:

$$\begin{aligned} W_{\|} &= 3.8 \times C_{x^2-y^2}\,(4C_{xy} + C_{xz})\ \mathrm{ms}^{-1} \\ W_{\perp} &= 3.8 \times C_{x^2-y^2} C_{xz}\ \mathrm{ms}^{-1} \end{aligned} \tag{8}$$

The coefficients $C_{x^2-y^2}$, C_{xy} and C_{xz} denote the percentage of admixture of the Cu $3d_{x^2-y^2}$, $3d_{xy}$ and $3d_{xz}$ orbital respectively. The result of the analysis is that the orbital relaxation of the nuclear spins can be appreciable because of the orbital degeneracy but that the absolute value depends strongly on the exact amount of admixture of the Cu $3d_{xz}$ and $3d_{xy}$ orbitals to the Wannier functions at the Fermi surface which are not known theoretically to an accuracy, ($\sim$ 5%), needed to determine if the orbital relaxation may play a role or not. Nevertheless it is important to keep this additional relaxation channel in mind if one discusses the limitations on hyperfine Hamiltonian and dynamical structure factor obtained from experiments.

4 Reexamination of the MMP model

Up to now we have only discussed the coupling of the electronic spin and orbital degree of freedom to the nuclear spin. Now we would like to turn our attention to the quantity which is of most interest from the theoretical point of view, the dynamical structure factor.

Recently we [9] reexamined the results of a previous phenomenological analysis of the nuclear magnetic relaxation experiments in the $YBa_2Cu_3O_{6+\delta}$ materials [13]. In the original MMP paper we made the assumption that the only temperature dependent quantity is the antiferromagnetic correlation length and the energy scale is described by a dynamical critical exponent and used the simplest form for a peaked structure factor describing the magnetic correlations, a Lorentzian. In the extended analysis we dropped the assumption that the correlation length is the only quantity which is temperature dependent and allowed the strength of the antiferromagnetic peak also to vary with temperature. Also we considered two models for the q dependence of the structure factor - a Lorentzian and a Gaussian. The parameters describing the correlation functions are the antiferromagnetic correlation length ξ and the strength of the antiferromagnetic peak which we denote with β. The two models for the dynamical structure factor in the limit frequency ω going to zero are:

$$\chi''_{Lor}(q,\omega) = \frac{\pi\chi_0\hbar\omega}{\Gamma}\left[1+\beta\frac{\xi^4}{(1+\xi^2(q-Q)^2)^2}\right] \tag{9}$$

$$\chi''_{Gauss}(q,\omega) = \frac{\pi\chi_0\hbar\omega}{\Gamma}\left[1+\beta\xi^4 e^{-\xi^2(q-Q)^2}\right] \tag{10}$$

where ξ is the antiferromagnetic correlation length, β is the strength of the antiferromagnetic peak at the zone corner, $Q = (\pi/a, \pi/a)$, Γ is the "bare" energy scale of the spin fluctuations and χ_0 is the static susceptibility.

Within the "one component picture" the Knight shift for a field applied in the direction α, K_α, and the relaxation rates, W_α of the various nuclei, denoted by $a = 63, 17, 89$ for Cu, O and Y respectively, for a field applied in the α direction are given by:

$$\begin{aligned} {}^aK_\alpha &= \lim_{q\to 0} {}^aF_\alpha(q)\chi'(q,\omega=0) \\ {}^aW_\alpha &= \frac{1}{4\mu_B^2}\lim_{\omega\to 0}\frac{k_BT}{\omega}\sum_{q,\overline{\alpha}}\left[{}^aF_{\overline{\alpha}}\right]^2\chi''(q,\omega) \end{aligned} \tag{11}$$

where $\overline{\alpha}$ is the direction perpendicular to α. The form factors ${}^aF_\alpha(q)$ have the dimension of energy and are basically the Fourier transform of the spin hyperfine Hamiltonian and can be found in the literature, e.g. [9]. To obtain the limits on the correlation lengths we considered two limits in one case the correlation length is temperature dependent and the strength is not. In the second case the strength of the peak is temperature dependent and the correlation length is temperature independent.

We state the results for the correlation lengths in Table (1).

	$YBa_2Cu_3O_7$		$YBa_2Cu_3O_{6.63}$	
T[K]	ξ_{Lor}	ξ_{Gauss}	ξ_{Lor}	ξ_{Gauss}
100	1.5 - 2.4	1.0 - 1.5	2.5 - 4.5	1.5 - 2.5
300	1.3 - 1.7	0.7 - 1.1	1.5 - 2.5	1.0 - 1.5

Table 1: correlation lengths from the NMR analysis

From these results we can conclude:

- The results for the correlation length are not compatible with the neutron scattering results by Rossat - Mignod [14].
- The neutron scattering experiments favor temperature independent correlation lengths.
- The experimental results by G. Aeppli and coworkers [15] seem to be more compatible with the correlation lengths required by NMR experiments.

The detailed analysis shows that the copper and oxygen relaxation time experiments are compatible with both assumptions - a temperature dependent correlation length or a temperature dependent strength of the antiferromagnetic peak - but that the crucial quantity which could determine which case it is is the ratio of the yttrium to the oxygen relaxation rate which was not measured in the same sample to a high enough accuracy up to now. For details of the analysis we refer the reader to our paper [9]. The original MMP analysis was intended to describe the spin correlations on a very low energy scale. A comparison of the nuclear magnetic experiments with the neutron scattering experiments requires an extrapolation to larger energies. For a thorough discussion we refer the reader to [9].

Recently Monthoux and Pines [16] put forward a theory of the normal and superconducting state of the CuO materials based on the notion that these materials are nearly antiferromagnetic Fermi liquids.

5 Dynamical structure factor for in the marginal Fermi liquid picture

The remaining physics question, as stated in the introduction, is off course where does one get the low energy physics scale from. The easiest way to obtain a low energy scale from a "high energy" model is to be close to a critical point in this class of models belong the "nested Fermi liquid", Hubbard RPA and dynamical phase separation models. Very close to the phase transition dynamical scaling should hold in that case the low energy scale is determined by the correlation length via $\omega \sim \Gamma \times (\xi/a)^{-z}$ where Γ is the energy scale determined by the "high energy physics", ξ is the correlation length and z is the dynamical scaling exponent. In this section we will circumnavigate the difficult question what is determining the low energy physics and answer the following related question: Assuming that the only effect of all the many body complications is to give rise to a single particle scattering rate which is proportional to the energy or temperature, whatever is less (which is the so called marginal Fermi liquid hypothesis) what can we know about the spin response. The original marginal Fermi liquid picture, [18], assumed that the selfenergy is nearly wavevector independent

$$\mathrm{Im}\Sigma \sim \max(\omega, T). \tag{12}$$

In the original MFL paper it was argued that the resulting response functions are a universal function of ω/T where ω is the frequency of the external field and T is the temperature and also weakly q dependent. The assumed factorizable form of the dynamical structure factor is not consistent with a different temperature dependence of the Cu and O relaxation rate. Neutron scattering experiments by Aeppli et al. [15] on the $\mathrm{LaSr_xCu_{2-x}O_4}$ for $x \approx 0.15$ showed that the dynamical structure factor is strongly q dependent even for the strongly doped material. Littlewood et al., [19] have proposed an extension of the original MFL picture. It is clear that in two dimensions the dynamical structure factor even for the noninteracting Fermi gas shows a strong q dependence at low frequencies. It is interesting to examine the predictions of the simple band structure of the CuO plane. Using a tight binding band structure one obtains an incommensurate peak structure with four posts which are a little bit shifted away from the zone corner, $Q = (\pi/a, \pi/a)$. Similar ideas were put forward by Levin et al. [17] What we found is that the position and magnitude seems to be compatible with a simple tight binding band structure picture for the CuO plane. The position and magnitude of the peaks will strongly depend on the doping. What is even more interesting is that if we include the marginal Fermi liquid form for the self energy we also can understand the temperature and energy dependence of the peak structure. In the regime which is accessible by the neutron scattering the main effect of the marginal Fermi liquid form of the quasiparticle propagator is the life time broadening which has the energy scale of the temperature.

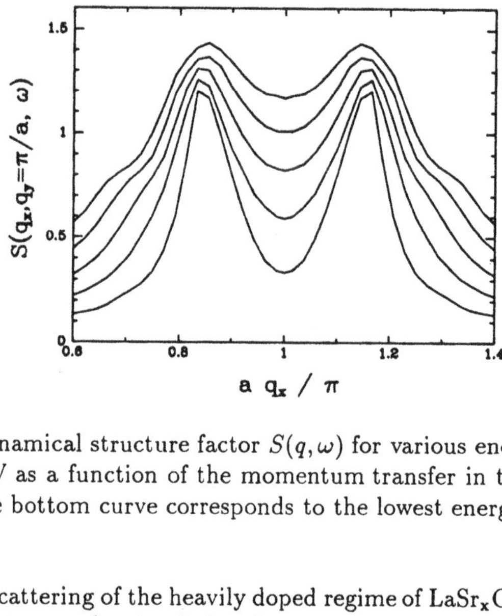

Figure 1: The dynamical structure factor $S(q,\omega)$ for various energy transfers, $\omega = 3$, 6, 12 and 15 meV as a function of the momentum transfer in the q_x direction. q_y is fixed to π/a. The bottom curve corresponds to the lowest energy transfer the top to the highest.

The neutron scattering of the heavily doped regime of $LaSr_xCu_{2-x}O_4$ can therefore be understood in terms of a simple band structure picture of marginal Fermi liquid quasiparticles. It is not clear if this is also true for the nuclear magnetic resonance experiments in the same system. Although the Cu nuclear magnetic relaxation rates seem to not so far off from what one would predict with this simple approach it is not clear if the oxygen NMR data can be understood within the context of the MFL picture.

6 Conclusions

By now it has become quite clear that temperature dependent antiferromagnetic correlations play an important role in the understanding of the magnetic properties of the CuO high temperature superconductors even in the strongly doped regime. The early interpretation of Millis, Monien and Pines [13] assumed a temperature dependent correlation length. The reexamination of the present nuclear magnetic resonance data by [9] demonstrates that one can not deduce from the nuclear magnetic relaxation measurements of the copper and oxygen rates alone that the correlation length is temperature dependent. We have proposed a simple model in which the temperature and frequency dependence of the dynamical structure factor in $LaSr_xCu_{2-x}O_4$, $x \approx 0.15$, can be understood in a simple tight binding band structure picture including lifetime effects.

It is a pleasure to thank my numerous collaborators. Many have worked in this interesting area of solid state physics - I apologize to those whose important work I did omit in this brief review. The research for this project was supported in part by an IBM fellowship and by the NSF Grant NSF-PHY-04035.

References

[1] P. W. Anderson, Science **235**, 1196 (1989)

[2] see for e.g. B. Keimer et al., Phys. Rev. Lett. **67**, 1930 (1991)

[3] For a review see, e.g. S. Chakravarty, "Spin Dynamics of Superconducting Cuprates", p. 198 in High Temperature Superconductivity: Proceedings of the Los Alamos Symposium - 1989, K. S. Bedell, D. Coffey, D. E. Meltzer, D. Pines and J. R. Schrieffer, eds. (Addison Wesley: Redwood City, CA) (1990).

[4] For a review see, e.g. D. J. Scalapino, "Results from Numerical Simulations of the 2D Hubbard Model", p. 314 and references therein, in High Temperature Superconductivity: Proceedings of the Los Alamos Symposium - 1989, K. S. Bedell, D. Coffey, D. E. Meltzer, D. Pines and J. R. Schrieffer, eds. (Addison Wesley: Redwood City, CA) (1990).

[5] F. Mila and T. M. Rice, Physica **C 157**, 561 (1989)

[6] B. S. Shastry, Phys. Rev. Lett. **63**, 1288 (1989)

[7] M. Takigawa, private communication

[8] see e.g. M. Schlüter and M. S. Hybertsen, Physica **C 162**, 583 (1989)

[9] A. J. Millis and H. Monien, Phys. Rev. B **B45**, 3059 (1992)

[10] M. Takigawa, J. L. Smith and W. L. Hults, Phys. Rev. **B44**, 7764 (1991)

[11] J. A. Martindale et al., Phys. Rev. Lett. **68**, 702 (1992)

[12] A. J. Millis and H. Monien, unpublished

[13] A. J. Millis, H. Monien and D. Pines, Phys. Rev. **B42**, 167 (1990)

[14] see e.g. J. Rossat-Mignod et al., Physica **C185**, 86 (1991)

[15] T. E. Mason, G. Aeppli, M.A. Mook, Phys. Rev. Lett. **68**, 1414 (1992)

[16] P. Monthoux and D. Pines, Phys. Rev. Lett. **69**, 961 (1992)

[17] J. P. Lu, Q. M. Si, J. H. Kim, K. Levin, Physica **C179**, 191 (1991)

[18] C. M. Varma et al., Phys. Rev. Lett. **63**, 1996 (1989)

[19] P. Littlewood, J. Zaanen, G. Aeppli and H. Monien, unpublished

OXYGEN ORDER AND SPIN STRUCTURE IN $YBa_2Cu_3O_x$ DEDUCED FROM COPPER NMR AND NQR

H. Lütgemeier*
IFF, Forschungszentrum Jülich, D–5170, Germany

I. Heinmaa
Institute of Chemical Physics and Biophysiscs, Tallinn, Estonia

ABSTRACT

The magnetic state and the arrangement of the oxygen ions in $YBa_2Cu_3O_x$, $6 < x < 7$, is investigated by analyzing the NQR spectra of copper. For this analysis it is important that the signals from the Cu(1) and Cu(2) sites can be distinguished in $GdBa_2Cu_3O_x$ due to the different influence of the magnetic Gd ions. At x near 6.5 the existence of the ortho–II structure with alternating filled and empty chains is evident, at higher x ortho–II and ortho–I with all chains filled coexist. The average chain length n is about 10 to 13 oxygen ions per chain in the region of the 60 K plateau and decreases fast at lower x. Superconductivity disappeares in samples with $n < 6$. No coexistence of magnetic order at the Cu(2) sites with superconductivity is found, but a slowing down of the magnetic fluctuations in samples with T_c below the 60 K plateau can be deduced from the disappearing Cu(2) NMR and NQR signals. In the antiferromagnetic region the spin structure is influenced by impurities like Fe, Al and Gd; but in any case the expectation value of the magnetic moments at the Cu(1) site in zero.

*The main part of this report has been presented at the Egyptian–German Workshop on HT_c Superconductivity at Fayonm and Shebin Al–Kome (1992) and will be published in Egyptian Journal of Solids.

1. Introduction

In $YBa_2Cu_3O_x$, $6 < x < 7$ (YBCO) and in most related compounds with Y replaced by rare earths like Gd, antiferromagnetic order of the copper ions in the Cu–O planes (Cu(2) sites) appears at an oxygen concentration x below about 6.4, with a Neel temperature which is about 400 K for x = 6.0. At x above 6.4 the samples become superconducting. The values of T_c increase with x showing a plateau in T_c of about 60 K in the region of $6.5 < x < 6.7$ and a second plateau with T_c of about 92 K for $x > 6.8$ [1]. Since in the structure of YBCO only the oxygen ions in the Cu(1) planes can be removed, these must be responsible for the transition from the antiferromagnetic insulator to the superconductor. Therefore the local arrangement of the oxygen ions in the Cu(1) planes is of great interest. A large amount of information about the structure has been obtained by diffraction methods, X–ray, neutron, and electron diffraction.

The structure of the antiferromagnet at x = 6.0 without oxygen in the Cu(1) plane is tetragonal. At a small increase of x the oxygen ions are distributed randomly on the sites between two Cu ions in this plane. At higher x, a transition to the orthorhombic structure of the superconductor at x = 7 appears, where only the O(1)–sites in b–direction, in the "chains", can be occupied. Since the value of Tc at intermediate x depends to a high degree on the thermal treatment of the samples it must be assumed that the arrangement of the oxygen ions on the O(1) sites is important for the superconductivity in oxygen depleted samples [1,2]. Different superstructures have been found by electron diffraction [1]. The structure at x = 6.5 with a unit cell of dimension 2a and every second chain either filled or empty is designated as ortho–II, and correspondingly the normal one at x = 7.0 as ortho–I. The problem of all diffraction methods is that local order can be detected only when the correlation length of the structure is long enough.

Since the electric field gradient (EFG) at a copper nucleus depends on the valence state of the ion itself and on the charge of the surrounding atoms, the copper NMR/NQR can be a direct method to study the surrounding of the Cu ions and by this the local order of oxygen in these materials. Unfortunately, the NQR spectra in oxygen deficient $YBCO_x$ change with x in a very complicated way [3,4,5]. Overlapping of the lines from the two copper isotopes (^{63}Cu, ^{65}Cu), from the Cu(1) and Cu(2) layers, and from the different configurations within the layers makes the analysis extremely difficult.

In this report we present the NQR and NMR spectra of copper in oxygen deficient $YBCO_x$ and $GdBCO_x$ with the aim to study the oxygen ordering in the lattice and the magnetic order of the Cu–ions in the antiferromagnetic state. We will use the advantage of the $GdBCO_x$ system where separating of the Cu(1) and Cu(2) NQR lines is possible [6,7]

2. NMR and NQR of copper

Natural copper contains two isotopes ^{63}Cu and ^{65}Cu, both with the nuclear spin I = 3/2. The magnetic dipole moments and the electric quadrupole moments differ by about 10%. Since the ^{63}Cu isotope of larger abundance (69%) reveals the smaller value of the gyromagnetic ratio, gamma, and the larger quadrupole moment, eQ, the origin of an interaction can be determined easily from the different behavior of both isotopes. The quadrupole moment couples to the electric field gradient tensor (EFG) determined by the charge distribution in the crystal, which can be expressed by the largest eigenvalue, eQ, and the asymmetry parameter, η [8]. Without a magnetic field, the degeneracy of the +–m levels is retained and for I=3/2 only one "pure NQR" line $|\pm 1/2\rangle \rightarrow |\pm 3/2\rangle$ can be observed at the frequency $\nu_{NQR} = \nu_Q (1+\eta^2/3)^{1/2}$ with the quadrupole frequency $\nu_Q = e^2qQ/2h$. The asymmetry parameter cannot be determined from the pure NQR spectrum for I = 3/2 since only one transition exists. This is possible in an applied magnetic field which leads at low field to the Zeeman splitting of the NQR lines or at high field to the NMR spectrum with quadrupole splitting. Since in both cases the splitting depends on the orientation of the magnetic field relative to the main axes of the EFG tensor, the spectra of powder samples become broad. In metallic samples only the region of the penetration depth of the r.f. field contributes to the signal. For this reason powder samples are used in most experiments and the orientation of the grains is of great importance.

In the case of antiferromagnetic order, the hyperfine interaction leads to a local magnetic field, the hyperfine field, at those Cu ions which carry the magnetic moments. The Zeeman splitting of the nuclear levels by this hyperfine field determines the frequency of the antiferromagnetic nuclear magnetic resonance (AFNMR) which can be observed without an applied magnetic field. For copper the corresponding frequencies are much larger than the NQR frequencies.

3. Experimental

All experiments presented here have been performed by the standard spin–echo technique at 4.2 or 1.2 K either by sweeping the frequency at zero field (NQR and AFNMR) or the magnetic field at fixed frequency (NMR). As the intensity of the echo signal is proportional to the square of the resonance frequency, the NQR spectra were normalized by dividing the echo intensities by the square of the frequency.

The NQR and NMR experiments with oxygen deficient samples of YBCO and GdBCO have been performed over a long period with samples of different preparation and origin. For the quantitative analysis discussed below we used three series of samples presented in Tab.1.

Slightly Gd–doped samples $Y_{0.999}Gd_{0.001}Ba_2Cu_3O_x$ (A1..A7) have been prepared by the equilibrium method at 650°C as described in ref.[9] and had been already used for Gd EPR experiments [9]. In order to get useful information from the EPR and NMR spectra, these samples had been uniaxially oriented in the magnetic field and fixed in epoxy. For comparison purposes and in order to investigate the influence of annealing, several $YBCO_x$ samples without Gd doping prepared by the quenching technique with subsequent annealing [10] were investigated, too (B1..B4). The extraction of oxygen in the series of $GdBCO_x$ samples (C1...C8) has been done at 810°C at controlled pressure [11]. The oxygen concentration x given in Tab.1 has been determined in different ways and the accuracy is difficult to estimate. Within each series of samples the increments of x are very reliable as was seen from the continuous development of all properties, especially from the x dependence of the EPR spectra which had been recorded at smaller steps of x [9]. As a result of the different preparation and calibration procedures, however, errors of x in the range of +–0.1 must be admitted in the comparison of results obtained for the different series of samples. The same holds for the comparison with data from the literature.

Tab.1:

Summary of samples investigated: oxygen concentration, x, T_c from susceptibility, relative intensities of the different components of the NQR signal of Cu(1), I2..I4, mean length of Cu–O chains, n (eq.3), and oxygen concentration in the Cu(1) plane, z (eq.1).

sample #	x	T_c K	I4* %	I4** %	I3 %	I2 %	n	z
A1	.6.0					100		0
A2	6.34		15	11	17	57	4.1	0.35
A3	6.43		27	9	14	50	6.2	0.43
A4	6.53	35	24	21	10	45	10.0	0.50
A5	6.59	56	8	37	8	47	12.3	0.49
A6	6.77	61	26	30	12	32	10.3	0.62
A7	7.0	90		100				1
B1	6.30	20	14	17	8	61	8.7	0.35
B2	6.54	56	7	26	8	59	9.2	0.37
B3	6.58	56	11	34	12	43	8.5	0.51
B4	6.74	70	8	48	10	34	12.2	0.61
C1	6.0					100		0
C2	6.35		7	12	36	45	2.1	0.37
C3	6.39		9	8	45	38	1.8	0.40
C4	6.50	20	9	35	19	37	5.6	0.53
C5	6.55	50	8	32	11	49	8.3	0.46
C6	6.65	52	10	36	12	42	8.7	0.52
C7	6.75	58	13	52	11	24	12.8	0.70
C8	7.0	92		100				1

A1..A7 $YBCO_x$ 0.1% Gd slowly cooled from 650°C [9]
B1,B2,B4 $YBCO_x$ annealed about 10 days 380°C [10]
B3 $YBCO_x$ as–quenched [10]
C1..C8 $GdBCO_x$ slowly cooled from 800°C [11]

4. Copper NQR Spectra in $YBCO_x$

Simple NQR spectra of copper are found in YBCO only for high and low oxygen concentration, x. In the stoichiometric compound with x near 7 the spectrum consists of two pairs of the $^{63}Cu/^{65}Cu$ lines at 31.5/29.1 MHz from the Cu(2) site and at 22.1/20.5 MHz assigned to the Cu(1) site, Cu(1)4, with four oxygen neighbors [6,12]. In the following we mention only the frequencies of the most abundant ^{63}Cu isotope. The lines of ^{65}Cu are shifted to lower frequency by the ratio of the nuclear quadrupole moments, $(^{65}eQ/^{63}eQ) = 0.923$, and the intensities are smaller by the ratio of the natural abundances, 0.45.

The assignment of the lines to the different sites was determined from the intensity ratio of the lines [12] (since one formula unit contains one Cu(1) and two Cu(2) sites), by symmetry arguments since only the Cu(2) sites can reveal the nearly axial symmetry of the EFG along the c–axis found in the NMR spectra for the line at 31.5 MHz [6], and by the influence of Gd ions on both lines [6]. In $GdBa_2Cu_3O_x$ the electron spins of the Gd ions order antiferromagnetically at about 2.2 K. Above this temperature the fluctuations of the magnetic moments induce strong relaxation of the nearby Cu(2) nuclei but leave the Cu(1) sites undistorted. Below the Neel temperature, however, the ordered moments influence both sites only weakly [7].

With decreasing x, $6.9 < x < 7.0$ the Cu(1) and Cu(2) lines become broader due to the creation of oxygen vacancies in the Cu(1)–O layer. The increase of the linewidth was found to be in agreement with a random distribution of the defects [13]. At $x < 6.9$ the spectra develop into a more complicated structure [3,4,5] which is the topic of our investigation.

In the antiferromagnetic compound with low oxygen concentration the narrow lines are found again. For x = 6.0 only the NQR lines of Cu(1) with two oxygen neighbours, Cu(1)2, exist (30.1 MHz). At slightly increased x, a second NQR line at 24.0 MHz appears which has been attributed to the Cu(1) sites with one oxygen neighbour in the chain, $Cu(1)_3$ [6]. This line was observed in a very broad region of the oxygen concentration $6.1 < x < 6.8$ [4]. The NQR spectrum of the Cu(2) sites is missing in the antiferromagnetic region since these sites carry the ordered magnetic moments leading to the hyperfine field at the copper nucleus. From the AFNMR spectrum in the region of 90 MHz [14] a quadrupole splitting corresponding to a frequency of 23.2 MHz is deduced which was found also above the Neel temperature by NMR in an external field only slightly shifted to 23.8 MHz [15].

5. Cu(1) Spectra

We have used the $GdBCO_x$ samples in order to separate the signals of the Cu(1) sites from the Cu(2) site spectrum. At 4.2 K the NQR signal (Fig.1) contains only the Cu(1) signal since the Cu(2) nuclei are distorted severely by the fluctuations of the nearby magnetic moments of the Gd ions. At 1.2 K below the Neel temperature of the Gd spins, however, the full spectrum of both layers is observed. It is evident that the Cu(1) spectrum contains the sharp lines whereas ths Cu(2) spectrum is so broad, that the contributions of the two isotopes are not separated. The comparison of the spectra at both temperatures for lower x shows that no signal of the Cu(2) layers exists below 25 MHz and no Cu(2) resonance can be observed in samples with low x showing T_c values below the region of the 60 K plateau.

Fig. 1: NQR spectrum of $GdBa_2Cu_3O_{6.75}$ (sample # C7) at 4.2 and 1.2 K.

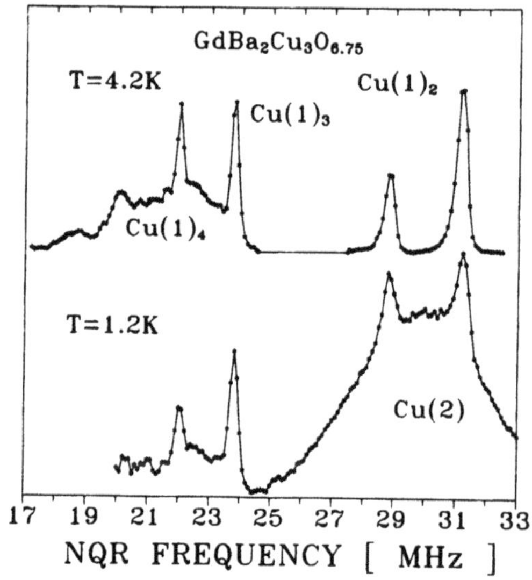

Fig. 2: NQR spectra of the Cu(1) sites in $GdBa_2Cu_3O_x$ measured at 4.2 K for samples C1, C2, C5, C7, and C8. Each value of the EFG leads to a doublet for the ^{63}Cu and the less abundant ^{65}Cu isotopes. The doublets with the ^{63}Cu line between 30 and 31.5 MHz belong to $Cu(1)_2$,those at 24 and 22.5 MHz to $Cu(1)_3$ and $Cu(1)_4$, respectively. Whereas the lines of $Cu(1)_3$ and $Cu(1)_2$ are relatively sharp at all x, $Cu(1)_4$ broadens severely for $x < 7$.

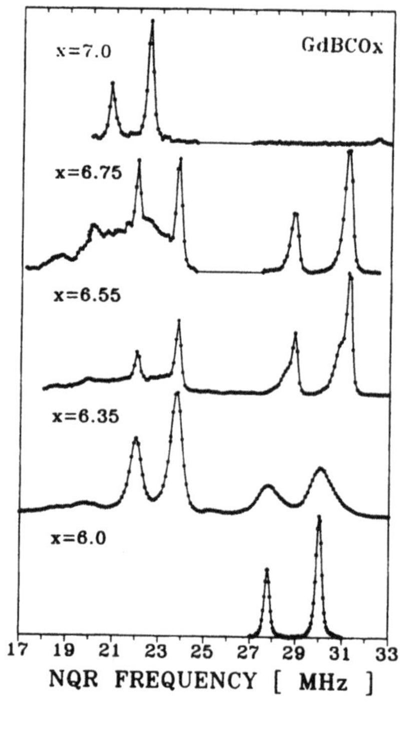

The Cu(1) spectrum in Fig.1 contains two narrow lines for ^{63}Cu at 31.2 and 23.8 MHz. Since they are near to the corresponding lines found at low x, we attribute these lines to the twofold and threefold coordinated Cu(1) sites, $Cu(1)_2$ and $Cu(1)_3$, respectively. Correspondingly, the broader structure below 22.5 MHz is assigned to the $Cu(1)_4$ site. In most samples we found also a line at about 20 MHz. Since the transverse ralaxation times of all these lines are similar (about 0.2 ms at 4.2 K) we assume that the structure at about 20 MHz must belong to the $Cu(1)_4$ site, too, most probably to sites in short chain segments.

The characteristic NQR spectra of Cu(1) in $GdBCO_x$ (Fig. 2.) show the change of the spectra with the oxygen concentrations x. The corresponding spectra of $YBCO_x$ compounds are quite similar with only a small shift of the frequencies, but mixed with the signal of the Cu(2) layers. In the high frequency region the spectra consist in the samples with x between 6.6 and 6.8 of a single line of ^{63}Cu at the frequency 31.2 MHz in $GdBCO_x$ and 31.5 MHz in $YBCO_x$. At x below about 6.55 two more lines appear and the spectra exhibit three lines A, B, and C at the frequencies 31.2, 30.9 and 30.2 MHz, respectively, in the case of GdBCO, and at 31.5, 30.9 and 30.4 MHz for YBCO (Fig. 3). With further decreasing x the spectra become broader and the fine structure disappears.

Fig. 3: NQR spectrum of the Cu(1) sites with twofold oxygen coordination in $YBa_2Cu_3O_{6.53}$ sample A4) with explanation of sites A between two full chains, B and C with one or two neighbouring empty chains, respectively.

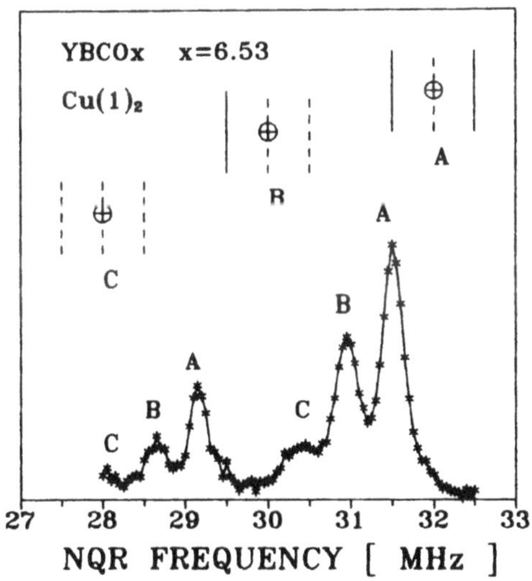

Different authors [3,4,5] have recorded similar data about the change of the NQR spectra around 30 MHz in this range of x, but with different explanations. The comparison of YBCO and GdBCO proves that all three lines belong to the Cu(1) sites. The threefold spectrum has been found at different nominal values of the oxygen concentration x, e.g. in a sample with $x = 6.4$ and $T_c = 40$ K studied earlier by neutron diffraction [16]. In all cases the lines A, B, and C were found together in one sample only if it was superconducting with T_c below 50 K.

Three different configurations of the twofold coordinated $Cu(1)_2$ site can be found easily in a model which assumes that the oxygen ions are ordered with either filled or empty chains. We assign the single line A, the only one seen in the spectra of the compounds with $x > 6.6$, to the $Cu(1)_2$ sites between two intact chains as shown in Fig. 3. Then the line B must be ascribed to the $Cu(1)_2$ sites between an intact and an empty one, and the line C to the $Cu(1)_2$ sites between two empty chains. Since the frequency of line C is only 0.2 MHz larger than the frequency of $Cu(1)_2$ at $x = 6.0$, our assignment is confirmed.

The evolution of the NQR spectra at lower oxygen concentration is seen most clearly in the measurements at 77 K by Serikov et al. [3] where with decreasing x from 6.44 to 6.13 the main intensity moves from the line A through line B to the line C. We attribute the broadening observed at $x < 6.5$ in our measurement at 4.2 K to the onset of the localization of the electronic spins at the Cu(2) sites. In the AF structure of pure YBCO with $x = 6.0$ the magnetic interactions of the Cu(1) sites with the two neighbouring Cu(2) planes are canceled exactly by symmetry [7]. If at increasing x the number of defects in the AF structure increases as is seen in the broadening of the AFNMR spectra at 90 MHz [16], the dipolar and transferred hyperfine fields are no longer canceled at the Cu(1) sites and cause the observed broadening of the NQR lines. This effect must disappear above the Neel temperature of the Cu(2) planes. Since Tn becomes low near $x = 6.4$, the spectra recorded at higher temperature [3,5] are not influenced by the disorder of the antiferromagnetic state.

6. Correlation time of the Cu(2) spins

The sequence of NQR spectra at decreasing x shows the development of the magnetic state at the Cu(2) sites in three steps: i) In the region of the 60 K plateau of T_c and at higher x the rate of the spin fluctuations is so large that the NQR lines at the Cu(1) and Cu(2) sites are not distorted. The influence of the fluctuations is important, however, for the spin–lattice relaxation. ii) The spin fluctuations become slower at the lower end of the 60 K plateau and cause the disappearing of the Cu(2) signal due to the fast nuclear relaxation, but do not yet distort the Cu(1) signal. iii) The onset of magnetic order in the Cu(2) layer becomes evident from the broadening of the Cu(1) NQR lines in samples with x reduced so much that they are no longer superconducting. Another indication of the onset of magnetism is found in the T_2 measurements of $GdBCO_x$, where the Cu(1) sites can be studied separately. Here we found a nonexponential magnetization decay of the Cu(1) lines in the samples $x = 6.50$ and $x = 6.39$ with a superposition of fast and slow relaxation rates. This inhomogeneous relaxation at the Cu(1) sites can be explained assuming an additional relaxation path due to slowly fluctuating spins distributed inhomogeneously in the Cu(2) layers.

7. Intensity distribution of the Cu(1) lines

The best proof of the site assignment of the NQR lines in the systems $YBCO_x$ and $GdBCO_x$ can be a reasonable intensity distribution of the Cu(1) lines belonging to the sites with two, three, and four nearest oxygen neighbours. In order to get a quantitative estimate of the intensity distribution of the Cu(1) lines we simulated the normalized experimental spectra by pairs of the $^{63}Cu/^{65}Cu$ lines. The relative intensities of $Cu(1)_2$ (I2), $Cu(1)_3$ (I3), and both components of $Cu(1)_4$ (I4* and I4**, see section 5) for all samples evaluated are given in Tab.1.

The variation of the intensity for the three configurations at increasing x in the series A of $YBCO_x$ of uniform preparation and in series C of $GdBCO_x$ reveal the general trend of increasing $I4 = I4^* + I4^{**}$ and decreasing I2 and I3. The reliability of the relative intensities can be checked by the oxygen concentration z in the Cu(1) layer which is determined by

$$z = I4 + I3/2, \quad I4 = I4^* + I4^{**}. \qquad (1)$$

The relation $z = x-6$ is fulfilled in nearly all samples within the limit of accuracy ($\pm$ 0.1) by which x and the intensities are determined (Tab.1), but the generally too small values of z show that the intensity I2 has been overestimated systematically at $x > 6.5$. The reason may be connected with problems of separating the $Cu(1)_2$ from the Cu(2) signal and with the correction for T_2. It is evident that the component at 20 MHz with the intensity $I4^*$ must belong really to the fourfold coordinated Cu(1), otherwise the value of z would be even smaller.

A random occupation of the O(1) sites in the Cu(1) chains by oxygen would lead to relative intensities depending on z as follows:

$$I2=(1-z)^2, \quad I3 = 2z(1-z) \quad \text{and} \quad I4 = z^2 . \qquad (2)$$

This is evidently not the case in Tab.1. The intensity I3 is too small by more than a factor 2 for all samples with x near 6.5. This is a first clear indication for the development of a non–random structure with long chain fragments in the region x = 6.5.

From the relative intensities we can estimate the average length of the chain fragments. The average number of oxygen atoms, n, in one chain fragment is simply determined by the ratio of the number of copper atoms within the chains, I4, and the number of the atoms terminating the chains, I3, by

$$n = 1 + 2(I4/I3) \qquad (3)$$

Fig. 4: Mean length of oxygen chains in $YBa_2Cu_3O_x$ and $GdBa_2Cu_3O_x$ versus x.

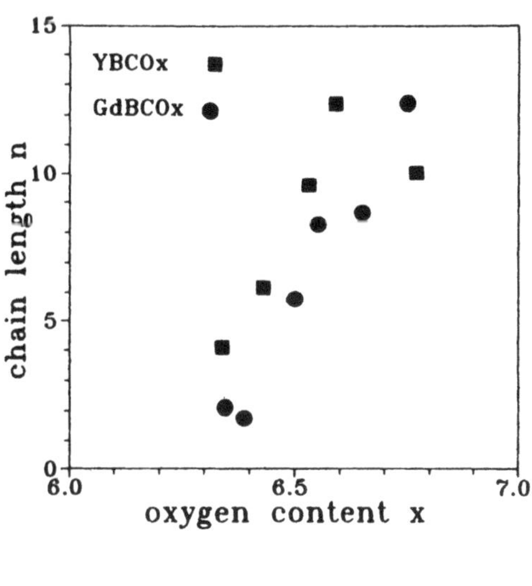

The values of n increase with x in $GdBCO_x$ and $YBCO_x$ as shown in Tab. 1 and in Fig. 4. The smaller values of n in $GdBCO_x$ at the same x compared with $YBCO_x$ can be caused by the different preparation method in the two systems, but in any case a strong increase of n is seen when x approaches 6.5. This is a clear evidence for the development of the ortho–II structure with filled and empty chains. Superconductivity appears in the samples with $n > 6$. The largest values of the chain length between 10 and 13 are found for $x > 6.5$.

8. Cu(2) spectra

The NQR spectra containing of the ^{63}Cu and ^{65}Cu lines of Cu(2) and $Cu(1)_2$ (Fig. 1 for 4.2 K) are too complicated for the analysis of the Cu(2) resonances. We found, that the changes in the local environment of the Cu(2) sites in $YBCO_x$ due to the oxygen depletion can be followed better in the field sweep NMR spectra. We have measured these spectra of the uniaxially oriented samples with the crystallographic c–axis either parallel or perpendicular to the external magnetic field. The latter case is suitable also for the usual, previously not oriented powder samples which turn out to be perfectly oriented with the c–axis perpendicular to H in the superconducting state [6]. The spectrum, given in Fig.5 shows the full field sweep

Fig. 5: NMR spectrum of $YBa_2Cu_3O_{6.54}$ (sample B2) at 4.2 K measured at 45.5 MHz, with the magnetic field applied perpendicular to the c–axis.

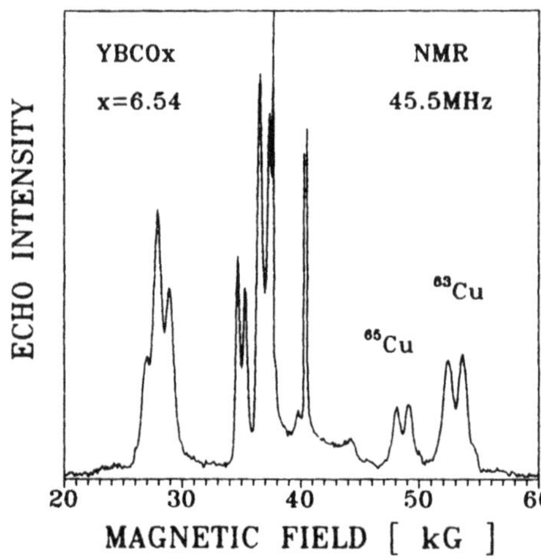

spectrum of the sample B2. Here the self–orientation of the superconducting grains in the magnetic field is seen from the well resolved lines. The spectrum shows the resonances from the two copper isotopes clearly separated in the high field satellite region as a result of the higher gamma value and a smaller quadrupolar shift of the isotope ^{65}Cu, but overlapping in the low field satellite. The most surprising result in this spectrum is the splitting of the Cu resonance with two lines of equal intensity and slightly different quadrupole coupling. In our NQR spectra the two lines are seen in the fit of the broad Cu(2) spectrum below the $Cu(1)_2$ signal with NQR frequencies of 30.8 and 27.7 MHz in agreement with results of Warren et al. for a sample with a single Cu isotope [18] and Shimizu et al. [19]. From the Cu(1) sites only the $Cu(1)_2$ contribution is seen as a shoulder at the high field side of the doublet, whereas the resonances of $Cu(1)_4$ and $Cu(1)_3$ sites are smeared over a large region of the spectrum as expected due to the strong asymmetry of the EFG tensor of these sites [6].

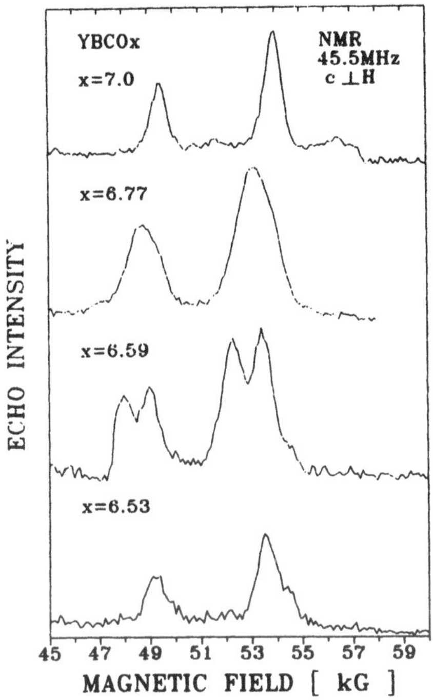

Fig. 6:The high–field satellite in the NMR spectrum of $YBa_2Cu_3O_x$ at different x measured at 40 MHz for the magnetic field perpendicular to the c–axis (samples A4, A5, A6, A7).

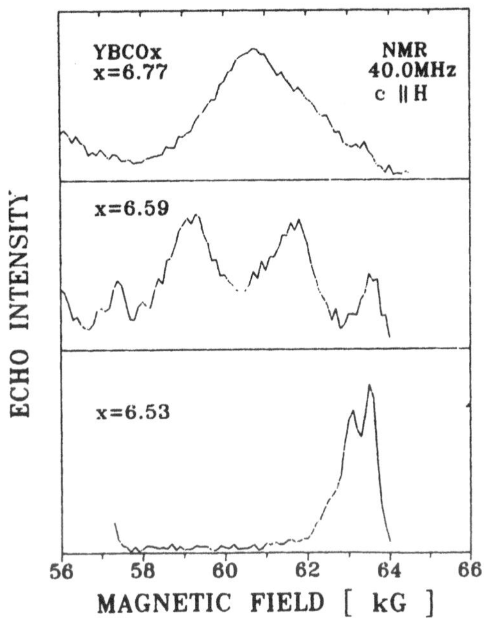

Fig. 7: The high–field satellite in the NMR spectrum of $YBa_2Cu_3O_x$ at different x measured at 40 MHz for the magnetic field applied parallel to the c–axis (samples A4, A5, A6)

The change of the NMR spectra in the region of the high–field satellite at decreasing x for grain oriented powder samples is shown for the magnetic field perpendicular and parallel to the c–axis in Figs. 6 and 7, respectively. The spectra for the perpendicular field reveal a continuous shift of the Cu(2) resonances to lower fields corresponding to lower NQR frequencies at x decreasing from 7.0 to 6.77 and the splitting around $x = 6.59$. The single line for $x = 6.53$ shifted again to higher field belongs to the Cu(1) resonances according to the interpretation of the spectra measured for the field parallel to the c–axis (Fig.7). Here the spectrum of the $x = 6.59$ compound shows again the splitting of the Cu(2) resonance. Another line is found at higher field which is the only component remaining at $x = 6.53$. The splitting of this line into three components in the NQR spectrum of Fig.3 identifies these as the contributions of $Cu(1)_2$ (A,B,C). Thus this spectrum proves again that the resonance of the Cu(2) sites disappears in superconducting samples of low x with T_C values below the 60 K plateau. The main advantage of the NMR spectra in this case is the separation of the Cu(2) and $Cu(1)_2$ contributions by different values of the orbital Knight shift at the two sites [19].

The splitting of the Cu(2) line is found in all samples with $x < 6.7$ but disappears at larger x. We did not find any indication for a continuous transition between these spectra like a change of the relative intensities of both peaks or a line shift, only the intensity between the peaks changes and is lower in the annealed sample B2 than in the quenched one B3. Thus it seems that the spectra at larger x consist of the superposition of the split ($x = 6.55$) and the unsplit ($x = 6.8$) components. This is most obvious in a spectrum published recently by Takigawa et al. for $x = 6.63$ [20].

9. Order of Oxygen in the Cu(1) Layers

The evaluation of the chain length n from the different components of the Cu(1) spectrum has already proved, that a random occupation of the O(1) sites in the Cu(1) plane does not exist in our samples. The spectra of the Cu(1) and Cu(2) sites in samples with x slightly higher than 6.5 can be described perfectly in the model of the ortho–II structure with every second chain filled: The two Cu(2) lines show two different Cu(2) sites to exist connected with "full" and "empty" chains, respectively. Here we assume that the line at higher field which means with the larger EFG, corresponds to the Cu(2) sites near the full chains, since it agrees nearly with the line for $x = 7$. In the same region the spectrum of $Cu(1)_2$ contains only the line A which we assigned to the sites in "empty chains" between "full" chains. As expected in the case of the pure ortho–II structure, this line represents nearly one half of all Cu(1) sites (Tab.1).

The Cu(1) resonance reveals the details about the correlation length in this structure. The average length of the "full" chains is rather short, about 10 oxygen ions per chain as shown by the large intensity I3 of the terminating $Cu(1)_3$ ions. This means that the number of defects in this structure is high and the coherence length can be small. Since all information we can gain from counting the different sites, concerns only the arrangement of the oxygen ions in the Cu(1) plane and not the sequence of the planes along the c–axis, we cannot prove the existence of a three dimensional order.

Above and below $x = 6.5$ the filling of the Cu(1) layer with the oxygen ions can be modeled in different ways. One group of models is based on the ortho–II structure with every other chain either empty or partly filled for $x < 6.5$ and either full or partly filled at $x > 6.5$. Within the partly filled chains different arrangements have been proposed, e.g. a random distribution [21], the condition of maximal distance between the oxygen ions [22], or ordered superstructures at multiples of $x{-}6 = 0.125$ [23]. These models show different dependence of the chain length on x (Fig.8). But they agree in the general behavior that after the divergence at $x = 6.5$, n becomes smaller again to about 6 at $x = 6.7$ by the formation of the new chains. At larger x, n does not differ much from the values expected for the random distribution.

Fig. 8: Mean length n of chains calculated for different models of the arrangement of the oxygen ions on the O(1) sites in the Cu(1) plane
1: random distribution (diluted ortho–I)
2: ortho–II with random filling of the "diluted" chains [21]
3: ortho–II with maximal distance in the "diluted" chains [22]
4: ordered structures based on ortho–II with larger unit cell [23].

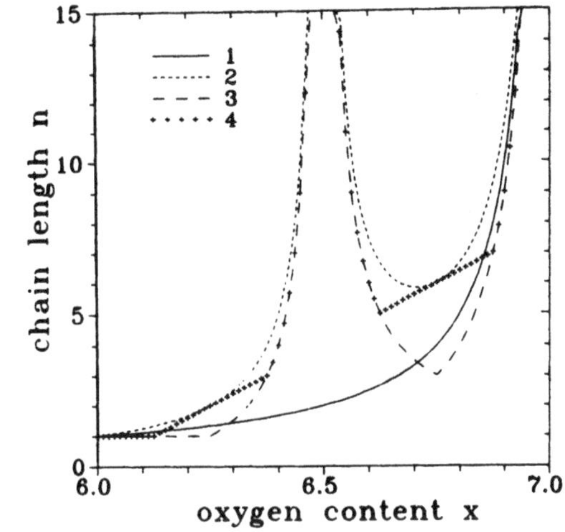

The experimental data in Tab.1 and Fig. 4 show no decrease of n for $x > 6.5$ which means that the oxygen ions entering the ortho–II structure form chains of the same length as in pure ortho–II. This may happen in two ways: the first is the formation of domains of ortho–I in coexistance with ortho–II [24], the second means ordered sequences of full and empty chains like ortho–III with one empty and two filled chains [25]. This second model can be excluded from the Cu(2) spectra since in this case the intensity of the line corresponding to the filled chains should increase at the cost of the other line. The spectra, however, show the ortho–II lines together with the broadened ortho–I line corresponding to the coexistence of the pure ortho–II structure with the oxygen depleted ortho–I of about $x = 6.8$.

In the range below the "60 K plateau" only the Cu(1) signal can be observed at least at the low temperatures of our experiment, 4.2 and 1.2 K. Here two features are evident: the decrease of the mean chain length n (Fig.4) calculated from the relative intensities of the $Cu(1)_3$ and $Cu(1)_4$ components in the spectrum, and the appearance of the lines B and C of $Cu(1)_2$ besides line A (Fig.3). As found already by Serikov et al. [3] the superconductivity disappears in the samples where the line A is missing in the NQR spectrum. Assuming an ideal arrangement of the oxygen in full and empty chains, one can see that for a structure with every third chain occupied ($x = 6.33$) only the line B would be seen, for every fifth chain occupied ($x = 6.2$) the lines B and C with equal intensities. In our not superconducting samples A2 and C2 ($x = 6.35$), the intensity of the $Cu(1)_2$ resonances is already shifted to the line C. This means, that the empty chains belong mainly to clusters with $x < 6.2$ and a considerable amount of the oxygen is arranged in clusters of larger x. In contrast, the NQR spectrum of the weakly superconducting sample B1 produced by long annealing ($x = 6.3$, $T_c = 20$ K) shows 25% intensity of the $Cu(1)_2$ resonances in the line A and 60% in B and, consistent with that, a large mean value of the chain length n. Thus we find that the main effect of the annealing in this region of the oxygen concentration is to resolve the clusters with large x, and to create long chains in the whole sample. Accordingly the condition for the onset of superconductivity is that the filled chain fragments with a length of about 6 are attached to filled fragments at the next nearest chain, that means the existence of extended regions of the ortho–II structure. In agreement with this result, domains of the ortho–II structure have been found also by neutron scattering in a superconducting single crystal with $T_c = 38$ K and x = 6.4 [26].

In a non superconducting crystal with x = 6.35 a different oxygen structure containing only isolated oxygen ions has been found by neutron [27] and X–ray [28] diffraction. This structure would lead to an intensity for the $Cu(1)_3$ sites, I3, of 70 %. As we found maximal values for I3 of only 45 % (Tab.1), this arrangement cannot be the dominant part in any of our samples. The intensity of the superstructure reflexes in both diffraction experiments, however, is small and corresponds to only 12 % of the total crystal[28]. Therefore the results from diffraction and NQR are not really contradictory.

10. The Antiferromagnetic State of $YBa_2Cu_3O_x$

In the antiferromagnetic state the hyperfine field of 7.95 T at the Cu(2) sites [14] is an indication of the magnetic order at these sites. It is independent of the oxygen concentration [17]; but above x = 6.2 the intensity of the AFNMR spectrum decreases due to the distortion of the antiferromagnetic coherence. The antiferromagnetic spin structure of pure YBCO is determined by three AF exchange

Fig. 9: The two different AF structures of Cu in 123 compounds with different stacking sequences of the magnetic Cu(2) layers.
A: structure of pure samples without a local magnetic field at Cu(1) sites.
B: structure in Fe doped samples allowing a dipolar and transferred hyperfine field at Cu(1).

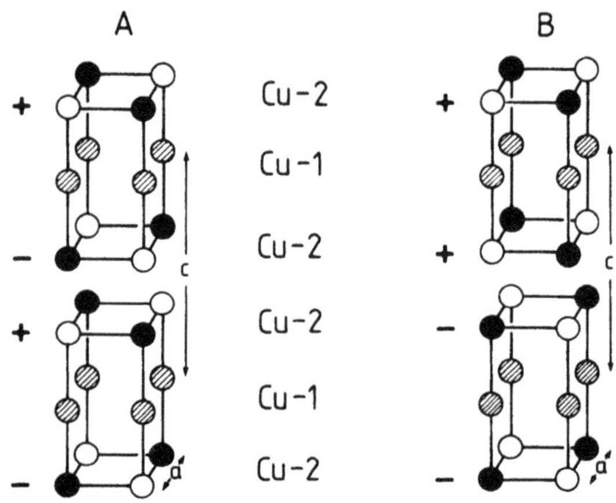

coupling constants [29]: the coupling of the nearest neighbor Cu(2) spin within the layers, J = 1700 K, between the two planes of a bilayer, $J_b > 10^{-2}J$, and a very weak one between the bilayers, $J' = 10^{-5}J$. By this symmetry (Fig. 9a) the local fields from the Cu(2) ions, either the dipolar or the transferred hyperfine fields, disappear at the Cu(1) sites as is seen from the narrow NQR lines of Cu(1) with two or three oxygen neighbors, $Cu(1)_2$ and $Cu(1)_3$, respectively (Fig. 2). From the linewidth follows an upper limit for a local field of less than 0.05 T at x = 6.3 which decreases to 0.02 T at x = 6.0.

The magnetic structure becomes different at low temperature, if the material is doped by about 1% of Fe. In this case a splitting of the $Cu(1)_2$ line [30] appears at 4.2 K caused by a field of 0.16 T perpendicular to the c–axis (Fig. 10). We have found the same splitting also for other trivalent ions, Al and Ga [7] which mainly enter the Cu(1) sites, but not for the divalent ions Zn and Ni, which probably occupy only the Cu(2) sites. Kumagai et al. [31] reported a different influence of impurities: split lines for magnetic Fe, Co, and Ni ions and unsplit lines for Al and Zn.

Fig. 10: NQR spectrum of the Cu(1) site in $YBa_2(Cu_{1-x}Fe_x)_3O_6$ with x = 0.016 Fe at 4.2 K showing the Zeeman splitting by a field of 0.16 T.

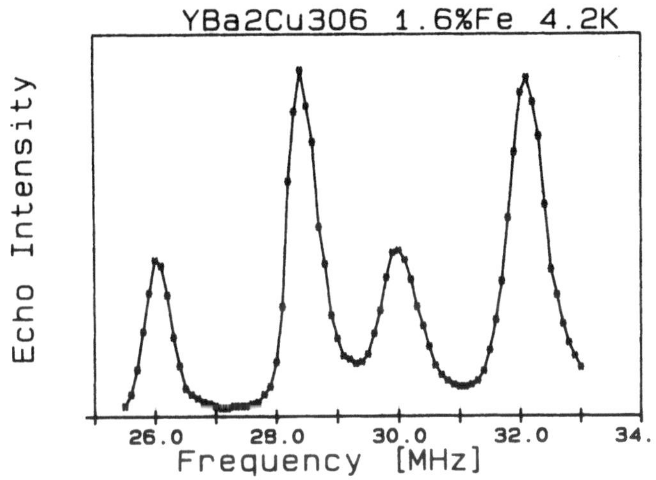

The reason of the local field at the Cu(1) ions is a changed stacking sequence of the AF Cu(2) layers which has been found also by neutron diffraction in single crystals of YBCO [32] and NdBCO [33]. In the new structure the magnetic moments of the two Cu(2) ions neighboring a Cu(1) ion are directed parallel (Fig. 9b) that means the stacking sequence is changed from +o–+o– to +o+–o–, where "+" and "–" represent the directions of the Cu(2) spins and "o" the zero moment of Cu(1). In this way the AF unit cell is doubled in c–direction as seen by neutron diffraction, and the field contributions from the Cu(2) moments do no longer cancel each other at the Cu(1) sites. The ordered Cu(2) magnetic moment of 0.7 μ_B leads to the dipolar field of only 0.02 T (the earlier estimate [7,30] was too large) and the main contribution to the local field must be a transferred hyperfine field of about 0.1 T mediated by each apex oxygen ion. The transferred hyperfine field by the interaction of two Cu(2) neighbors in the layer has been estimated to about 2 T [19]. This means, that the hyperfine interaction Cu(1)–O–Cu(2) is only one order of magnitude smaller than the in plane Cu(2)–O–Cu(2) interaction. The ratio of interlayer to intralayer hyperfine interaction is asthonishingly large compared with the small ratio of the corresponding components of the exchange interaction [29].

The comparison of the AFNMR spectra of the two structures shows the same hyperfine field [34] which means that the change of the stacking sequence has no influence on the ordered magnetic moments of the Cu(2) sites. The transition between both structures has been studied for doping with Fe (Fig. 11). At increasing temperature the local field decreases continously without a coexistence of split and unsplit components in the spectrum, and disappears far below the Neel temperature of about 400 K [30]. Thus the transition at increasing temperature can be described in the easiest way as a turning of the staggered magnetization of the two planes neighboring the Cu(1) layer from parallel to antiparallel orientation with an intermediate canted spin structure.

The reason for the change of the spin structure is still unclear. According to our measurements a magnetic effect can be excluded since also the nonmagnetic ions Al and Ga show the same influence. One explanation may be that the very weak antiferromagnetic interaction in the chain Cu(2)–O–Cu(1)–O–Cu(2) is modified by a trivalent ion on the Cu(1) site into a stronger ferromagnetic one. Since the interaction within the Cu(2) planes is much stronger, a small number of ferromagnetic links can turn the staggered magnetization in the whole structure. In the case of the very low doping by 0.2 % Fe (Fig. 11) the ferromagnetic interaction is not strong enough and the canted state with a lower local field persists down to 1.2 K.

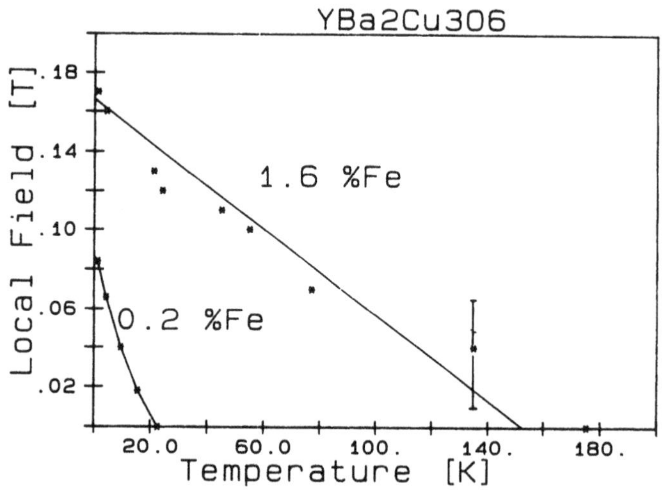

Fig. 11: Temperature dependence of the Zeeman splitting in $YBa_2(Cu_{1-x}Fe_x)_3O_6$ for x = 0.016 and 0.002.

The unsplit lines in pure AF YBCO and GdBCO with x > 6.0 (Fig. 2) show that the oxygen in the chain sites and the neighboring Cu(1) ions do not change the magnetic structure. The situation of NbBCO, however, is still unclear. In this case we have found only one sample with x = 6.0 showing the "normal" unsplit Cu(1) spectrum, whereas four other samples revealed the split spectrum like the Fe doped samples and thus confirm the result from neutron diffraction of the different groundstate of NdBCO [33]. As it seems to be possible that, depending on the conditions of preparation, Nd partly replaces the Ba ions, this may be another way to introduce ferromagnetic couplings into the YBCO structure. In any case, however, the small local field at the Cu(1) sites proves that the expectation value of the magnetic moments at Cu(1) is extremely small in contrast to the value of about 0.4 μ_B deduced from the neutron spectra [32]. Assuming an hyperfine coupling constant of 10 T/μ_B and the full field found at Cu(1) as an on–site hyperfine field, the upper limit of a static magnetic moment at Cu(1) would be only 0.02 μ_B.

11. Summary

We have discussed the change of the magnetic state of the Cu(2) ions and of the arrangement of the oxygen ions in the Cu(1) plane in YBCOx when the oxygen concentration x increases from 6 to 7. The maim information comes from the fact, that Cu(1) ions with different oxygen concentration can be distinguished by the EFG values generated by the surrounding ions. The second source of information is the NQR signal of the Cu(2) sites which is influenced by the oxygen ions in the Cu(1) plane, too. In our measurements at 4.2 K and 1.2 K the NMR and NQR resonance of Cu(2) can be observed only in samples with x > 6.5 i.e.in samples where T_c is in the region of the 60 K plateau or higher.

For the $3d^9$ state of the Cu^{2+} ions at the Cu(2) sites a magnetic moment is expected. This moment is evident in the antiferromagnetic state at low x [14]. A fluctuating magnetic moment with strong antiferromagnetic correlations is assumed generally to explain the temperature dependence of the spin–lattice relaxation rate of the Cu(2) sites of samples in the 60K plateau and at larger x [20]. When the Cu(2) signal disappears in samples with lower Tc this must be due to the slowing down of the magnetic fluctuations, but without freezing in of the Cu(2) spins in a spin–glas or antiferromagnetic state. This is proved by the narrow linewidth of the NQR lines of Cu(1) in all samples which reveal the superconductivity. Frozen spins are detected by the broadening of the Cu(1) line only in samples which do not become superconducting. The development of the antiferromagnetically ordered state with defects is evident from the broadened AFNMR spectrum of the Cu(2) sublattice at about x = 6.3 [17]. The NQR and AFNMR spectra become narrow near x = 6.0 without a change of the hyperfine field at the Cu(2) sites. This means that introducing a small amount of oxygen into the Cu(1) plane distorts only the magnetic order in the Cu(2) plane without changing the localized magnetic moment.

The NQR lines of all configurations of the Cu(1) sites reveal no magnetic splitting and prove that the time average of the magnetic moment of the Cu(1) sites is zero at any oxygen concentration. A small local magnetic field is induced at the Cu(1) sites, however, in antiferromagnetic samples doped with some trivalent ions like Fe, Al, or Gd which modify the stacking sequence of the antiferromagnetically ordered Cu(2) planes at low temperature [7].

The dependence of the relative intensities of the different configurations of the Cu(1) ions on the oxygen concentration x proves a non–random distribution of the oxygen ions on the chain sites and the existence of the ortho–II structure near x = 6.5 with alternatively filled and empty chains. In the region of the 60 K plateau the average length of these chains is about 10 to 13 oxygen ions per chain. At higher x the coexistance of the ortho–II with the ortho–I structure containing only filled chains is the best description of our results. At low x superconductivity disappears when the chain length is below 6 and no extended domains of the ortho–II structure exist.

12. Acknowledgement

We are indebted to many colleagues who supplied the samples used for this investigation, especially to A. Janossy and L. Bottyan, Budapest, G. Krabbes, Dresden, and M. Buchgeister, Bonn. Financial support by the BMFT through the DAAD is acknowledged. I.H. likes to thank Prof. Zinn for the hospitality he received at the KFA.

13. References

1. R.J. Cava, A.W. Hewat, E.A. Hewat, B. Batlogg, M. Marezio, K.M. Rabe, J.J. Krajewski, W.F. Peck Jr. and L.W. Rupp Jr., Physica **C 165** (1990) 419.
2. H. Claus, S. Yang, A.P. Paulikas, J.W. Downey and B.W. Veal, Physica C **171** (1990) 205.
3. V.V. Serikov, A.M. Bogdanovich, S.V. Verkhovskii, Yu.I. Zhdanov, B.A. Aleksashin, K.N. Mikhalev, V.L. Kozhevnikov, and S.M. Chesnitskii, JETP Lett., **47** (1988) 534.
4. H. Yasuoka, T. Shimizu, T. Imai and S. Sasaki, Hyperfine Int. **49** (1989) 167.
5. A.J. Vega, W.E. Farneth, E.M. McCarron and R.K. Bordia, Phys. Rev. **B39** (1989) 2322.
6. H. Lütgemeier, Physica **C 153–155** (1988) 95.
7. H. Lütgemeier and I. Heinmaa, Proc. of XXVI Zakopane School on Physics, Zakopane, 1991, eds. J. Stanek and A.T.Pedziwiatr, (World Scientific, 1991, p.264).
8. T.P. Das and E.L. Hahn, Solid State Physics Suppl.1, F. Seitz and D. Turnbull eds., (Academic Press, New York 1958).
9. A. Janossy, A. Rockenbauer, S. Pekker,G. Oszlonyi, G. Faigel and L. Korecz, Physica **C 171** (1990) 457, S. Pekker, A. Janossy, and A. Rockenbauer, Physica **C 181** (1991) 11.
10. G. Krabbes, W. Bieger, J. Thomas, P. Verges and H. Lütgemeier, Physica **C190** (1992), to be published.
11. M. Buchgeister, P. Herzog, S.M. Hosseini, K. Kopitzki and D.Wagener, Physica **C178** (1991) 105.
12. M. Mali, D. Brinkmann, L. Pauli, J. Roos and H. Zimmermann, Phys. Lett. **A124** (1987) 112.
13. H. Schiefer, M. Mali, J. Roos, H. Zimmermann, D. Brinkmann, S. Rusiecki and E. Kaldis, Physica **C 162–164** (1989) 171.
14. P. Mendels and H. Alloul, Physica **C 156** (1988) 355; H. Yasuoka, T. Shimizu, Y. Ueda, and K. Kosuge, J.Phys.Soc.Jpn **57** (1988) 2659; Y. Yamada, K. Ishida, Y. Kitaoka, K. Asayama, H. Takagi, H. Iwabuchi, and S. Uchida, J.Phys.Soc.Jpn **57** (1988) 2633.
15. M. Mali, I. Mangelschots, H. Zimmermann and D. Brinkmann, Physica C **175** (1991) 581.
16. B. Rupp, P. Fischer, E. Pörschke, R.R. Arons, and P. Meuffels, Physica C **559** (1988).
17. M. Matsumura, H. Yamagata, Y. Yamada, K. Ishida, Y. Kitaoka, K. Asayama, H. Takagi, H. Iwabuchi, and S. Uchida, J.Phys.Soc.Jpn **57** (1988) 3297; P. Mendels, H. Alloul, J.F. Marucco, J. Arabski, and G. Collin, Physica **C 171** (1990) 429.

18. W.W. Warren, Jr., R.E. Walstedt, G.F. Brennert, R.F. Bell, G.P. Espinosa and R.J. Cava, Physica C **162–164** (1989) 179.
19. T. Shimizu, H. Yasuoka, T. Tsuda, K. Koga and Y. Ueda, Bull. Magn. Res. **12** (1990) 39.
20. M. Takigawa, A.P. Reyes, P.C. Hammel, J.D. Thompson, R.H. Heffner, Z. Fisk, and K.C. Ott, Phys.Rev.B **43** (1991) 247.
21. B.W. Veal and A.P. Paulikas, Physica C **194** (1991) 321.
22. J. Zaanen, A.T. Paxton, O. Jepsen and O.K. Andersen, Phys. Rev. Letters, **60** (1988) 2685.
A. Latge, E.V. Anda and J.L. Moran–Lopez, Phys. Rev.B **42** (1990) 4288).
23. M.A. Alario–Franko, C. Chaillout, J.J. Capponi, J. Chenavas and M. Marezio, Physica C **156** (1988) 455.
24. C.F. Poulsen, N.H. Andersen, J.V. Andersen, H. Bohr and O.G. Mouritsen, Nature **349** (1991) 594.
25. G. Ceder, M. Asta and D.de Fontaine, Physica C **177** (1991) 106.
26. Th. Zeiske, R. Sonntag, D. Hohlwein, N.H. Andersen, and Th. Wolf, Nature **353** (1991) 542.
27. R. Sonntag, D. Hohlwein, T. Bröckel, and G. Collin, Phys.Rev.Lett. **66** (1991) 1497.
28. Th. Zeiske, D. Hohlwein, R. Sonntag, F. Kubanek, and G. Collin, Z.Phys. B 11 (1992) 15.
29. P. Burlet, this conference; J. Rossat–Mignod, L.P. Reynault, M.J. Jurgens, C. Vettier, P. Burlet, J.Y. Henry, and G. Lapertot, Physica **C162–164** (1989) 1269.
30. R.A. Brand, Ch. Sauer, P.M. Meuffels, Hyperfine Int. **55** (1990) 1229; H. Lütgemeier, Hyperfine Int. **61** (1990) 1051.
31. K. Kumagai, T. Takatsuka and A. Yammaka, JMMM **104–107** (1992) 577.
32. H. Kadowakai, M. Nishi, Y. Yamada, H. Takeya, H. Takei, S.M. Shapiro, and G. Shirane, Phys. Rev. **B37** (1988) 7932.
33. W.H. Li, J.W. Lynn, and Z. Fisk, Phys. Rev. **B41** (1990) 4098.
34. L. Bottyan, H. Lütgemeier, J. Dengler, S. Pekker, A. Rockenbauer, A. Janossy and D.L. Nagy, Electronic Properties of High–T_C Superconductors, eds. H. Kuzmany, M. Mehring and J. Fink (Springer 1990, p. 230).

STATIC AND DYNAMIC MAGNETIC PROPERTIES OF Ba, Cu AND O IN $YBa_2Cu_4O_8$ AND $Y_2Ba_4Cu_7O_{15.1}$

I. MANGELSCHOTS
Physik-Institut, Universität Zürich, Schönberggasse 9, CH-8001 Zurich, Switzerland and IBM Research Division, Zurich Research Laboratory, CH-8803 Rüschlikon, Switzerland

M. MALI, J. ROOS, R. STERN, M. BANKAY, A. LOMBARDI AND D. BRINKMANN
Physik-Institut, Universität Zürich, Schönberggasse 9, CH-8001 Zurich, Switzerland

We present recent results of ^{17}O, $^{63,65}Cu$ and $^{135,137}Ba$ nuclear magnetic (NMR) and nuclear quadrupolar resonance (NQR) studies in the normal-conducting state of the 82 K superconductor $YBa_2Cu_4O_8$ and the 93 K superconductor $Y_2Ba_4Cu_7O_{15.1}$.

The electric field gradient tensors of the four O sites and the Ba site in $YBa_2Cu_4O_8$ agree well with those obtained from linearized augmented plane wave calculations. The NQR frequency at the Ba site exhibits a similar T-behavior as that observed in non-cubic metals.

The T-dependence of the plane Cu and O magnetic shift tensors, together with the T-behavior of the spin-lattice relaxation rate ($1/T_1$) at the plane Cu, suggest the opening of a pseudo-spin gap well above T_c. The very different T-dependence of $1/T_1$ between the plane O and Cu sites points to the reduced role of the antiferromagnetic correlated spin fluctuations in the relaxation of the plane O. In contrast to all O and Cu sites, whose $1/T_1$ are of magnetic origin, $1/T_1$ at Ba is of predominantly quadrupolar nature between 85 K and 400 K.

The $^{63,65}Cu$ NQR spectrum in $Y_2Ba_4Cu_7O_{15.1}$ suggests that the CuO_2 planes in the 1-2-3 and 1-2-4 block have a different doping level, and $1/T_1$ measurements at both plane Cu sites provide direct evidence for interplanar coupling.

1. INTRODUCTION

The understanding of the electron spin fluctuations and their possible interplay with superconductivity in the complex high-T_c copper oxides have attracted a lot of interest since both aspects are of great importance for an adequate description of these materials. NMR and NQR have played a vital role in the study of the low-frequency spectrum of the spin fluctuations by probing the wave-vector (q) and frequency (ω) dependent dynamical spin susceptibility, $\chi(q,\omega)$. Relevant information can be gained from the temperature dependence and the anisotropy of both the magnetic shift tensor (K) and $1/T_1$ at ***different atomic*** sites. K measures the total uniform static susceptibility, and $1/T_1$ provides information about the q sum of the imaginary part of $\chi(q,\omega_0)$ with ω_0 the nuclear Larmor frequency. On the other hand, neutron scattering experiments probe the q and ω dependence of the imaginary part of $\chi(q,\omega)$ and normally occur at much higher energies than NMR ($E^{neutrsc} \approx 20$ meV, $E^{NMR} \approx 10^{-4}$ meV).

In addition to information about the spin dynamics in the high-T_c compounds, NQR can also give insight into the electronic structure of these compounds since the nuclear quadrupole frequency, ν_Q, being proportional to the nuclear quadrupole coupling constant, is directly related to the electric field gradient (EFG) at the corresponding nuclear site. Our NMR and NQR investigations have focussed on the $YBa_2Cu_4O_8$ (1-2-4) compound [1] because of its high oxygen stability, which leads to well-defined samples with respect to their oxygen content. Their structural homogeneity is reflected by their very narrow $^{63,65}Cu$ NQR resonance lines [2]. As will be discussed below, the lower T_c of the 1-2-4 compound (T_c = 82 K), compared to that of $YBa_2Cu_3O_7$ (1-2-3-7), T_c = 92 K, can be qualitatively accounted for by its lower doping level. The crystal structure of 1-2-4, determined by X–ray [3] and neutron diffraction [4] studies, closely resembles that of 1-2-3-7. The unit cell consists of two 1-2-3-7 unit cells joined chain-to-chain while the second cell is displaced by $b/2$ along the b axis. Hence, the only difference to the crystallographic structure of 1-2-3-7 is the double CuO chain in 1-2-4.

In this paper, we report on the T-dependence of K and $1/T_1$ at the plane ^{63}Cu and ^{17}O sites, together with the T-behavior of ν_Q and $1/T_1$ at the Ba site in 1-2-4. Linearized augmented plane wave calculations can successfully describe the EFG tensors of the Ba site, the double-chain Cu site and all four O sites in the 1-2-4 material, whereas the theoretical EFG tensor of the plane Cu site deviates by a factor of three from the experimental one.

Recently we began $^{63,65}Cu$ NQR measurements in the $Y_2Ba_4Cu_7O_{15}$ (2-4-7) compound [5], which has been first synthesized under high O_2 pressure by Karpinski and Kaldis [6]. However, single-phase samples can also be obtained under normal O_2 pressure [7]. The 2-4-7 compound can be considered a *natural* superlattice, consisting of alternating 1-2-4 and 1-2-3 blocks with their double and single CuO chains, respectively [8]. T_c can be varied between 20 K and 95 K by changing the oxygen content; this is assumed to happen more easily in the single chain of the 1-2-3 block [9] than in the stable double chain of the 1-2-4 block. The similar lattice constants of pure 1-2-4 and 2-4-7 suggest that the rigid 1-2-4 block imposes its structure on that of the 2-4-7 compound, which is also reflected in the fact that the a-, b- and c-axes in 2-4-7 depend only slightly on the oxygen content. This might also be responsible for the oxygen disorder in the single chains of the 1-2-3 block. Neutron experiments [10] revealed that a certain number of single chains are lying in the a-direction which is perpendicular to the double chains. The question arises how long a chain fragment is within this zig-zag single chain structure and whether there is a relation between this length and the optimal superconducting properties of this material. By probing the EFG of a certain nucleus, NQR is a suitable tool for studying this problem because it is very sensitive to structural disordering. We discuss the $^{63,65}Cu$ NQR spectrum at 100 K and the T-dependence of $1/T_1$ at the plane Cu sites in the 93 K $Y_2Ba_4Cu_7O_{15.1}$ (2-4-7-15.1) superconductor.

2. EXPERIMENTAL

The $^{63,65}Cu$ and ^{17}O NMR investigations were carried out in a fixed external magnetic field B_0 = 5 or 9 T on c-axis oriented 1-2-4 powders (T_c = 82 K), which were synthesized under high oxygen pressure as described in Ref. [1]. The degree of alignment was checked by neutron diffraction, yielding a c-axis angular distribution of roughly ± 2 degrees. The procedure used for the ^{17}O enrichment is described in detail in Ref. [11]. The Ba NQR experiments were done on a ^{135,137}Ba-enriched 1-2-4 powder ($T_c = 80$ K) prepared at ambient pressure [12]. The 2-4-7-15.1 sample with $T_c = 93$ K was synthesized under 20 bar O_2 [9]. The absolute oxygen content was determined from the weight change of the sample during reduction in a H_2 atmosphere [13].

Both the NMR and NQR measurements are performed using a conventional pulsed NMR spectrometer. For the optimization of the signal-to-noise ratio, all spin-echo signals are detected with the Four-Phase-Spin-Echosequence ($[\pi/2]_{x/-x}$-τ-$[\pi]_{x/-x}$). The ^{63,65}Cu and ^{17}O resonance lines (FWHH(^{63}Cu2) = 75 kHz, FWHH(^{17}O) $\approx$ 25 kHz; T = 100 K; FWHH: full width at half height) in 1-2-4 are obtained by Fourier transformation of the spin echoes. The ^{135,137}Ba resonance lines (FWHH(^{137}Ba) =315 kHz, FWHH(^{135}Ba) = 180 kHz; T = 100 K) and the ^{63,65}Cu resonance spectrum in 2-4-7 are taken by time integration of the spin-echo signals while sweeping the spectrometer frequency in 50 kHz steps.

3. ELECTRIC FIELD GRADIENTS IN $YBa_2Cu_4O_8$

3.1. Theory

The EFG tensor is a ***ground-state*** property of a crystal that depends sensitively on the charge distribution in the material. It consists of two contributions: (i) the lattice part refers to the charge distribution of the lattice ions surrounding the nucleus, and (ii) the valence contribution results from electrons (holes) in incompletely filled electronic shells of the atom to which the nucleus belongs. Therefore, the principal components of the EFG tensor, $V_{\alpha\alpha}$, can be written as

$$V_{\alpha\alpha} = V_{\alpha\alpha}^{\mathrm{LAT}} + V_{\alpha\alpha}^{\mathrm{VAL}} \tag{1}$$

with $V_{\alpha\alpha}^{\mathrm{LAT}}$ and $V_{\alpha\alpha}^{\mathrm{VAL}}$ denoting the lattice and the valence part, respectively. The following convention for $V_{\alpha\alpha}$ is used: $\mid V_{XX} \mid \leq \mid V_{YY} \mid \leq \mid V_{ZZ} \mid$ with X, Y and Z the principal axes of the EFG, which lie along the crystalline axes in 1-2-4. The asymmetry parameter of the EFG, η, is defined as $\eta = (V_{XX} - V_{YY})/V_{ZZ}$. In the 1-2-4 compound, Ba — being a ion^{+2} — has a closed electronic shell, and therefore $V_{\alpha\alpha}$ is fully determined by the lattice part. In the case of the Cu sites, both having an open Cu3d shell, the valence as well as the lattice parts contribute to $V_{\alpha\alpha}$. V_{ZZ}^{VAL} can then be written in terms of the number of holes, n_h, in the different Cu3d orbitals:

$$V_{ZZ}^{\mathrm{VAL}} = \mathrm{A} < r^{-3} >_{3d} \left[n_h(3d_{3z^2-r^2}) - n_h(3d_{x^2-y^2}) - n_h(3d_{xy}) + \frac{1}{2} n_h(3d_{xz}) - \frac{1}{2} n_h(3d_{yz}) \right] \tag{2}$$

where A is a constant. For a nucleus with $I = 3/2$, as is the case for Ba and Cu, the following relation between ν_Q and $V_{\alpha\alpha}$ is valid:

$$\nu_Q = \frac{eQV_{ZZ}}{2h} \sqrt{1 + \frac{1}{3}\eta^2} \tag{3}$$

Thus, ν_Q directly depends on the EFG tensor and therefore reflects the charge distribution around the nucleus.

3.2. Electric Field Gradients at the Ba and the Four O Sites

The EFG tensors for the four crystallographically inequivalent O sites in 1-2-4 were determined by exact diagonalization of the Hamiltonian using the ^{17}O NMR spectra ($B_0 \parallel c$, $B_0 \perp c$) at $T = 100$ K. This procedure has been described extensively in [11] and the resulting EFG tensors for the two plane oxygen, O2 and O3, the double-chain oxygen, O4, and the apical oxygen, O1, are displayed in Table 1, together with the EFG results obtained by linearized augmented plane wave calculations in combination with the local-density approximation [14]. Note that the identification of the principal axes of the in-plane components of the experimental EFG tensor ($B_0 \perp c$) is based on one of the main results of these theoretical calculations, namely, that V_{ZZ} of all O sites lies along the CuO bond axis; i.e. V_{ZZ} is parallel to the crystallographic a-, b- and c-axis for O2, (O3, O4) and O1, respectively. Good qualitative agreement between the experimental and calculated EFGs, with typical deviations of 12%, is found for all four O sites. In 1-2-4, the components of the EFG tensors of O2, O3 and O1 are slightly smaller than in 1-2-3-7; for the O4 site, however, a significant difference between the two compounds appears, which most likely reflects the additional Cu1 neighboring the O4 site in the double CuO chain in 1-2-4. This drastically changes the charge distribution around the O4 position and consequently η.

Table 1: Comparison between the experimental and calculated EFG components (in 10^{21} Vm^{-2}) for all oxygen sites in $YBa_2Cu_4O_8$.

Position		V_{aa}	V_{bb}	V_{cc}	η
O1	experiment	−3.976	−6.626	10.6	0.252
	theory	−5.0	−6.5	11.5	0.1
O2	experiment	9.993	−6.06	−3.932	0.213
	theory	11.2	−6.7	−4.5	0.2
O3	experiment	−6.12	9.968	−3.848	0.228
	theory	−6.6	11.1	−4.5	0.2
O4	experiment	−9.170	9.807	0.637	0.868
	theory	−10.7	10.7	0	1.0

Good agreement between the experimental and theoretical ν_Q is also found for the Ba site, as is shown in Table 2 together with the corresponding experimental EFG data in 1-2-3-7 [15] In 1-2-4 a significantly smaller V_{aa} at Ba is obtained than in 1-2-3-7 [15] reflecting that the charge "concentration" towards the direction of the apical O in 1-2-4 is much less than that in

Table 2: Comparison between the calculated ν_Q (in MHz) obtained from the theoretical EFG tensor (in 10^{-21} Vm^{-2}), and the experimental ν_Q at the Ba site in $YBa_2Cu_4O_8$, together with the experimental EFG tensor and ν_Q at Ba in $YBa_2Cu_3O_7$.

$YBa_2Cu_4O_8$						
V_{aa}^{theor}	V_{bb}^{theor}	V_{cc}^{theor}	V_{zz}^{theor}	η^{theor}	ν_Q^{theor}	ν_Q^{exp} $T = 150$ K
-6.1	-2.3	8.4	8.4	0.5	28.4	31.6
$YBa_2Cu_3O_7$						
V_{aa}^{exp}	V_{bb}^{exp}	V_{cc}^{exp}	V_{zz}^{exp}	η^{exp}		ν_Q^{exp} $T = 4.2$ K
-8.35	-0.35	8.70	8.70	0.92		33.3

the 1-2-3-7 compound. This might be correlated with the fact that in 1-2-4 the CuO_2 planes are less polarized by the apex oxygen [16]. This result suggests that Ba is a promising candidate for observing the electronic charge transfer, which leads to an increase in T_c under pressure. Figure 1 shows that the temperature dependence of ν_Q at the Ba site in both 1-2-4 and 1-2-3-7 [15] can be well described by

$$\nu_Q(T) = \nu_Q(0)(1 - \alpha T^{3/2}), \qquad (4)$$

with α= 6.5 10^{-6} K$^{-3/2}$ for 1-2-4. This T-behavior is also applicable to non-cubic metals, in which the lattice potential is screened by the conduction electrons [17].

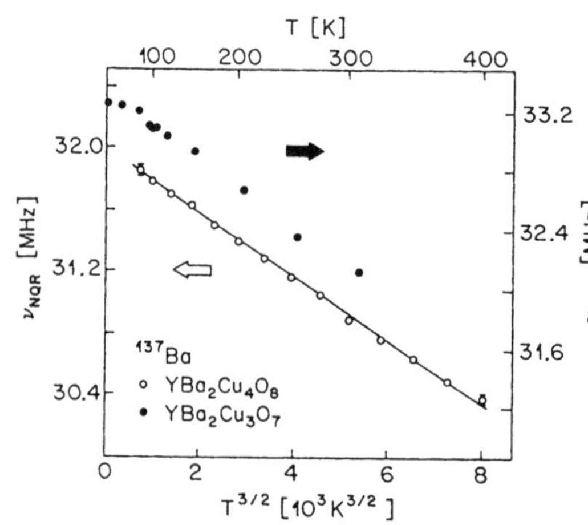

Figure 1. Temperature dependence of $\nu_Q(T)$ at ^{137}Ba in 1-2-4 and 1-2-3-7 [15]. The solid line represents the fit to $\nu_Q(T) = \nu_Q(0)(1 - \alpha T)^{3/2}$.

4. MAGNETIC SHIFT

The magnetic shift, K, is defined as the frequency displacement of the NMR line with respect to the frequency in a diamagnetic reference sample. The K tensor is described within the frame of the principal axes of the EFG tensor [11]. The following notations are used for the principal components of the K tensor at the O(2,3) and O4 sites: K_{cc} describes the tensor component perpendicular to the CuO_2 planes ($B_0 \| c$), whereas the components for $B_0 \perp c$ are given by $K_{\|}$ and $K_{\perp}$ for $B_0 \parallel$ CuO-bond axis ($B_0 \parallel Z$) and $B_0 \perp$ CuO-bond axis, respectively. The K tensors at the O(2,3) and O4 sites were determined for all temperatures from the central and satellite lines, using second-order perturbation theory [11].

The T-behavior of the K tensor at the O(2,3) and O4 sites in the normal-conducting phase, obtained in $B_0 = 5.173$ T, is shown in Fig. 2(a) and (b), respectively. Unlike the strong T-dependence of the K tensor at the O(2,3) sites (Fig. 2(a)), the K tensors at the O1 (not shown) and O4 sites (Fig. 2(b)) display almost no T-dependence between 85 and 350 K. The result at the O4 site is consistent with the T-independent K tensor at the Cu1 site [18]. The validity of the Korringa law at the Cu1 site (T_1TK^2 = const) together with the Pauli-like behavior of $K(T)$ at O4 in the normal state indicate that the conducting charge carriers in the double chains behave as Fermi-liquid quasiparticles. However, the rapid drop of K_{cc} at T_c on the O4 site suggests that singlet pairing also takes place in the double chains of the 1-2-4 compound [19].

All components of the K tensor at the O(2,3) site (Fig. 2(a)) significantly diminish with decreasing T. This strong T-behavior is also observed in O-deficient $YBa_2Cu_3O_{6+x}$ ($x \approx 0.05$, 0.08) (1-2-3-(6+x)) single crystals [20],[21] and oriented powders [22],[23], in contrast to the T-independent K tensor in 1-2-3-7 [21],[24], and is a typical feature of not-optimally doped high-T_c materials. Generally, the observed K_{ii} consists of a spin part and an orbital part that are proportional to the spin and orbital susceptibilities, respectively. The latter contribution is T-independent:

$$K_{ii}(T) = K_{ii}^{spin}(T) + K_{ii}^{orb} \tag{5}$$

$$K_{ii}^{spin}(T) = A_{ii}^{spin}\chi^{spin}(T) \tag{6}$$

$$K_{ii}^{orb} = A_{ii}^{orb}\chi_{ii}^{orb} \tag{7}$$

where A is the hyperfine coupling. Using the K_{ii}^{orb} values from 1-2-3-6.6 [24] and 1-2-3-7 [22], it is possible to calculate the $K_{ii}^{spin}(T)$ arising from the spin density in the oxygen $2s$ and $2p$ states. Furthermore, according to Fig. 2(a), $K_{\perp}$ and K_{cc} follow the same T-dependence, indicating that K^{spin} at the plane O site is uniaxial. Consequently, the separate contributions of the $2s$ and $2p$ states to K_{ii}^{spin} are obtained from K_{iso}^{spin} and K_{ax}^{spin}, respectively:

$$K_{iso}^{spin} = \frac{K_{\|}^{spin} + K_{\perp}^{spin} + K_{cc}^{spin}}{3} \tag{8}$$

$$K_{ax}^{spin} = \frac{K_{\parallel}^{spin} - K_{\perp}^{spin}}{3} \tag{9}$$

Figure 3 demonstrates that various components of the ^{17}O and ^{63}Cu K^{spin} tensor ($^{17}K_{ax}^{spin}$, $^{17}K_{iso}^{spin}$, $^{17}K_{cc}^{spin}$, $^{63}K_{ab}^{spin}$) exhibit a common T-behavior. Hence, all these components are coupled to the same T-dependent spin susceptibility. The $^{17}K_{iso}^{spin}$ probing χ_{2s}^{spin} probably originates from the polarization of the O2s states by the Cu3d spin density, reflected by $^{63}K_{ab}^{spin}$, due to the strong hybridization of the O2s states and the Cu3d states.

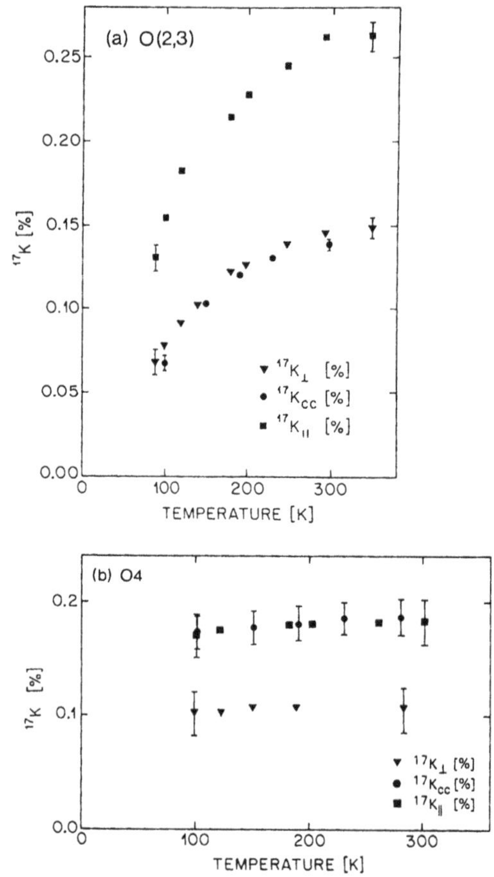

Figure 2. Temperature dependence of the principal components of the K tensor at the O(2,3) and O4 sites in 1-2-4 at B_0 = 5.173 T [11]. Note that the principal axes of the EFG tensor, whose Z-axis is along the CuO-bond axis, are taken as local frame for the K tensor. **(a)** Data for the plane O(2,3) site, from [11]; **(b)** data for the chain O4 site.

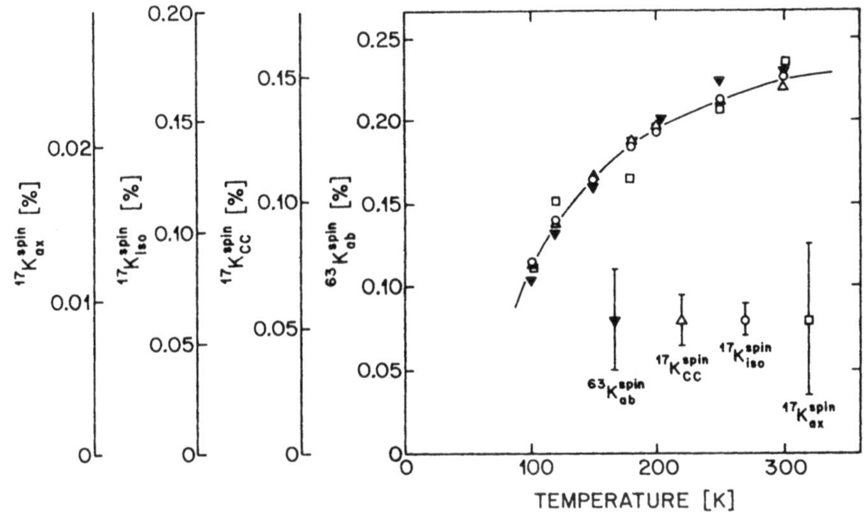

Figure 3. Various magnetic shift components of O(2,3) and Cu2 in 1-2-4 [11] are plotted against the temperature (with different vertical scales). From [11].

This explains the similar T-behavior of $^{17}K^{spin}_{iso}$ and $^{63}K^{spin}_{ab}$. On the other hand, $^{17}K^{spin}_{ax}$ measures χ^{spin}_{2p}, which arises from two dipolar interactions: (i) with the Cu_{3d} electronic spin-density due to the covalency and overlap between the O2p and the Cu3$d_{x^2-y^2}$ orbitals, and (ii) with the O_{2p} hole spin-density. Assuming that there exists a hole spin-susceptibility so that contribution (ii) of $^{17}K^{spin}_{ax}$ is not negligible compared to contribution (i), then the similar T-dependences of χ^{spin}_{2p} ($^{17}K^{spin}_{ax}$) and χ^{spin}_{3d} ($^{63}K^{spin}_{ab}$) suggest that the spin of the O2p hole forms a single entity with the Cu3d spin system. This result supports the one-spin component model and is further confirmed by the fact that the ratio χ_{2p}/χ_{3d} is independent of the hole concentration [11]. The origin of the strong T-dependence of χ^{spin}_{2p} and χ^{spin}_{3d} is still under discussion. It has been mentioned several times that a spin gap, whose existence is experimentally evidenced by NMR and neutron scattering experiments (see Section 5), might also be responsible for the observed T-behavior of χ^{spin}. However, at the moment there is no quantitative understanding for this result.

5. SPIN-LATTICE RELAXATION

Measurements of $1/T_1$ provide useful information about the nature of the low-frequency spin fluctuations. With the Moriya expression [25], $(T_1T)^{-1}$ can be directly related to the dynamical spin susceptibility:

$$\frac{1}{T_1T} = \frac{\gamma_n^2 k_B}{2\mu_B^2}\sum_q |A_q|^2 \frac{\mathrm{Im}\chi(q,\omega_0)}{\omega_0}$$
$$A_q = \sum_i A_i \exp(i\vec{q}\cdot\vec{r}_i)\,, \tag{10}$$

where γ_n is the nuclear gyromagnetic ratio, A_i is the hyperfine coupling between the nuclear and electron spin at r_i, and ω_0 is the nuclear Larmor frequency. Figure 4 shows the normal-state T-dependence of $(T_1T)^{-1}$ at the plane ^{17}O and ^{63}Cu sites, denoted $^{17}(T_1T)^{-1}$ and $^{63}(T_1T)^{-1}$, respectively, for $B_0 \parallel c$ in 1-2-4. The $^{17}(T_1T)^{-1}$ and $^{63}(T_1T)^{-1}$ measurements are made on the high-frequency quadrupole satellite to ensure that only the signals of the O(2,3) and Cu2 sites, respectively were detected. Unlike the T-independent $(T_1T)^{-1}$ in metals, both $^{17}(T_1T)^{-1}$ and $^{63}(T_1T)^{-1}$ in 1-2-4 display a strong T-dependence in the normal-conducting state: $^{17}(T_1T)^{-1}$ is a monotone function of T, whereas $^{63}(T_1T)^{-1}$ first increases until $T \approx 130$ K and then rapidly decreases as T is reduced. It is interesting that in the same T-regime $^{63}(T_1T)^{-1}$ exceeds $^{17}(T_1T)^{-1}$ by as much as a factor of 20. The relaxation mechanisms for both the O(2,3) and Cu2 nuclei are of magnetic origin.

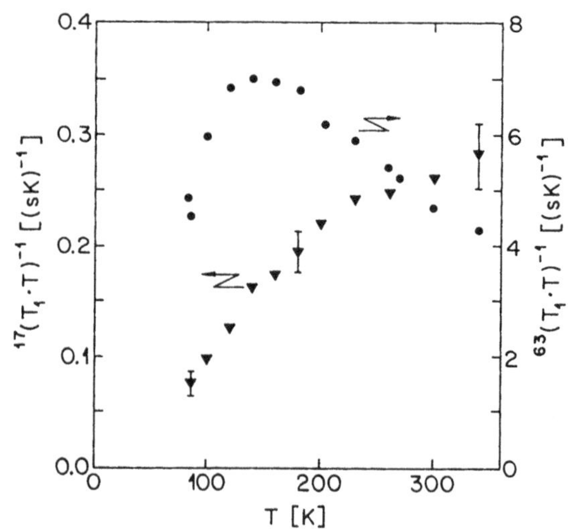

Figure 4. Temperature dependence of $(T_1T)^{-1}$ at Cu2 and O(2,3) in 1-2-4, with $B_0 \parallel c$ ($B_0 =$ 5.173 T).

The markedly different T-behavior of $(T_1T)^{-1}$ at Cu2, compared to that measured at O(2,3) above T_c, can be qualitatively understood in terms of the *one*-spin component model with a ***single*** dynamic susceptibility, $\chi(q,\omega)$, arising from *one* spin degree of freedom as proposed in Refs. [26]-[29]. The various nuclei scan different q-regions of the same $\chi(q,\omega)$ at ω_0, which accounts for the different T-behavior of $(T_1T)^{-1}$ for the Cu2 and O(2,3) sites. The Cu nucleus samples the spin fluctuations of the entire q-space, and its $(T_1T)^{-1}$ is greatly enhanced by the antiferromagnetic (AF) correlated spin fluctuations, which are peaked at $Q_{\mathrm{AF}} = (\frac{\pi}{a}, \frac{\pi}{a})$, as found from neutron scattering experiments [30],[31]. On the other hand, $(T_1T)^{-1}$ at O(2,3) is determined by the fluctuations at $q \neq Q_{\mathrm{AF}}$, owing to the fact that the AF-correlated spin fluctuations are filtered out at the geometrically symmetric O site.

The $(T_1T)^{-1}$ at Cu2 in 1-2-4, which is dominated by the AF fluctuations at Q_{AF}, exhibits a maximum well above T_c, namely at $T \approx 130$ K, as is illustrated by Fig. 4. This has also been observed in oxygen-deficient 1-2-3-(6+x) materials (x = 0.5 [21], x = 0.6 [24]), but not in 1-2-3-7 [21], where $^{63}(T_1T)^{-1}$ starts to decrease close to T_c. Neutron scattering experiments in 1-2-3-6.5 at $q \approx Q_{\mathrm{AF}}$ [31] indicate that the decrease of $^{63}(T_1T)^{-1}$ above T_c is due to the opening of a pseudo gap in the magnetic excitations. Through this gap, a transfer of low to high energy occurs in the spectral weight of the magnetic excitations with decreasing T. The energy of this pseudo-spin gap strongly depends on the doping level of the material [31]. The opening of the pseudo-spin gap ***above*** T_c, expressed by the maximum in $^{63}(T_1T)^{-1}$ ***above*** T_c, is like a fingerprint for the not-optimally doped (underdoped) high-T_c materials (such as 1-2-4, 1-2-3-(6+x) (x < 0.94), 2-4-7-15.1 (see Section 6.2), $La_{2-x}Sr_xCuO_4$ [32]). The physical origin of this pseudo-spin gap is still under debate. Within the Hubbard [33] and the $t-J$ model [34], the spin gap is caused by the nesting property of the band structure, whereas according to [35] it is a direct consequence of the narrow conduction band. In the revised phenomenological model developed by Millis *et al.* [29] (MMP), the pseudo-spin gap follows from the T-behavior of the energy of the spin fluctuations around Q_{AF}.

As discussed, there is a good agreement between the neutron scattering results and ^{63}Cu NMR data. However, recently neutron scattering experiments have shown that the AF-correlated fluctuations involve ***only*** the ***in-plane*** components [31]. Thus, the out-of-plane spin components are not AF correlated. In the framework of the one-spin component model, this result — together with the experimentally determined anisotropy of $^{63}T_1^{-1}$ — has a direct influence on the ratio of direct hyperfine coupling of the nuclear Cu spin with its on-site Cu3d spin (A_{ab}) and the transferred hyperfine coupling with the nearest-neighboring Cu3d spin (B). From this ratio and the measured magnetic shift components, the macroscopic susceptibility can be calculated, which can thus be compared with the experimental one. For spin fluctuations that are AF correlated in both the in-plane and out-of-plane directions, the calculated and experimental macroscopic susceptibilities agree quantitatively well [36], whereas less good agreement (by a factor of two) has been obtained in the case of only in-plane AF-correlated spin fluctuations.

We now turn to the $(T_1T)^{-1}$ results at the O(2,3) site in 1-2-4. In Fig. 5 the normal-state behavior of $^{17}(T_1T)^{-1}$ for $B_0 \parallel c$ is shown for different 1-2-4 samples and for 1-2-3-(6+x) [23] compounds with various oxygen contents (x). In all samples, $^{17}(T_1T)^{-1}$ shows a monotonous decrease with decreasing T in the normal state. However, the $^{17}(T_1T)^{-1}$ in our 1-2-4 sample, synthesized under high pressure with T_c = 82 K, displays a much stronger T-dependence than in the 1-2-4 material prepared with the HIP method and which has a T_c = 74 K [19].

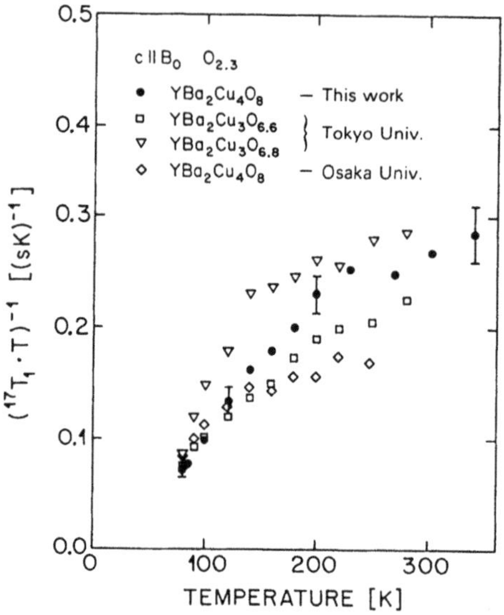

Figure 5. Comparison of the T-behavior of $(T_1T)^{-1}$ at O(2,3) with $B_0 \parallel c$ in 1-2-4 and 1-2-3-(6+x) [23].

It seems that for a fixed temperature, $^{17}(T_1T)^{-1}$ increases in going from 1-2-3-6.6 and 1-2-4 to 1-2-3-6.8.

Another important result is demonstrated in Fig. 6, which shows that in the 1-2-4 compound, $^{17}(T_1T)^{-1}$ follows almost the same T-behavior as $K^{\rm spin}_{cc}$. This result, T_1TK = const., has also been found in 1-2-3-(6+x) (x = 0.5 [21], [23], x=0.63 [24] and x=0.8 [23]), and seems to be a typical feature of the not-optimally doped materials, since in 1-2-3-7 the Korringa law, T_1TK^2 = const., is obeyed.

As mentioned above, $^{17}(T_1T)^{-1}$ is expected to be primarily determined by the uncorrelated spin fluctuations at $q \simeq 0$. Assuming a Lorentzian frequency spectrum with width Γ, Eq. (10) can be written for O(2,3) as

$$\frac{1}{T_1T} \propto \sum_{q\simeq} \frac{{\rm Im}\chi(q,\omega_0)}{\omega_0} \simeq \frac{\chi(q \simeq 0)(T)}{\Gamma_{q\sim 0}}. \tag{11}$$

Since $K^{\rm spin}$ is proportional to $\chi^{\rm spin}$ [Eq. (8)], Fig. 6 implies that $^{17}(T_1T)^{-1}$ and $\chi^{\rm spin}$ have a common T-dependence, which based on Eq. (11) indicates that $\Gamma_{q\simeq 0}$ in 1-2-4 is T-independent. The magnitude of $^{17}(T_1TK^{\rm spin}_{cc})^{-1}$ in 1-2-4 is similar to that obtained in 1-2-3-6.63 [24], which is in contrast to [19] where $\Gamma_{q\simeq 0}$ in 1-2-4 is found to be twice as large as the value in 1-2-3-6.5.

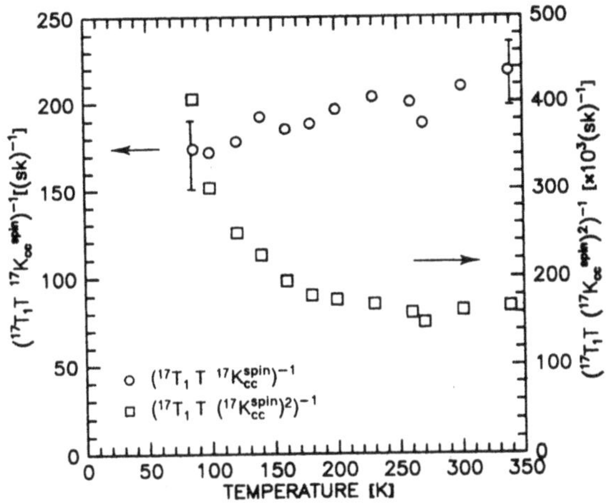

Figure 6. Circles and squares: $^{17}(T_1T)^{-1}$ divided by $^{17}K_{cc}^{\mathrm{spin}}$ and $(^{17}K_{cc}^{\mathrm{spin}})^2$, respectively, plotted against temperature.

Figure 7 plots the T-dependence of the $^{17}T_1/^{63}T_1$ ratio for $B_0 \parallel c$ in 1-2-4, together with related data in 1-2-3-6.5 [21]. Similar results are also obtained in 1-2-3-6.63 [24]. The $^{17}T_1/^{63}T_1$ ratios in the 1-2-4 and 1-2-3-6.5 [21] compounds display a much stronger T-dependence than that in the 1-2-3-7 material [21], in which $^{17}T_1/^{63}T_1$ is in overall smaller compared to 1-2-4 and 1-2-3-6.5.

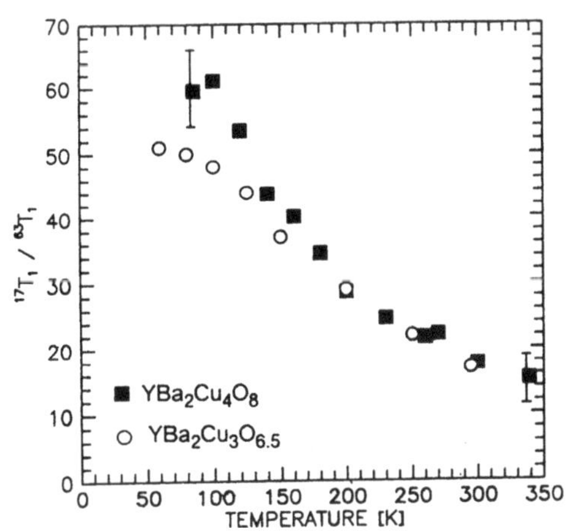

Figure 7. Temperature dependence of the ratio $^{17}T_1/^{63}T_1$ for ■: 1-2-4 ($B_0 \parallel c$) and ○: 1-2-3-6.5 ($B_0 \parallel c$ for Cu2 and $\perp$ CuO-bond axis for O(2,3) [21]).

In the MMP model, ${}^{17}T_1/{}^{63}T_1$ is a measure for the AF fluctuations. This model predicts that for large AF-correlation lengths ξ, ${}^{17}T_1/{}^{63}T_1$ is proportional to ξ^2. Thus the analysis of the results in Fig. 7 ***in the framework of the MMP model*** suggests that the 1-2-4 and 1-2-3-6.5 materials have a similar ξ, which is considerably larger than that in the 1-2-3-7 compound [21] and which strongly depends on T in the normal-conducting state. This is in contrast to the neutron-scattering experiments [31], which up to 250 K yield a T-independent ξ in all high-T_c systems. According to MMP, this discrepancy is due to the fact that the present neutron measurements have to be performed at such high energies that it is no longer possible to see a T-dependent AF-correlation length [29].

Recently, we have also measured the T-behavior of T_1^{-1} at the Ba site in the 1-2-4 compound. The relaxation mechanism of the Ba nucleus is predominantly of quadrupolar character ($\approx$ 80–90 %) in a wide T-range (85 K–400 K), in contrast to 1-2-3-7 [15], where it is predominantly of magnetic nature. In preliminary measurements of T_1^{-1} at ^{137}Ba in 1-2-4, a maximum in $T_1^{-1}(T)$ was found at $T \approx 270$ K. Unfortunately, this maximum could not be reproduced. Instead, it is found that ${}^{137}T_1^{-1}$ follows a T^2 law from 85 K up to 400 K, suggesting that phonons might lie at the origin of the Ba relaxation in 1-2-4.

6. ^{63,65}Cu NQR RESULTS IN $Y_2Ba_4Cu_7O_{15.1}$

6.1. The ^{63,65}Cu NQR spectrum

Figure 8 shows the ^{63,65}Cu resonance spectrum in 2-4-7-15.1 at T = 100 K. Resonance lines from four crystallographically inequivalent Cu sites are observed: ***two plane*** Cu resonance lines arising from the plane Cu in the 1-2-3 and 1-2-4 blocks, and ***two chain*** Cu resonance signals coming from the Cu in the single and double chain in the 1-2-3 and 1-2-4 blocks, respectively. As is clearly seen in Table 3, the ν_Q of the plane Cu in the 1-2-3 and 1-2-4 blocks differ only slightly from those of the "pure" 1-2-3 and 1-2-4 [2] structures. Based on Eqs. (2) and (3), we conclude that the CuO_2 planes of the 1-2-3 and 1-2-4 blocks experience a different doping level: in the former the CuO_2 plane is more heavily doped than in the latter, which is similar in the pure materials [11]. This result further indicates that the charge transfer between the 1-2-3 and 1-2-4 layers in 2-4-7-15.1 is negligible. Measurements of the T-dependence of the magnetic shift at both plane Cu sites and the Y site [5] can provide further information about this topic. Both plane Cu signals are broadened with respect to those of pure materials.

The resonance lines of the single and double-chain Cu sites have also nearly the same ν_Q as in the pure materials. The double-chain Cu signal is not much broadened with respect to that in the bulk 1-2-4 material, proving that the double chain in the 2-4-7-15.1 compound is well formed. On the other hand, the single-chain line is much broader than in the bulk 1-2-3 material. Many samples that are either oxygen-deficient or have a non-optimized T_c^{onset} and a broad superconducting transition display single-chain signals, which can be spread out over several MHz [37]. This large broadening most probably arises from the presence of single-chain Cu sites with different EFGs due to their different oxygen coordinations, which results in a distribution of ν_Q [38]. This might be a consequence of the zig-zag single-chain structure along the *a*- and *b*-axes [10], which prevents the formation of long single-chain fragments in one direction and might negatively affect the superconducting properties.

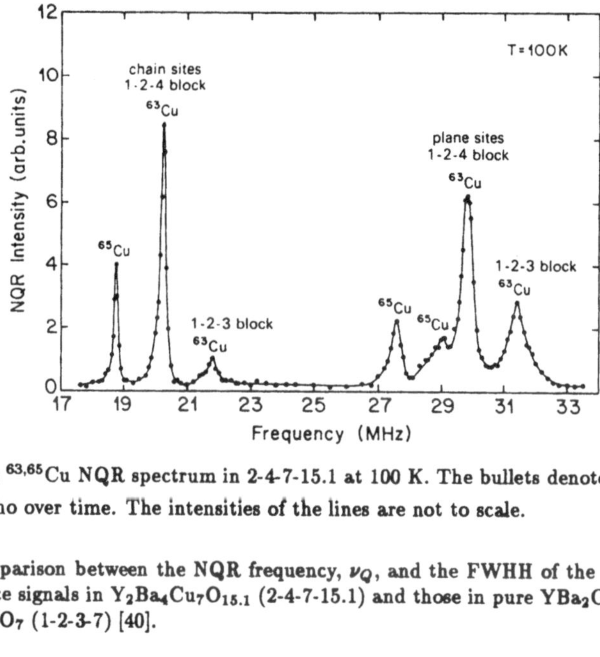

Figure 8. The ^{63,65}Cu NQR spectrum in 2-4-7-15.1 at 100 K. The bullets denote the integration of the spin-echo over time. The intensities of the lines are not to scale.

Table 3: Comparison between the NQR frequency, ν_Q, and the FWHH of the plane and chain ^{63}Cu resonance signals in $Y_2Ba_4Cu_7O_{15.1}$ (2-4-7-15.1) and those in pure $YBa_2Cu_4O_8$ (1-2-4) [2] and $YBa_2Cu_3O_7$ (1-2-3-7) [40].

		ν_q (MHz)		FWHH (kHz)	
T = 100 K		PLANE Cu2	CHAIN Cu1	PLANE Cu2	CHAIN Cu1
2-4-7-15.1	1-2-3 block	31.40 ± 0.08	21.80 ± 0.05	750 ± 80	450 ± 60
	1-2-4 block	29.80 ± 0.05	20.35 ± 0.04	430 ± 60	180 ± 30
1-2-3-7		31.19 ± 0.01	22.16 ± 0.01	200	90
1-2-4		29.700 ± 0.06	20.00 ± 0.03	180 ± 10	80± 10

6.2. The Spin-Lattice Relaxation at the Plane Cu in $Y_2Ba_4Cu_7O_{15.1}$

Figure 9 depicts the T-dependence of $(T_1T)^{-1}$ at the plane Cu in the 1-2-3 and 1-2-4 blocks in 2-4-7-15.1, while the inset shows the T-behavior of $(T_1T)^{-1}$ in the case of 1-2-3-7 and 1-2-4 [18]. According to Fig. 9, $(T_1T)^{-1}$ and thus the imaginary part of the spin susceptibility [Eq. (10)] of the plane Cu in both the 1-2-3 and 1-2-4 blocks start to decrease above T_c (around T = 130 K), as is the case for the bulk-oxygen-deficient 1-2-3-(6+x) (x < 0.94) material [21],[20] and the 1-2-4 compound [18]. This indicates that the 2-4-7-15.1 material, with $T_c = 93$ K, is in the "underdoped" regime, suggesting that its T_c^{optimal} has not yet been reached, so that T_c can be further increased either by O doping or by applying pressure. It is interesting to note that $(T_1T)^{-1}$ of the plane Cu in the 1-2-3 block is lower than that in the 1-2-4 block. This might be a consequence of the higher doping level of the CuO_2 plane in the 1-2-3 block than that in the 1-2-4 block, since T_1^{-1} at the plane Cu is primarily caused by the AF-correlated spin fluctuations at ω_0, which are found to be progressively suppressed with increasing carrier concentration [39].

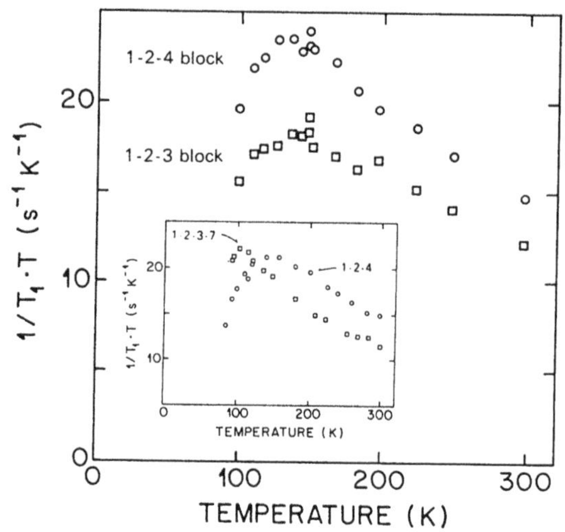

Figure 9. Temperature dependence of $(T_1T)^{-1}$ of the plane Cu site in the 1-2-4 and 1-2-3 blocks in 2-4-7-15.1 measured by NQR. The inset shows related data in the pure 1-2-4 [18] and 1-2-3-7 compounds.

When plotting $1/T_1$ of the plane Cu in the 1-2-4 block versus $1/T_1$ of the plane Cu in the 1-2-3 block (Fig. 10), it is clearly seen that the plane Cu relaxation rates of the 1-2-4 and 1-2-3 blocks are proportional to each other in the T-range from 100 to 300 K, implying a common spin dynamics in the different blocks and hence a coupling of these layers.

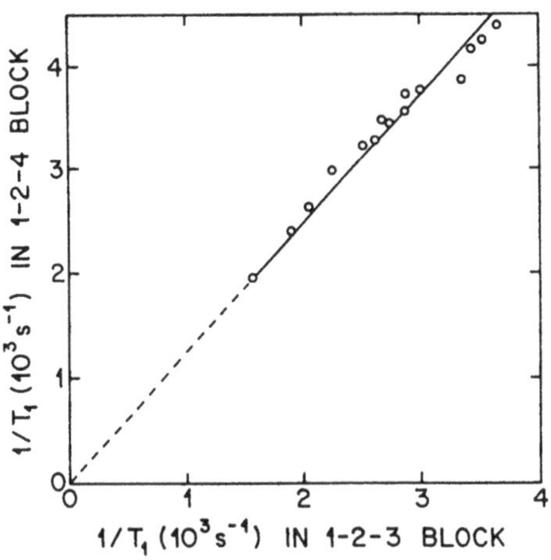

Figure 10. $1/T_1$ at the plane Cu in the 1-2-4 block vs. $1/T_1$ of the plane Cu in 1-2-3 block with the temperature as implicit parameter.

7. SUMMARY

The experimental EFG tensors of the Ba, the double chain Cu and the four O sites in $YBa_2Cu_4O_8$ are in a good quantitative agreement with those obtained by the linearized augmented plane wave calculations. However, a large discrepancy between experiment and theory is observed for the EFG tensor of the plane Cu site, which is most probably due to the inadequate treatment of the exchange-correlation effects at this site.

Strong similarities in the normal-state T-dependence of both K and $(T_1T)^{-1}$ at the planar Cu and O sites are observed between the $YBa_2Cu_4O_8$ and the substoichiometric $YBa_2Cu_3O_{6+x}$. As they do in $YBa_2Cu_3O_{6.6}$, all K components at the planar Cu and O sites in $YBa_2Cu_4O_8$ strongly decrease with decreasing T. The similar T-dependence of K at the plane Cu and O sites supports the picture that there is one spin component in the CuO_2 planes, provided a hole spin susceptibility exists. The strong T-behavior of K in the normal state possibly reflects the opening of a spin gap well above T_c, which is also observed in the T-dependence of $(T_1T)^{-1}$ at the plane Cu site and which is directly detected in neutron scattering experiments.

The very different T-dependence of $(T_1T)^{-1}$ at the plane Cu from that at the plane O site can be qualitatively explained within the one-spin component model, assuming that the AFM-correlated spin fluctuations, dominating the plane Cu $(T_1T)^{-1}$, are canceled at the planar O site. Furthermore, (T_1TK) at the plane O site in $YBa_2Cu_4O_8$ is found to be constant in the normal state with a value almost equal to that in $YBa_2Cu_3O_{6.6}$, indicating that the energy of the uncorrelated quasiparticles is similar in both compounds.

From the Cu NQR spectrum in $Y_2Ba_4Cu_7O_{15.1}$, it is inferred that the CuO_2 plane in the 1-2-3 block has a higher doping level than in the 1-2-4 block. Interplanar coupling is directly evidenced from the one-to-one correlation in the T-dependence of $(T_1T)^{-1}$ at the plane Cu in the 1-2-3 and 1-2-4 block.

Acknowledgements

We would like to thank K. A. Müller and G. Benedek for organizing this workshop and for the opportunity to contribute to it.

We greatly acknowledge the group of E. Kaldis (ETH Zurich) and the group of J. Muller (University of Geneva) for providing the 1-2-4 and 2-4-7 samples, respectively. We further thank A. Yakubowskii for the ^{135,137}Ba-enriched 1-2-4 samples.

This work has been financially supported by the "Schweizerische Nationalfonds."

References

[1] J. Karpinski, E. Kaldis, E. Jilek, S. Rusiecki and B. Bucher, *Nature* **336** (1988), 660.

[2] H. Zimmermann, M. Mali, D. Brinkmann, J. Karpinski, E. Kaldis and S. Rusiecki, *Physica C* **159** (1989), 681.

[3] P. Marsh, R. M. Fleming, M. L. Mandich, A. M. Desantolo, J. Kwok, M. Hong and L. J. Martinez-Miranda, *Nature* **334** (1988), 141.

[4] P. Fisher, J. Karpinski, E. Kaldis, E. Jilek and S. Rusiecki, *Solid State Commun* **69** (1989), 531; E. Kaldis, P. Fisher, A. W. Hewat, E. A. Hewat, J. Karpinski and S. Rusiecki, *Physica C* **159** (1989), 668.

[5] R. Stern, M. Mali, J. Roos, D. Brinkmann, J. Muller and Th. Graf, to be published.

[6] J. Karpinski and E. Kaldis, *Nature* **331** (1988), 242.

[7] K. Koyama, A. Junod, T. Graf, G. Triscone and J. Muller, *Physica C* **185-189** (1991) 461.

[8] P. Bordet, C. Chaillout, J. Chenavas, J. Hodeau, M. Marezio, J. Karpinski and E. Kaldis, *Nature* **334** (1988), 596.

[9] J.Y. Genoud, T. Graf, G. Triscone, A. Junod and J. Muller, *Physica C* **192** (1992), 137.

[10] A. W. Hewat, P. Fischer, E. Kaldis, E. A. Hewat, E. Jilek, J. Karpinski and S. Rusiecki, *J. Less-Common Met.* **164–165** (1990), 39.

[11] I. Mangelschots, M. Mali, J. Roos, D. Brinkmann, S. Rusiecki, J. Karpinski and E. Kaldis, *Physica C* **194** (1992), 277.

[12] A. Lombardi, M. Mali, J. Roos, D. Brinkmann and A. Yakubovskii, to be published.

[13] T. Graf, G. Triscone and J. Muller, *J. Less-Common Met.* **159** (1990), 349.

[14] C. Ambrosch-Draxl, P. Blaha and K. Schwarz, *Phys. Rev. B* **44** (1991), 5141.

[15] A. Yakubovskii, A. Egorov and H. Lütgemeier, *Applied Magnetic Resonance* (in press).

[16] A. R. Bishop, R. L. Martin, K. A. Müller and Z. Tesanovic, ***Z. Phys. B: Cond. Matter* 76** (1989), 17.

[17] K. Nishiyama, F. Dimmling, Th. Kornrumpf and D. Riegel, ***Phys. Rev. Lett.* 37** (1976), 357.

[18] H. Zimmermann, M. Mali, I. Mangelschots, J. Roos, L. Pauli, D. Brinkmann, J. Karpinski, S. Rusiecki and E. Kaldis, ***J. Less-Common Met.* 164-165** (1990), 138.

[19] G.q. Zheng, Y. Kitaoka, K. Asayama, Y. Kodama and Y. Yamada, ***Physica C* 193** (1992), 154.

[20] Y. Berthier: these proceedings.

[21] C. Berthier, Y. Berthier, P. Butaud, W. G. Clark, J. A. Gillet, M. Horvatic, P. Ségransan, J. and Y. Henry, ***Applied Magnetic Resonance*** (in press).

[22] M. Takigawa, P. C. Hammel, R. F. Heffner, Z. Fisk, K. C. Ott and J. D. Thompson, ***Phys. Rev. Lett.* 63** (1989), 1865.

[23] Y. Yoshinari, H. Yasuoka, Y. Ueda, K. Koga and K. Kosuge, ***J. Phys. Soc. Jpn.* 59** (1990), 3698.

[24] M. Takigawa, A. P. Reyes, P. C. Hammel, J. D. Thompson, R. F. Heffner, Z. Fisk and K. C. Ott, ***Phys. Rev. B* 43** (1991), 247.

[25] T. Moriya, ***J. Phys. Soc. Jpn.* 18** (1963), 516.

[26] B. S. Shastry, ***Phys. Rev. Lett.* 63** (1989), 1288.

[27] F. Mila and T. M. Rice, ***Physica C* 157** (1989), 561.

[28] A. J. Millis, H. Monien and D. Pines, ***Phys. Rev. B* 42** (1990), 167

[29] A. J. Millis and H. Monien, ***Phys. Rev. B* 45** (1992), 3059.

[30] J. M. Tranquada and G. Shirane, in *Dynamics of Magnetic Fluctuations in High-Temperature Superconductors*, eds. G. Reiter, P. Horsch and G. C. Psaltakis (Plenum Press, New York, 1991).

[31] J. Rossat-Mignod, L. P. Regnault, C. Vettier, P. Bourges, P. Burlet, J. Bossy, J. Y. Henry and G. Lapertot, ***Proc. "Int'l. Conf. Neutron Scattering"***, Oxford, UK (1991), in press.

[32] Y. Kitaoka, K. Ishida, S. Ohsugi, K. Fujiwara and K. Asayama, ***Physica C* 185-189** (1991), 98.

[33] N. Bulut and D. J. Scalapino, preprint.

[34] D. R. Grempel and M. Lavagna, preprint.

[35] V. V. Moshchalkov and J. Rossat-Mignod, preprint.

[36] H. Zimmermann; Ph. D. Thesis, University of Zurich, Switzerland (1991).

[37] I. Mangelschots, M. Mali, J. Roos, D. Brinkmann, J. Karpinski, S. Rusiecki and E. Kaldis, presented at the Spring Meeting of the Swiss Physical Society (1991).

[38] H. Lütgemeier, these proceedings.

[39] G. Shirane, ***Physica C* 185-189** (1991), 80.

[40] H. Schriefer, M. Mali, J. Roos, H. Zimmermann, D. Brinkmann, S. Rusiecki and E. Kaldis, ***Physica C* 162-164** (1989), 171.

POSITIONAL SPLITTING OF APEX OXYGEN AND NONLINEAR EXCITATIONS IN CUPRATES

HELMUT ESCHRIG, STEFAN-LUDWIG DRECHSLER, AND JIRI MALEK
IFW Dresden e. V., Helmholtzstrasse 20, D-O-8027 Dresden, Germany

ABSTRACT

Structures of cuprate superconductors are stacking sequences of copper oxide layers and block layers, the latter doping the former. The undoped copper oxide layers are antiferromagnetic charge transfer insulators where the charge transfer gap at the Fermi level is between oxygen *p*-states and the upper Hubbard band of copper *d*-states. The doped copper oxide layers are strange metals. There is evidence for quasi-onedimensionality in many of the block layers. A model scenario is developed with the exchange of electrons between a conduction band of the copper oxide layer and low-lying non-linear polaronic excitations in a Peierls gap of a block layer.

1. Introduction

Many of the block layers of cuprate superconductors seem to contain an element of (quasi-)onedimensionality, the chains of the YBaCuO system forming the most pronounced example. The actual structures of Bi and Tl systems show splitting of Bi–O and Tl–O distances, respectively, which are due to a slight ion-radii missfit with the lattice, and which produce quasi-onedimensionality in these structures [1]. Moreover, an ordering of Tl-vacancies in one dimension seems to influence superconductivity [2]. As found recently [3], the La-system is superconducting only in the orthorhombic phase and not in the tetragonal one. Orthorhombicity again tends towards quasi-onedimensionality of the planes.

In [4], a microscopic scenario of s-channel superconductivity was proposed, based on the existence of a Peierls phase in the block layers sandwiched between CuO_2-planes, and on bipolaronic charge transfer excitations from the planes into the Peierls phase. On the basis of results of model calculations, this scenario is discussed in the following from a more general point of view. It is capable of explaining also the strange lifetime behaviour of the conduction electrons near the Fermi level and hence of explaining why the cuprates are strange metals above T_c. In section 2 the model is developed, and in section 3 its essential properties are discussed.

2. A model chain

As a model case, the $CuO_3^{-3-2\delta}$-chain of $YBa_2Cu_3O_7$ is considered. If the Cu-3*d* shell and the O-2*p* shells were filled, the charge per formula unit would be −5. As the cations Y and Ba are fully ionized giving a charge +7 per formula unit in $YBa_2Cu_3O_7$, the charge of the two CuO_2-planes and the CuO_3-chain must in total be −7. Assuming as usual a charge $-2+\delta$ per formula unit of the CuO_2-plane, where δ is the hole doping level, $-3-2\delta$ is obtained for the charge per formula unit of the CuO_3-chain. With the Cu-3*d*-O-2*p* band complex completely filled, CuO_3 would carry a charge -5. This means that the uppermost band of the Cu-3*d*-O-2*p* band complex of the chain contains 2δ electrons per unit cell for both spin directions.

The chain consists of alternating O ions and CuO_2 radicals, the latter oriented in c-direction with apical oxygens above and below the central Cu ion. The uppermost band considered above is the antibonding band formed of $p\sigma$-states at the chain oxygen sites (sites A) and of antibonding molecular $dp\sigma$-states at the CuO_2 sites (sites B). The corresponding Hamiltonian matrix consists of two on-site energies ε_A and ε_B and one transfer integral t between sites A and B (see Fig.1).

One-dimensional metallic systems are subject to Peierls transitions, i.e. the electron-lattice interaction cannot be neglected. We consider two kinds of chain deformation coupling to the electron states of interest: symmetric (stretching) displacements $u = z_2 - z_1$ of the two apical O ions, and longitudinal displacements y of the chain sites. A subscript m denotes the site number in the chain, and $v_m = y_{m+1} - y_m$ is the stretching of the link on the right of the mth chain site. The resulting Hamiltonian is

$$H = \sum_{m,s} \varepsilon_m c^\dagger_{m,s} c_{m,s} - t_{m,m+1}(c^\dagger_{m+1,s} c_{m,s} + h.c.) + \frac{K_u}{2} \sum_{m\in A} u_m^2 + \frac{K_v}{2} \sum_m v_m^2, \quad (1)$$

where the electronic matrix elements are expanded in powers of displacement coordinates according to

$$\varepsilon_m = \begin{cases} \varepsilon_A - \alpha_A u_m - \beta_A(v_m + v_{m-1}) & m \in A \\ \varepsilon_B - \alpha_B(u_{m+1} + u_{m-1}) - \beta_B(v_m + v_{m-1}) & m \in B, \end{cases} \quad (2)$$

$$t_{m,m+1} = \begin{cases} -[t_0 - \gamma_1 u_m - \gamma_2 v_m] & m \in A \\ t_0 - \gamma_1 u_{m+1} - \gamma_2 v_m & m \in B. \end{cases} \quad (3)$$

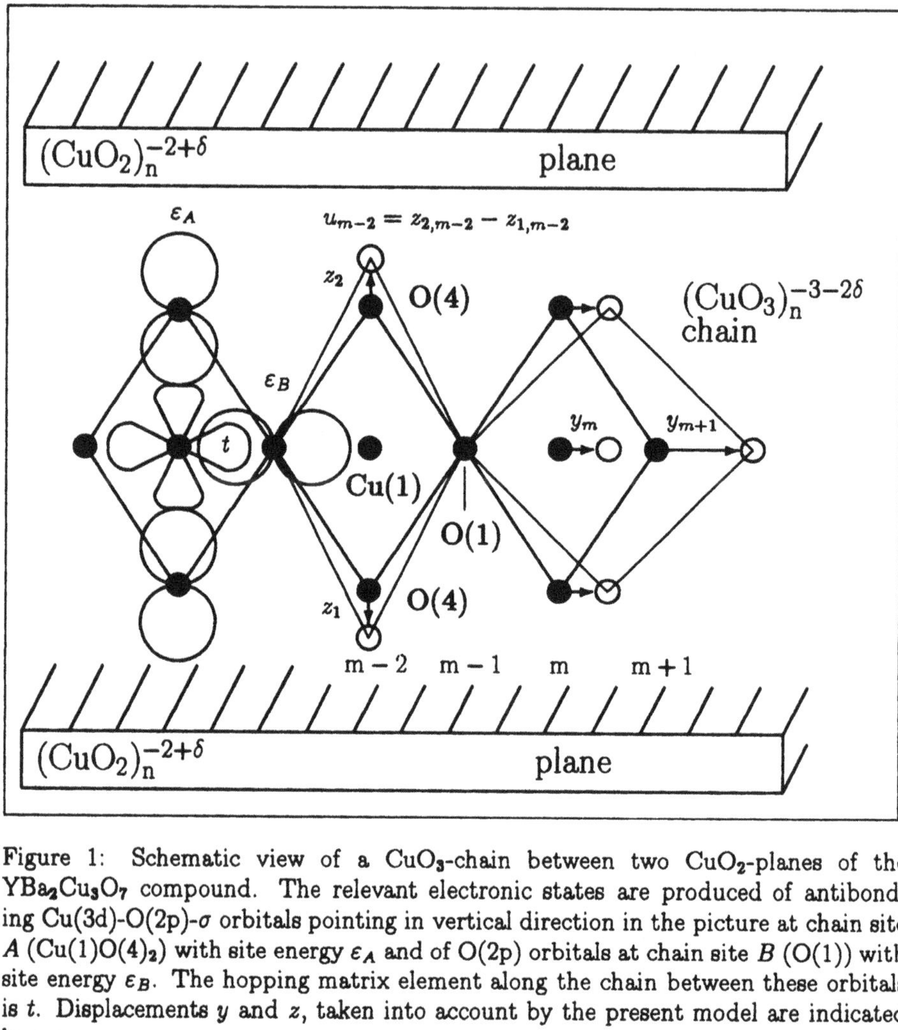

Figure 1: Schematic view of a CuO_3-chain between two CuO_2-planes of the $YBa_2Cu_3O_7$ compound. The relevant electronic states are produced of antibonding Cu(3d)-O(2p)-σ orbitals pointing in vertical direction in the picture at chain site A (Cu(1)O(4)$_2$) with site energy ε_A and of O(2p) orbitals at chain site B (O(1)) with site energy ε_B. The hopping matrix element along the chain between these orbitals is t. Displacements y and z, taken into account by the present model are indicated by arrows.

Only first order terms are kept. From self-consistent LDA electronic structure calculations we find

$$\varepsilon_A = -3.59\,\text{eV}, \quad \varepsilon_B = -5.30\,\text{eV}, \quad t_0 = 1.67\,\text{eV} \tag{4}$$

$$\alpha_A = -0.816\,\text{eV/Å}, \quad \alpha_B = -0.408\,\text{eV/Å}, \quad \gamma_1 = 1.32\,\text{eV/Å}. \tag{5}$$

We have no calculated values for $\beta_{A,B}$ and γ_2, we put

$$\beta_{A,B} = 0.4 \ldots 2\,\mathrm{eV/Å}, \quad \gamma_2 = \gamma_1. \tag{6}$$

On-site correlations not included in the LDA, especially at the Cu-site, can modify these parameters, especially α_A. Therefore we have also done calculations with varied values.

The K-values may be extracted from the experimental frequencies of the corresponding phonon modes to be

$$K_u = 7.78\,\mathrm{eV/Å^2}, \quad K_v = 4.87\,\mathrm{eV/Å^2}. \tag{7}$$

Note that the Hamiltonian (1) neglects the vibrational kinetic energy and keeps only the electronic energy and the static deformation energy.

A single-electron wavefunction for the quantum state μ on the chain may be expanded with respect to site wavefunctions $\phi_m(\mathbf{r})$ as

$$\Psi_\mu(\mathbf{r}) = \sum_m \psi_\mu(m)\phi_m(\mathbf{r}). \tag{8}$$

With the density matrix

$$P_{m,m'} = \sum_\mu^{occ.} \psi_\mu^*(m)\psi_\mu(m') \tag{9}$$

constructed from the wavefunction coefficients, the total energy may be expressed as

$$E_{tot} = \sum_m [\epsilon_m P_{m,m} - t_{m,m+1}(P_{m,m+1} + P_{m+1,m})] + \frac{1}{2}K_u \sum_k u_{2k}^2 + \frac{1}{2}K_v \sum_m v_m^2. \tag{10}$$

Varying this expression with respect to the $\psi_\mu^*(m)$, the u_{2k}, and the v_m, one obtains a closed non-linear equation set for these quantities, which, for a finite chain, can be solved iteratively. Doing so we impose additionally the constraints

$$\sum_m u_m = 0, \qquad \sum_m v_m = 0. \tag{11}$$

The first one implies some renormalization of the equilibrium bondlength between Cu and the apical O. The second one simply fixes the total chain length.

It may be reasonable to impose even stronger constraints on the v_m than simply fixing the total chain length. Since the chains are embedded in the cationic lattice frame, they cannot quite freely stretch locally. We have also done calculations with the y_m of the A-sites fixed, i.e., $v_{m-1} + v_m = 0, \quad m \in A$.

3. Commensurability and low-lying excitations

The situation for the excitation spectrum of the model chain is schematically sketched on Fig.2. If the band filling ratio δ is a ratio of small integers, say 1/2, 1/3,

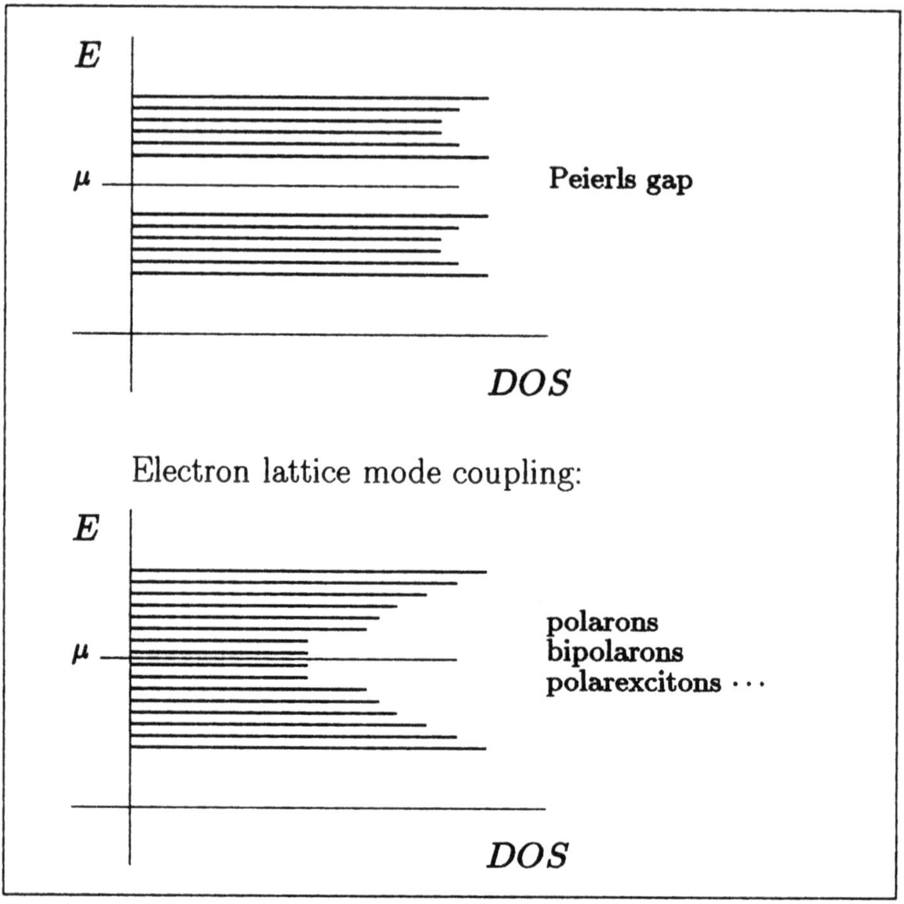

Figure 2: Nonlinear excitations closing the Peierls gap as explained in the text. The density of states vs. energy around Fermi level μ is sketched.

or $1/4$, and β is not too large, the variation of (10) yields a homogeneous Peierls state with a commensurate lattice deformation and an electronic gap at the Fermi level of typically 2 eV (upper part of Fig.2).

Experimentally, the band filling ratio δ of the antibonding chain band is not precisely known. As explained at the beginning of the preceding section, it is equal to the hole doping level of the CuO_2-planes. There are indications that the latter changes stepwise, e.g. from the stepwise change of the superconducting transition temperature with doping.

Table 1: Ground state properties and formation energies of polaronic states.
The formation energies are given with respect to a Fermi level assumed in the middle of the Peierls gap. The following abbreviations are used:
ν - band filling ratio, Δu (Δv) - amplitudes of the apical (inchain) component of the superstructure, E_{gap} - Peierls gap, $P_{\pm}$ - $\pm$ charged polaron, $P_{\pm 2}$ - $\pm$ charged bipolaron, P_0 - polarexciton, B_0 - biexciton.
^{x}) A-sites fixed, *) $\alpha_A = 0$, **) $\alpha_A = 0$, ***) $\alpha_A = 0.6$eV/Å, ****) $\alpha_A = 0.8$eV/Å

ν	$\beta_{A,B}$ [eV/Å]	Δu [Å]	Δv [Å]	E_{gap} [eV]	$E_{P_{-2}}$ [eV]	E_{P_-} [eV]	E_{P_0} [eV]	P_+ [eV]	$E_{P_{+2}}$ [eV]	E_{B_0} [eV]
1/2^{x})	1.0	0.11	0.24	1.77	1.023	0.76	0.96	0.907	1.1	-
1/2	1.0	0.11	0.24	1.77	-0.403	0.37	0.94	0.63	0.704	0.116
1/2*)	0.4	0.14	0.15	1.16	-0.136	0.312	0.68	0.453	0.554	0.325
1/3	1.5	0.11	0.49	2.36	0.096	0.64	1.11	0.219	-0.212	0...0.1
1/3**)	1.0	0.16	0.34	1.61	0.008	0.563	0.877	0.280	0.162	0.03
1/4	1.75	0.11	0.69	2.68	0.819	1.04	1.33	-	-0.665	0...0.154
1/4***)	1.2	0.12	0.40	1.27	0.309	0.51	-	0.027	-0.121	0.054
1/5****)	0.8	0.25	0.41	1.51	0.625	0.644	0.402	-0.153	-0.481	0.088

Commensurability of δ yields cusps in the total energy of the chain at those band filling ratios. Since the width of the gap is of the order of the bandwidth of the conduction band of the planes, it is natural to assume that δ is close to one of the above given values. If not, the convexity of the total energy of the chain, caused by the cusps, would lead to a phase separation with neighbouring rational δ-values, the mixed-phase total energy being the concave enveloppe of the pure-phase one. (Thermodynamic stability demands the total energy to be a concave function of particle number.)

Note, however, that the Peierls gap is a fictitious quantity: it would be the gap for electronic excitations if the lattice were fixed at the commensurate Peierls distorted state.

Adding one ore two electrons or holes to this state, or exciting an electron in the density matrix (9) leads to inhomogeneous lattice deformations and to a large variety of autolocalized excited states within the Peierls gap (lower part of Fig.2). Using the model parameters of the last section, a rich Zoo of low-lying excitations, polarons, bipolarons, both of electron and hole type, polarexcitons, etc. etc. is found within the Peierls gap. They are localized typically over three lattice sites. Table 1 presents a selection of typical results for homogeneous Peierls deformation amplitudes, gap energies, and formation energies of nonlinear excitations.

Since the Fermi level within the Peierls gap is determined by the position and filling of the conduction band of the CuO_2-planes, it is not unlikely that a nonzero density of states of the above considered non-linear excitations, both of bosonic and fermionic type, is present right on the Fermi level. Such a situation opens possibilities to explain the 'strangeness' of the metallic state of the CuO_2-planes, with bipolarons for instance forming heavy scattering centres for conduction holes, as well as of moderately exotic coupling mechanisms capable of leading to superconductivity. The occupation number of bipolaronic and other bosonic low energy excitations is proportional to temperature, and hence a contribution to the scattering rate of conduction electrons proportional to temperature is expected [5]. Hall and thermopower data and marginal Fermi liquid behaviour may be interpreted similiarly [6,7].

Recent structural data seem to give evidence of a splitting of the apical oxygen position [8] in c-direction in $YBa_2Cu_3O_7$ from EXAFS, and a modulation of this position in the ab-plane [9] from inelastic neutron scattering, respectively. An y-displacement of O(1) atoms was deduced from thermal diffuse electron scattering [10]. Disorder or modulations of the O(4) position can also explain the unusually large broadening (splitting) of the Cu(2) NQR line [11]: The O(4) position has a strong influence on the ratio of $3d_{z^2}$ vs. $3d_{x^2-y^2}$ occupation and hence on the quadrupolar field at Cu(2).

4. Acknowledgements

Encouraging discussions of one of the authors (H. E.) with K. A. Müller at a number of occassions are gratefully acknowledged.

5. References

[1] H. Eschrig in: S. Maekawa and M. Sato (eds.), Physics of high-temperature superconductors, Springer-Verlag Berlin Heidelberg 1992, p. 45.

[2] A. W. Hewat, E. A. Hewat, P. Bordet, J. Capponi, C. Chaillout, J. Chenevas, J.-L. Hodeau, and M. Marezio, Proc. Int. Sem. HTSC Dubna 89, World Scientific, Singapore 1990, p. 24.

[3] H. Takagi, R. J. Cava, M. Marezio, B. Batlogg, J. J. Krajewski, W. F. Peck, Jr., P. Bordet, and D. E. Cox, Phys. Rev. Lett. **68**, 3777 (1992).

[4] H. Eschrig and S.-L. Drechsler, Physica C**173**, 80 (1991); H. Eschrig, S.-L. Drechsler, and J. Malek, Physica C**185–189**, 3597 (1991).

[5] Y. Bar-Yam, Phys. Rev. **B43**, 359, 2601 (1991).

[6] R. I. Shekhter, G. Wendin, and I. O. Kulik, Phys. Rev. **B44**, 10215 (1991).

[7] B. R. Alascio, R. Allub, C. R. Proetto, and C. Ventura, Solid St. Commun. **77**, 949 (1991).

[8] J. Mustre de Leon, S. D. Conradson, I. Batistic, A. R. Bishop, I. D. Raistrick, M. C. Aronson, and F. H. Garzon, Phys. Rev. **B45**, 2447 (1992).

[9] M. Arai, K. Yamada, Y.Hidaka, S. Itoh, Z. A. Bowden, A. D. Taylor, and Y. Endoh, Phys. Rev. Lett. **69**, 359 (1992).

[10] Y. Koyama and Y. Hasebe, Phys. Rev. **B37**, 5831 (1988).

[11] H. Lütgemeier and I. Heinmaa, these procedings.

Cooper pair formation by distortive electron-lattice coupling

V. Hizhnyakov[+], E. Sigmund and A.Bill

Institut für Theoretische Physik, Universität Stuttgart,
Pfaffenwaldring 57, 7000 Stuttgart 80, BRD
[+] *Institute of Physics, Estonian Academy of Sciences,*
Riia 142, Tartu, Estonia

ABSTRACT

It is shown that Cooper pairs with zero total momentum can be formed as a result of static lattice distortions. The corresponding electron-lattice interaction leads to a non-retarded attraction of charge carriers. In the superconducting state a BCS like gap appears, which instead of the phonon energy is mainly characterized by the Fermi energy.

1. Introduction.

The electron-phonon interaction studied so far in the framework of conventional BCS superconductivity leads to an attraction only for charge carriers with energies near the Fermi energy [1,3]. Usually this property is connected with phonon retarding effects. As we will show below, electron-lattice interaction can induce an attraction between charge-carriers independently of their energy. This attraction arises from static distortions of the lattice caused by the particles. It is non-retarded since phonon propagation effects are not involved. We will show that two different contributions to this type of interaction can be distinguished. One is independent of nuclear masses whereas the other depends on them. These couplings have vanishing, respectively non-vanishing interaction functions at q=0. The non-vanishing part of the interaction at $q \to 0$ is not only caused by short range covalent couplings, but also by non-totally screened Madelung interactions. These last interactions are relatively strong in high-T_c superconductors (see e.g. [4-6]). In the following we pay special attention to these at q=0 non-vanishing interactions.

2. Electron-lattice coupling.

The system is described by the hamiltonian

$$H = H_e + H_L + H_{e-L} \tag{1}$$

where

$$H_e = \sum_k \varepsilon(k)\, c_k^+ c_k \tag{2}$$

is the particle (electron or hole) hamiltonian, c_k^+ (c_k) the creation (destruction) operators of the particle with wave vector k (for simplicity spin indices are omitted).

$$H_L = T + U = \sum_{\nu q} h\omega_{\nu q}(b_{\nu q}^+ b_{\nu q} + \frac{1}{2}) \tag{3}$$

is the hamiltonian of the lattice vibrations, $b_{\nu q}^+$ ($b_{\nu q}$) the creation (destruction) operator of the phonon branch ν with wave vector q and frequency $\omega_{\nu q}$. T describes the kinetic energy and

$$U = \frac{1}{2}\sum_{rr'}\sum_{nn'} \Phi\binom{n-n'}{r,r'}\, x\binom{n}{r} x\binom{n'}{r'} \tag{4}$$

the potential energy of the vibrational system. $x\binom{n}{r}$ are Cartesian coordinates of the atom r in the unit cell n (r includes the number of the atom in the cell and the Cartesian component x, y or z). H_{e-L} is the electron-lattice interaction, which is experienced by an added particle.

First we consider only short-range electron-lattice interactions and restrict ourselves to the linear approximation in the lattice coordinates. For the many atomic systems this interaction can always be divided into two parts:

(1) the first part ($H^{(1)}_{e-L}$) takes into account the relative ionic shifts of the atoms inside the unitcell (unit cell distortion) to which the particle is added,

(2) the second part ($H^{(2)}_{e-L}$) takes into account the shifts of the atoms belonging to all other unit cells.

We assume the interactions to be diagonal in the electronic operators:

$$H^{(i)}_{e-L} = W_i \sum_n c^+_n c_n Q^{(i)}_n \qquad (i=1,2) \tag{5}$$

c^+_n, c_n are operators for electrons (holes) at site n , $W_{1,2}$ is the interaction constant. Spin indices are omitted. The configurational coordinate $Q^{(1)}_n$ is a linear combination of the coordinates of the atoms satisfying the condition

$$U + W_1 Q^{(1)}_n = \text{const} + \frac{1}{2} \sum_{rr'} \sum_{n'n''} \Phi\begin{pmatrix} n'-n'' \\ r \;\; r' \end{pmatrix} \left(x\begin{pmatrix} n' \\ r \end{pmatrix} - \delta_{n,n'} x^{(0)}_r \right) \left(x\begin{pmatrix} n'' \\ r' \end{pmatrix} - \delta_{n,n''} x^{(0)}_{r'} \right) \tag{6}$$

where $x^{(0)}_r$ is the change of the equilibrium position the atom r in the unit cell n caused by the particle situated in the cell (i.e. $c^+_n c_{n'} = \delta_{nn'}$). This condition is a consequence of the appearence of the shift of the atoms in the elementary cell n if a particle is added to the cell. It leads to

$$Q^{(1)}_n = \sum_{rr'} \sum_{n'} a(r) \Phi\begin{pmatrix} n-n' \\ r,r' \end{pmatrix} x\begin{pmatrix} n' \\ r' \end{pmatrix} \tag{7}$$

where $a(r) = -W_1^{-1} x^{(0)}_r$. From (6) and (7) one sees that $Q^{(1)}_n$ can be presented as a linear combination of the derivatives of the potential energy

$$Q_n^{(1)} = (\nabla_n U) \tag{7a}$$

where

$$\nabla_n = \sum_r a(r) \,\partial/\partial x\binom{n}{r} \tag{8}$$

Consequently, the interaction $H^{(1)}_{e-L}$ can be presented in the form

$$H^{(1)}_{e-L} = W_1 \sum_n c_n^+ c_n \left[\nabla_n , H_L\right] \tag{9}$$

and it actually describes the electron-lattice coupling associated with the distortion of the unit cell by the particle added to the cell.

At this step we can generalize our considerations in order to take non-totally screened long-range Madelung interactions into account. This is done by replacing expr. (8) by

$$\nabla_n = \sum_{r,\delta} a(r,\delta) \,\partial/\partial x\binom{n+\delta}{r}$$

where n+δ numbers the unit cells, r the atoms in the cells and a(r,δ) describes the shifts of these atoms when a particle is added to the cell n. In k-representation

$$\nabla_n = -\left(\sum_{\nu q} g_1(\nu,q)\, b^+_{\nu q}\, e^{iqn} - \text{h.c.}\right) \tag{8a}$$

$$c_n^+ = N^{-1/2} \sum_k c_k^+ e^{ikn},$$

where

$$g_1(\nu,q) = N^{-1/2} \sum_{r,\delta} e^{iq\delta} (M_r \omega_{\nu q}/2h)^{1/2} a(r,\delta)\, \xi_r(\nu,q) \tag{10}$$

N is the number of elementary cells in the crystal, $\xi_r(\nu,q)$ the phonon polarization vector at the atom r. M_r is the mass of the atom r. Now the shape of the distortive interaction in k-representation can easily be found from exprs. (3), (8a) and (9):

$$H^{(1)}_{e-L} = -W_1 \sum_\nu \sum_{k,q} g_1(\nu,q)\, c^+_k c_{k+q} \Big[b^+_{\nu q}, \sum_{\nu',q'} h\omega_{\nu'q'}\, b^+_{\nu'q'} b_{\nu'q'} \Big] + \text{h.c.}$$

$$= W_1 \sum_\nu \sum_{k,q} h\omega_{\nu q}\, g_1(\nu,q)\, b^+_{\nu q} c^+_k c_{k+q} + \text{h.c.} \tag{11}$$

One can see that the considered interaction does not vanish in the $q\to 0$ limit, since $g_1(\nu,0)\neq 0$. Due to the mentioned property of $H^{(1)}_{e-L}$ the interaction with optical branches for q=0

$$H^{(1)}_{e-L,O} = \sum_\nu h\omega_{\nu O}\, g_1(\nu,0)(b^+_{\nu O} + b_{\nu O}) \sum_k c^+_k c_k$$

gives a finite contribution to the energy of the system. This term leads to the static distortion of the elementary cells of the whole lattice due to the presence of the charge-carriers.

A simple example of this type short-range electron-lattice interaction is given by the motion of the charge-carriers only along one type of atoms (r_o) in the elementary cells, if these atoms are located in non centro- symmetrical positions (this was studied in [8]). In this case, if a particle occurs on such an atom a force may appear that moves the atom out of its equilibrium position. Then in expression (6) $x^{(O)}_r = x_O \delta_{rr_o}$ and we get [8]

$$H^{(1)}_{e-L} = -x_O \sum_n c^+_n c_n \frac{\partial U}{\partial x\binom{n}{r_o}}$$

As was pointed out above long-range Madelung interactions also can lead to the electron-lattice coupling of the considered type.

The second interaction

$$H^{(2)}_{e-L} = W_2 \sum_\nu \sum_{k,q} h\omega_{\nu q}\, g_2(\nu,q)\, b^+_{\nu q} c^+_k c_{k+q} + \text{h.c.}$$

can also be brought into a commutator form similar to eq. (9), but unlike the first interaction ∇_n depends here on the nuclear masses. Furthermore g_2 vanishes for q=0.

Depending on the physical situation the first ($\sim W_1$) or the second part ($\sim W_2$) of the electron lattice interaction may dominate. In the following we consider both interactions together and thus drop the indexes (1) and (2).

3. Unitary transformations.

By means of unitary transformations the linear electron-lattice coupling can be excluded and replaced by a series of electron-electron and nonlinear electron-lattice coupling terms. Usually this is done in k-representation by a Fröhlich type transformation [9]. For the distortive interactions however, a local transformation, which totally or partly removes H_{e-L} is more suitable. After performing such a transformation, in a second step, the residual electron-lattice coupling and the interactions appearing due to the first transformation, if $\varepsilon(k)\neq 0$, are then removed by a Fröhlich type transformation. In this way the electron-electron interaction leading to Cooper pair formation reaches its simplest form.

a) First unitary transformation

The first unitary transformation is of the form

$$\tilde{H} = e^{S_1} H e^{-S_1} \tag{12}$$

where

$$S_1 = -\frac{1}{2} W \sum_n c_n c_n \nabla_n = \frac{1}{2} W \sum_{\nu q} g(\nu,q)\, c_k^+ c_{k+q} b_{\nu q}^+ - \text{h.c.} \tag{13}$$

We would like to draw the attention of the reader to the unusual (for this type of transformation) factor 1/2 entering expression (13) and which plays an important role in our derivations.

We perform the transformation up to terms of the order W^2. In this approximation

$$\widetilde{H} \approx H + [S_1, H] + \frac{1}{2}[S_1,[S_1,H]] \tag{14}$$

The commutators entering this expressions equal

$$[S_1, H_L] = -\frac{1}{2} H_{e-L}$$

$$[S_1, H_e] = -\frac{1}{2} W \sum_{\nu} \sum_{kq} g(\nu,q) \Big(\varepsilon(k+q) - \varepsilon(k)\Big) c_k^+ c_{k+q} b_{\nu q}^+ + \text{h.c.}$$

$$[S_1, H_{e-L}] = -2[S_1,[S_1, H_L]] = -\frac{1}{2} \sum_{nn'} V_{nn'} c_n^+ c_n c_{n'}^+ c_{n'}$$

$$[S_1,[S_1,H_e]] = \frac{1}{4} W^2 \sum_{kk_1k'} \sum \Big(\varepsilon(k) + \varepsilon(k') - 2\varepsilon(k_1)\Big) a_k^+ a_{k'} \nabla_{k_1-k} \nabla_{k'-k_1}$$

where

$$V_{nn'} = W^2 \sum_{rr'} \sum_{\delta,\delta'} a(r,\delta)\, a(r',\delta)\, \Phi\begin{pmatrix} n-n'+\delta-\delta' \\ r,r' \end{pmatrix}$$

$$\nabla_k = N^{-1} \sum_n e^{-ikn} \nabla_n$$

Taking into account the above relations we get

$$\widetilde{H} \approx H_e + H_L + H_{e-e} + \widetilde{H}_{e-L} + \frac{1}{2}[S_1,[S_1,H_e]] \tag{14a}$$

where

$$H_{e-e} = -\frac{3}{8} \sum_{nn'} V_{nn'} c_n^+ c_n c_{n'}^+ c_{n'} \tag{15}$$

is a negative Hubbard-type electron-electron interaction, which comes from the static lattice distortion caused by the particle.

$$\widetilde{H}_{e-L} = \sum_{\nu} \sum_{kq} \widetilde{g}_{\nu}(k,q)\, b_{\nu q}^+ c_k^+ c_{k+q} + \text{h.c.} \tag{16}$$

is the transformed electron-phonon interaction, with

$$\tilde{g}_\nu(k,q) = \frac{1}{2} W g(\nu,q)[\varepsilon(k+q) - \varepsilon(k) + h\omega_{\nu q}] \tag{17}$$

We stress the fact here that the electron and phonon energies enter expression (17) with the same weight which is a consequence of the choice of the factor 1/2 in expr. (13).

b) Second unitary transformation

In a second step the residual linear electron-lattice interaction (16) can be removed by a Fröhlich type transformation [9]

$$\tilde{\tilde{H}} = e^{S_2} \tilde{H} e^{-S_2} \tag{18}$$

where

$$S_2 = -\left(\sum_{kq} \sum_{\nu} \frac{\tilde{g}_\nu(k,q)}{(\varepsilon(k+q) - \varepsilon(k) - h\omega_{\nu q})} c_k^+ c_{k+q} b_{\nu q}^+ - \text{h.c.} \right) \tag{19}$$

The transformed hamiltonian up to terms of the order W^2 reads

$$\tilde{\tilde{H}} \approx H_e + H_L + H' + H_{e-L}^{(2)} - \frac{3}{8} \sum_{k,k',k''} \tilde{W}^2(k-k') c_k^+ c_{k'} c_{k''}^+ c_{k-k'+k''} \tag{20}$$

where

$$\tilde{W}^2(q) = W^2 \sum_\nu h\omega_{\nu q} |g(\nu,q)|^2 \tag{21}$$

$$H' = \frac{1}{2} \sum_{kk'} \sum_{q\nu} \left[\frac{\tilde{g}_\nu^*(k,q) \tilde{g}_\nu(k',q)}{(\varepsilon(k+q) - \varepsilon(k) - h\omega_{\nu q})} c_k^+ c_{k+q} c_{k'+q}^+ c_{k'} + \text{h.c.} \right] \tag{22}$$

$$H_{e-L}^{(2)} = -\frac{W^2}{2} \sum\sum_{kq\nu} |g(\nu,q)|^2 \frac{(\varepsilon(k+q) - \varepsilon(k))^3 + 3(h\omega_{\nu q})^2 (\varepsilon(k+q) - \varepsilon(k))}{(\varepsilon(k+q) - \varepsilon(k))^2 - (h\omega_{\nu q})^2} \times c_k^+ c_k \left(b_{\nu k}^+ b_{\nu k} + \frac{1}{2} \right) \tag{23}$$

Terms proportional to $(c^+_k c_{k'} b^+_{\nu q}(b_{\nu,k+q-k'}+b^+_{\nu,k'-k-q})+h.c.)$ for $k \neq k'$ are not taken into account since they contribute to the particle-particle interaction only in higher order of the expansion.

4. Cooper pair interaction.

From the hamiltonian $\tilde{\tilde{H}}$ the most interesting part for superconductivity, is obtained by setting $k'=-(k+q)$ in the terms $\sim (c^+_k c_{k+q} c^+_{k'+q} c_{k'} + h.c.)$. This term corresponds to Cooper pairs with zero total momentum. We assume the electrons (holes) to form singlet pairs, i.e. the spins are ordered antiparallel. For the considered $k'=-(k+q)$ we have

$$\tilde{g}^*_\nu(k,q)\tilde{g}_\nu(-(k+q),q) =$$
$$-\frac{1}{4} W^2 |g(\nu,q)|^2 \left(\varepsilon(k+q)-\varepsilon(k)+h\omega_{\nu q}\right)\left(\varepsilon(k+q)-\varepsilon(k)-h\omega_{\nu q}\right)$$

One sees that in expr. (22) the denominator cancels with the last multiplier. This cancellation occurs because of the same weight of the electron and phonon energies in $\tilde{g}_\nu$ (see expr. (17)), which is a consequence of the factor 1/2 in the first transformation as was already pointed out. As a result the corresponding particle-particle interaction takes the same form as the interaction, which comes from the static lattice distortion caused by the added particles (see (15) and the last term of (20)). The sum of both terms gives the following effective interaction between the particles of the Cooper pair, with zero total momentum

$$H_{SC} = -\sum_{kk'} \tilde{W}^2(k-k') c^+_k c^+_{-k} c_{-k'} c_{k'} \tag{24}$$

It arises from the distortion of the lattice caused by charge carriers forming Cooper pair. Consequently the interaction leading to Cooper pairs in the considered approximation is of distortive nature. It is analogous to that in the negative Hubbard case [10], being attractive for all k-values of the partices. Furthermore, for $W_2=0$ it is independent of nuclear masses. This is in contrast to the ordinary Fröhlich type consideration, which imply an attraction only for particles with energies satisfying the relation $|\varepsilon(k+q) - \varepsilon(k)| < h\omega_{\nu q}$, and which depends on nuclear masses.

5. Superconducting gap and T_c.

Starting from the reduced interaction H_{SC} the superconducting gap and the critical temperature T_c can be calculated. The gap equation reads [1,2]

$$\widetilde{W}_1^2(0)\int_0^B \frac{d\varepsilon\ \rho(\varepsilon)\,\mathrm{th}\big((\varepsilon-E_F)/2k_BT\big)}{\sqrt{(\varepsilon-E_F)^2+\Delta_0^2}} = 1 \tag{25}$$

Here ρ is the density of states and Δ_0 is the gap.

In the cuprate systems superconductivity takes place in the two-dimensional CuO_2 planes. Therefore, at high hole concentrations the effective dimensionality is close to two. In this case the density of states can be approximated by the constant $\rho=\rho(E_F)$. This leads to ($T=0$)

$$\Delta_0 = 2(BE_F)^{1/2}\exp\Big(-\frac{1}{w}\Big)$$

where

$$w = 2\widetilde{W}^2(0)\rho. \tag{26}$$

For lower hole concentrations the effective dimensionality d of the system can be smaller, because of the percolative regime the systems comes into [11] . Using for the density of states the expression $\rho(\varepsilon)\sim \varepsilon^{(d-2)/2}$ the solution of the gap equation, for $1 \le d \le 2$, can be approximated by the following formula ($T=0$)[12]

$$\Delta_0 \approx 2E_e \exp\Big(-\frac{1}{w}\Big) \tag{27}$$

where $E_e = E_F^{1-\alpha} B^{\alpha}$, $\alpha \approx (d-1)/2$. The value of the transition temperature is given by the BCS relation $2\Delta_0 = 3.5\ k_B T_c^{(0)}$. Consequently, already in the regime of weak couplings T_c is proportional to the electronic energy E_e, which can reach much higher values than $\hbar\omega$. The exponent in (26) and consequently T_c increase with softening of the system. In the case of a distortive interaction ($W_2=0$) only, all parameters entering the gap equation do not depend on nuclear masses, and therefore no isotope effect occurs. This effect appears if the second type electron-lattice interaction ($\sim W_2$) is present.

6. Discussion

a) The above considered Cooper pairing induced by distortive electron-lattice interaction can be generalized taking into account an arbitrary number of electron (hole) bands, an arbitrary "size" of the site (e.g. Wannier) states and transitive electron-lattice interactions . The corresponding electron-lattice Hamiltonian then reads

$$H_{e-L} = \sum_{n\delta} \sum_{ii'} W_{ii',\delta}\, c^+_{in} c_{i'n+\delta} \left[\nabla_n \binom{\delta}{ii'}, H_L \right] \tag{28}$$

where i and i' count the electronic bands

$$\nabla_n \binom{\delta}{ii'} = -N^{-1/2} \sum_{\nu q} \left(g_{ii',\delta}(\nu,q)\, b^+_{\nu q}\, e^{iqn} - \text{h.c.} \right) \tag{29}$$

is the generalized shift operator, and

$$c^+_{in} = N^{-1/2} \sum_k \varphi^*_i(k)\, e^{ikn} c^+_{ik}, \tag{30}$$

is the Wannier particle creation operator for the band i. $W_{ii',\delta}$ and $g_{ii',\delta}(\nu,q)$ are interaction parameters. Applying unitary transformations analogous to (13) and (19) one obtains the Cooper pair interaction in the form

$$H_{sc} = \sum_{kq} \sum_{ii'} \sum_{i_1 i_1'} \sum_{\nu} \hbar\omega_{\nu q} W_{k+q}\binom{ii'}{\nu q} W^*_{-k}\binom{i_1 i'_1}{\nu q} \times c^+_{ik} c^+_{i_1(-k)} c_{i_1'(-k-q)} c_{i'(k+q)} + \text{h.c.} \tag{31}$$

where

$$W_k\binom{ii'}{\nu q} = \sum_{\delta} e^{-ik\delta} W_{ii',\delta} \varphi^*_i(k)\varphi_{i'}(k+q) \tag{32}$$

This type of Cooper pair interaction has a rather general form allowing superconductivity with a number of different order parameters (see e.g. [13,14]). For example, if the terms $W_k\binom{ii'}{\nu q}$ for $i \neq i'$ are dominating then superconductivity is mainly caused by vibronic mixing of different bands, i.e. by Jahn-Teller or pseudo Jahn-Teller interactions.

In the perovskite cuprates an example of this interaction is the vibronic mixing of $d_{x^2-y^2}$ and $d_{3z^2-r^2}$ orbitals of the Cu ions in the CuO_2 planes by different vibrational modes (including the apex oxygen vibrations) polarized in the direction perpendicular to the planes. The importance of this specific vibronic interaction for high-T_c superconductivity lies in the fact that through it a superconducting coupling between charge carriers of different CuO_2 planes is induced. This means that the third dimension is switched on and T_c takes - via a logarithmic factor— a non-zero value (in the strictly two-dimensional case T_c would be zero). Especially important in this context are the non-totally screened Madelung interactions, which are long-range and can, therefore, be rather strong between CuO_2 planes. The essential mixing of $d_{x^2-y^2}$ and $d_{3z^2-r^2}$ Cu orbitals was recently found by di Castro et al [15]. The Jahn-Teller (pseudo Jahn-Teller) coupling as a reason for high-T_c superconductivity was already discussed by Bednorz and Müller [16].

b) In this article the distortive mechanism of Cooper pair formation is considered in the weak coupling limit. The related range of interaction parameters can be estimated from the condition that the renormalization of the phonon frequencies has to be small. In the following we study this condition for the actual case of low band filling i.e. when the hole (electron) concentration c_0 is small ($c_0 \ll 1$) leading to a small Fermi energy $E_F \ll B$ and to small values of the wave vectors k for occupied states: $k \le k_F \ll \pi/a_0$. a_0 is the lattice constant.

In the one band case the renormalization of the phonon frequency $\omega_{\nu q}$ for $q > k_F$, $\varepsilon_q \ge h\omega_q$ is of the order (see (23))

$$\Delta\omega_{\nu q} \sim (W_1^2/2h) \sum_k |g_1(\nu,q)\varphi(k)\varphi(k+q)|^2 \varepsilon_q^{\cdot} c_k^+ c_k \sim W_1^2 E_{\nu q} c_0 \tag{33}$$

$E_{\nu q} \le \varepsilon_q$ is the characteristic energy of the electrons (holes) giving the main contribution to the phonon renormalization (we take here the factors $\varphi(k)$ into account like in (33)). The long wave phonons with $q \to 0$ are not renormalized in the normal state (in the superconducting state such a renormalization exists [6]).

The question now arises, if in the weak renormalization case $\Delta\omega_{\nu q}/\omega_{\nu q} \ll 1$ one still can obtain high T_c values. Taking for Eq.(27) the energy value $E_e \sim 1$ eV one finds $T_c \sim 100$K for the parameter $w \sim 0.2$. According to (26) and (37) the parameter w can be estimated

$$w \sim 2N_0 W_1^2 h\omega_0 \rho_0 \sim 4\left(\Delta\omega_0/\omega_0\right)\left((h^2\omega_0^2\rho_0 N_0)/(E_0 c_0)\right) \tag{34}$$

where N_0 is the number of interacting modes and $\Delta\omega_0, \omega_0$ and E_0 are the mean values of $\Delta\omega_q, \omega_q$ and $E_{\nu q}$. For a large number of interacting modes, $N_0 \sim 10$, small doping $c_0 \sim 0.1$ and long-range Madelung interactions the parameter in the bracket may be of the order of 1 or less. Then the value $w \sim 0.2$ required to reach $T_c \sim 100$K can be obtained for the case that the phonon frequency renormalization due to charge carriers is some percent only. This shows us that the distortive mechanism for Cooper pairing in crystals with large elementary cells and low Fermi energy can already in the weak coupling limit lead to a superconducting state with $T_c \sim 100$K.

7. Conclusion.

In conclusion, in this article we proposed a new type of k-space pairing interaction induced by the electron (hole)-lattice coupling. This type of interaction arises from distortions induced by charge carriers. Unlike the usual result for phononic pairing mechanism the distortive attraction between particles in Cooper pair occur independently of the k-vectors of the particles.

The interaction was worked out for the weak coupling case. In this case one gets a BCS like gap expression, which, however, instead of the phonon energy is characterized by a much larger energy, which exeeds the Fermi energy.

Two different contributions to the proposed distortional Cooper pairing interaction in crystals with many atoms in the unitary cell are distinguished. One does not depend on nuclear masses whereas the other depends on them. Therefore, an isotope effect although reduced still appears.

The results presented here gain experimental support. In the low doping regime a clear evidence is found experimentally showing that T_c increases with the Fermi energy E_F[17].

All the new high-T_c superconducting materials have complicated chemical structures with many atoms in the elementary cell. When doping the crystal structural distortions are observed. E.g. in the La and Y cuprates the out of plane shifts of the O ions in the CuO_2 planes are removed at high hole concentrations. Other contributions to the distortive type coupling, which are probably of importance for the high-T_c cuprates, may be connected with the apex oxygen atoms [18], the position of which is strongly governed by holes, and by all other ions situated between the planes, due to the non-totally screened Madelung interactions.

In addition, only small isotope effects are observed in the perovskite superconductors (see [14] and references therein).

These observations are in agreement with the model proposed here for the pairing mechanism for charge carriers which, as we have shown, already for the weak coupling case can cause superconducting phase transitions at high temperatures.

Acknowledgment

We want to thank K.A. Müller for stimulating discussions and for pointing us to the importance of apex oxygen and out off plane distortions in the high-T_c cuprates. We also acknowledge discussions and helpful remarks by M. Schneider. Financial support of the Bundesministerium für Forschung und Technologie and the Land Baden-Württemberg is acknowledged.

References

1. J. Bardeen, J.R. Schrieffer, *Phys.Rev.* **106** (1957) 162; **108** (1957)1175
2. J.R. Schrieffer, "*Theory of Superconductivity*", W.A. Benjamin, Inc., (New York, Amsterdam, 1964)
3. D.J. Scalapino, in "*Superconductivity*", ed. R.D. Parks, vol.1, (Dekker, NY,1969)
4. R.E. Cohen, W.E. Picket, L.L. Boyer, H. Krakauer, *Phys.Rev.Lett.* **60** (1988) 817
5. S. Barisic, K. Kupcic, I. Batistic, *Int. J. Mod. Phys.* **B3** (1989) 2051
6. R. Zeyher, *Z.Physik* **B80** (1990) 187
7. H. Gränicher, K.A. Müller, *Mat.Res.Bull.* **6** (1971) 977
8. V. Hizhnyakov, E. Sigmund, *Proceedings of the Estonian Acad.Sci., Physics, Mathematics*,**40** (1991) 337
9. H. Fröhlich, *Phys.Rev.* **79** (1950)845
10. R. Micnas, J. Ranninger, S. Robaszkiewicz, *Rev. Mod. Phys.* **62** (1990)113

11. V. Hizhnyakov, E. Sigmund, *Physica* **C156** (1988)655 , Materials Letters **9** (1990)425
12. A. Bill, M. Schneider, (1991) private communication
13. H. Suhl, B.T. Matthias, L.R. Walker, *Phys.Rev.Lett.* **3** (1959)552
14. V.A. Moskalenko, M.E. Palistrant, V.M. Vakalyuk, *Sov.Phys.Uspeki* **34**,(1991)717
15. C. Di Castro et al., (1992) in the same volume
16. J.G. Bednorz, K.A. Müller, *Z.Physik* **B64**,(1986)188
17. Y.J. Uemura et al *Phys.Rev.Lett.* **62**, 2317 (1989); *Phys.Rev.* **B38**,909 (1988); *Phys.Rev.Lett.* **66**, 2665 (1991)
18. K.A. Müller, *Z. Physik* **B80**, 193 (1990)

BIPOLARONIC CHARGE DENSITY WAVES POLARONIC SPIN DENSITY WAVES and HIGH Tc SUPERCONDUCTIVITY

Serge AUBRY*
CNLS, Los Alamos National Laboratory
Los Alamos, NM 87545, USA

ABSTRACT

At large enough electron phonon coupling, the existence of bipolaronic, polaronic and mixed states is rigourously proven for the adiabatic Holstein model at any dimension and any band filling. The ground-state is one of them which then prove the existence of insulating Bipolaronic Charge Density Waves. The role of the quantum lattice fluctuations is analysed and found to be negligible in that regime but to become essential in case of phonon softening then favouring the occurrence of superconductivity. When a strong Hubbard term is also present, the bipolarons break into polarons and the ground state is expected to be a polaronic spin density wave. If the repulsive Hubbard term is comparable to the electron-phonon coupling, the energy for breaking a bipolaron into two polarons can become small and we get instead of these two degenerate structures, a pair of polarons bounded by a spin resonance which we call "spin resonant bipolaron". This resonant bipolaron is still strongly bound, but the role of the quantum lattice fluctuations becomes now very important and yields a sharp broadening of the band width of this resonant bipolaron. Thus, the strong quantum character of these resonant bipolarons could prevent their localization into real space structures which could be insulating bipolaronic CDWs or polaronic SDWs, then favouring the formation of a superconducting coherent state with a possible high Tc.

1. Early Results and New Ideas: a New Microscopic Approach for High Tc Superconductors

In the continuation of some recent works[1-3] on the bipolaronic structures of the adiabatic Holstein model, we propose a new direction for the understanding of high Tc superconductors, but also nearby insulating materials which could exhibit either charge density waves or spin density waves. However, this theory is at an embryonic stage and needs much more quantitative developments.

Basically, electrons in solids are quantum particles with spin 1/2 (Fermions) submitted to only two types of interactions which both originate from Coulomb forces and which can potentially explain all the properties of the electrons in condensed matter. First they are coupled to the positions of the ions of the underlying lattice. This type of interaction in fact couples the electrons, to the vibration modes of the lattice. This interaction is thus described as an electron-phonon coupling. The second type of interaction is due to the Coulomb repulsion of the electrons one with each other.

It is well-known that the ground-state of the system sharply depends on the strength of these interactions. It is known that because of the electron phonon coupling, the ground-

* On leave of *Laboratoire Leon Brillouin, (Laboratoire commun CEA-CNRS) CEN Saclay 91191-Gif-sur-Yvette (France)*

state of the electronic system can be a superconductor (SC) and for quasi-one dimensional system, it might be a Peierls-Frohlich Charge Density Waves (CDW) (Note however that we conjecture that the Peierls-Frohlich CDWs with a gapless phason mode, are in fact unstable against true superconductivity). We have proven (and will describe in the following) a result valid for classes of models which are not only quasi one dimensional but at any dimension, which asserts that when the electron phonon coupling is large enough, the ground state cannot remain superconducting but becomes an insulating bipolaronic CDW (BCDW). When the electron electron coupling becomes more important, the ground-state of the system is changed and could become magnetic. Our results suggests that when the electron phonon coupling is also strong, the ground state could become a polaronic magnetic structure or polaronic spin density wave (PSDW).

Thus, for having a model with high Tc SC, increasing only the electron phonon coupling is not sufficient because beyond a certain critical point, superconductivity necessarily becomes unstable against the formation of a BCDW. This structure becomes insulating because the bipolaronic band width practically vanishes at large electron phonon coupling. If a strong electron-electron interaction is also present, one could get an insulating PSDW which should be insulating for the same reason. Our basic idea for interpreting the existence of high Tc SC is that at large electron phonon coupling, there exists an intermediate regime between these two phases (BCDW and PSDW) where our method of proof which can predict only insulating states, does not apply. In that region, the electron electron repulsion becomes such that the energy for breaking a bipolaron into two polarons is almost zero. There we get quantum resonance between different electronic states favouring the emergence of quantum states such as superconductivity. The internal structure of the bipolarons changes significantly. Instead of corresponding to a pair of electrons in the same state with opposite spins, the bipolaron becomes (mostly) a *quantum resonant* state between the two spin states $\uparrow\downarrow$ and $\downarrow\uparrow$ of a pair of polarons. This spin resonant bipolaron consists into two polarons bounded at some distance which depends on the strength of electron electron repulsion. Of course, the binding energy decays as a function of the electronic repulsion. This structure cannot be obtained within usual mean field approaches of electron Hamiltonians. An interesting experimental consequence is that this resonant bipolaron should involve spin fluctuations at an energy scale comparable to the binding energy of this pair.

This electron pairing only is not sufficient for having a high Tc. It is in fact associated with a sharp increase of the effect of the quantum lattice fluctuations (in the intermediate regime) resulting in a strong enhancement of the bipolaron band width. Then, if this band width energy becomes larger than the typical energies of the interacting potentials between bipolarons and polarons, the condensation of the electronic state into an insulating structure corresponding to the static ordering of localized bipolarons (BCDW) or of localized polarons (PSDW) could be avoided. Then the ground state has to be a coherent quantum state of delocalized resonant bipolarons (which can now be considered as Cooper pairs in an extended sense). While the electron electron interaction increases, the critical temperature of the resulting superconducting state is expected to reach a maximum at relatively high temperature when the (decaying) binding energy of a resonant bipolaron is still large while the bipolaron band width is close to its maximum.

A proper treatment of these interactions in realistic models is in fact a rather tough problem, if one wish to avoid the standard but somewhat uncontrolled approximations. It appears useful to simplify the real models as much as possible and then try a more rigourous treatment. Although these oversimplified models are likely not convenient for precise quantitative applications, the paradigms generated by studying these simple models, could be extendible to more realistic models with growing complexity. This is the reason why at the present time, we mainly focus on the Holstein-Hubbard model which is likely the "minimal" model involving both type of interactions and thus for which we can expect different structures for the ground-state which can either be superconductors, charge density waves or magnetic structures. Despite its simplicity, a deeper understanding of this simple model more than more complex models treated with rather loose approximations, should help for understanding the origin of high Tc superconductors and in particular for testing the above conjecture concerning the microscopic origin of the high Tc superconductors.

2- The Holstein-Hubbard model

2.1. Definitions

The hamiltonian of Holstein-Hubbard Model is the sum of four terms

$$H = H_k + H_p + H_{ep} + H_u \tag{1}$$

and can be defined on any discrete lattice at arbitrary dimension. The discreteness of this lattice turns out to be physically essential in our approach because it allows one to take into account the periodicity of the crystal which is not considered as uniform (jellium). In fact, this lattice is not necessarily periodic but can be chosen quasiperiodic or random or else if for example one wish to investigate quasi crystals or amorphous structures. It is useful to explicit the physical energy units of the model before to work with reduced dimensionless constant with an appropriate choice of the physical units.

The first part H_k is the standard Hamiltonian

$$H_k = -T \sum_{<i,j>,\sigma} c^+_{i,\sigma} c_{j,\sigma} \tag{2}$$

of a single band of non interacting electrons in a tight binding representation with non degenerate s states, written with the standard creation and annihilation Fermion operators $c^+_{i,\sigma}$ and $c_{i,\sigma}$ at site i. $<i,j>$ are neighbouring sites on the considered lattice and σ is the electron spin $\pm 1/2$ denoted $\uparrow$ or $\downarrow$. T is the transfer integral. Thus 2dT is the electronic band width for a d dimensional square lattice. The typical physical units for T are eV (electron-Volts).

The second part H_p of the Hamiltonian describes independent Einstein oscillators at site i. It is written with the standard creation and annihilation Boson operators $a_i^\dagger$ and a_i at site i and involves the phonon energy $\hbar\omega_0$. It can be also written with the position and momentum operators u_i and p_i. The typical physical units for $\hbar\omega_0$ are in general much smaller than the band width unit T and range around 10^{-2} eV.

$$H_p = \sum_i \hbar\,\omega_0 (a_i^+ a_i + \frac{1}{2}) = \sum_i (\frac{p_i^2}{2m} + \frac{1}{2} m\,\omega_0^2 u_i^2) \tag{3}$$

where m is the mass of the oscillator and $m\,\omega_0^2$ the stiffness constant. We have $[u_n,p_n] = i\hbar$, where [a,b] denotes the commutator of a and b.

The third part H_{ep} of the Hamiltonian couples linearly the electron density

$$n_i = c_{i,\uparrow}^+ c_{i,\uparrow} + c_{i,\downarrow}^+ c_{i,\downarrow} = n_{i,\uparrow} + n_{i,\downarrow} \tag{4a}$$

at site i to the position u_i of the oscillator, or equivalently to $(a_i^+ + a_i)$

$$H_{ep} = \lambda \sum_i n_i u_i = g \sum_i n_i (a_i^+ + a_i) \tag{4b}$$

Constants λ and g are related by $$g = \sqrt{\frac{\hbar}{2\,m\,\omega_0}}\;\lambda \tag{4c}$$

Finally, the fourth term H_u

$$H_u = \sum_i \upsilon\, n_{i,\uparrow} n_{i,\downarrow} = \sum_i \upsilon\, c_{i,\uparrow}^+ c_{i,\uparrow} c_{i,\downarrow}^+ c_{i,\downarrow} \tag{5}$$

with constant υ describes within the Hubbard approximation, the Coulomb repulsion between two electrons with opposite spins. The energy range of g and υ in different physical systems, can be very large which allows one to get different physical states such as superconductivity, CDWs or SDWs.

2.2. The anti-integrable limit[3]

For understanding some aspects of the strong electron-phonon limit ($g \gg \sqrt{T\hbar\omega_0}$), it is now useful to define the conjugate operators

$$\mathbf{u}_n = \frac{\hbar\,\omega_0}{4g} (a_n^+ + a_n) \qquad \text{and} \qquad \mathbf{p}_n = \frac{2g}{\hbar\,\omega_0}\, i\, (a_n^+ - a_n) \tag{6}$$

where $[\mathbf{u}_n,\mathbf{p}_n] = i$. By choosing $$E_0 = 8\, g^2/(\hbar\,\omega_0) \tag{7a}$$
as energy unit, our Hamiltonian H can be written as the sum of three terms

$$\mathbf{H} = \frac{H}{E_0} = H_{AI} + t\,H_K + \beta^2 H_Q \tag{7b}$$

where the main part $$H_{AI} = \sum_i \frac{1}{2} (\mathbf{u}_i^2 + \mathbf{n}_i \mathbf{u}_i) + U\, n_{i,\uparrow} n_{i,\downarrow} \tag{8a}$$

with $$U = \frac{\upsilon\,\hbar\,\omega_0}{8\,g^2} \tag{8b}$$

is called the "anti-integrable Hamiltonian". The appendix explains the origin of this denomination which comes out from mathematical analogies between this problem and the problem of chaos in dynamical systems[3-4]. It corresponds to the terms of the initial Hamiltonian which only involves the electron density operators $n_{i,\uparrow}$ and $n_{i,\downarrow}$ and the position operator $\mathbf{u}_i$. Since each of these operators commutes with H_{AI}, its eigenstates are readily obtained. These ones can be classified according to an arbitrary distribution of the electrons $n_{i,\uparrow} = 0$ or 1 and $n_{i,\downarrow} = 0$ or 1. Most of these distributions are chaotic. There are two perturbative terms in (7). H_K corresponds to the kinetic energy operator of the electrons

$$H_K = -\frac{1}{2} \sum_{<i,j>,\sigma} c^+_{i,\sigma} c_{j,\sigma} \tag{9a}$$

with the coefficient

$$t = \frac{T \hbar\omega_0}{4 g^2} \tag{9b}$$

H_Q corresponds to the kinetic energy of the atoms of the lattice

$$HQ = \frac{1}{2} \sum_i \mathbf{p}_i^2 \tag{10a}$$

with the coefficient

$$\beta = \frac{1}{8} \left(\frac{\hbar\omega_0}{g} \right)^2 \tag{10b}$$

(note here that our definition for β has been changed in that paper compared to ref.3) These dimensionless coefficients t and β measure the strength of the quantum fluctuations due to the electrons and to the lattice respectively. They can be chosen as independent parameters instead of the initial parameters. This change of variable is suitable when we have both $t \ll 1$ and $\beta \ll 1$.

In fact at the present stage, our exact results are only obtained for $U=0$ although we hope to extend them for $U \neq 0$. We will discuss later in the paper the role of this term. We assume that it is zero from nowon till we specify that it is not zero in section 6.

Our approach contrasts with most theories for electrons. In these theories, the kinetic energy of the electrons is considered to be the major term of the Hamiltonian for which the ground-state is a Fermi liquid. The problem turns out to understand the properties of the electronic system by perturbation of this Fermi liquid. We approach the same problem from the opposite limit (which we call anti-integrable) where the electrons form an insulating structure of (quasi) classical particles randomly localized on a lattice.

2.3. *The adiabatic and the Born-Oppenheimer approximation*

We obtain exact results for the perturbation of this anti-integrable Hamiltonian H_{AI} for small t when there is no Hubbard term ($U=0$) and at the adiabatic limit that is for $\beta=0$. Since there is several terms in the above Hamiltonian, there is of course several kinds of limits depending on their relative strength. The adiabatic approximation is often used in physics. It is obtained by dropping the atomic kinetic energy term H_Q of the atoms of the lattice, in the Hamiltonian. Since the position operator u_i now commutes with the Hamiltonian, it can be considered as a scalar variable. The electronic energy is

$$\Phi_{elect.}(\{\mathbf{u}_i\}) = \sum_\nu \sigma_\nu E_\nu(\{\mathbf{u}_i\}) \tag{11a}$$

where we define the occupation number σ_ν of the electronic state ν as 0,1/2 or 1 when it is unoccupied, occupied by one electron or by two electrons with opposite spins respectively. For the ground state, σ_ν is chosen according to the Fermi rule $\sigma_\nu=1$ for $E_\nu \leq E_F$ and $\sigma_\nu=0$

for $E_\nu \geq E_F$ where E_F is the Fermi energy. $\Phi_{elect.}(\{u_i\})$ is a function of the set of coordinates $\{u_i\}$ through the eigen energy equation for the electronic state ν

$$- t \, \Delta\Psi_i^\nu + u_i \, \Psi_i^\nu = E_\nu(\{u_i\}) \, \Psi_i^\nu \quad \text{where} \quad \sum_{j \text{ close to } i} \Psi_j^\nu = \Delta\Psi_i^\nu \tag{11b}$$

Thus the problem of finding the ground state and metastable states is equivalent to find the minima and the local minima of the sum

$$\Phi(\{u_i\}) = \Phi_{elast.}(\{u_i\}) + \Phi_{elect.}(\{u_i\}) \tag{11c}$$

where

$$\Phi_{elast.}(\{u_i\}) = \sum_i 1/2 \; u_i^2 \tag{11d}$$

In general, this explicit calculation cannot be done analytically but is affordable by numerical means. This approximation is in fact equivalent to a meanfield approximation on the electron phonon coupling which consists in assuming

$$u_i n_i \cong \langle u_i \rangle n_i + u_i \langle n_i \rangle - \langle u_i \rangle \langle n_i \rangle \tag{12}$$

The Born-Oppenheimer (BO) approximation is an improvement of the adiabatic approximation. It does not neglect the quantum lattice fluctuations but assumes that the electronic state is in adiabatic equilibrium with the atomic positions. Then the electron coordinates can be eliminate at least formally, and the atomic lattice is described by the effective Hamiltonian

$$H_{BO} = \frac{8g^2}{\hbar\omega_0} \left(\Phi(\{u_i\}) + \beta^2 \sum_i 1/2 \; p_i^2 \right) \tag{13a}$$

Since this Hamiltonian with a non linear potential, is still not analytically solvable, the effective potential energy is expanded at second order in the neighbourhood of an adiabatic solution $\underline{u}_i$ (which fulfils

$$\frac{\partial\Phi}{\partial u_i} = \underline{u}_i + \sum_\nu \sigma_\nu \, |\Psi_i^\nu(\{\underline{u}_i\})|^2) = 0 \tag{13b}$$

for all i) as a function of $\varepsilon_i = u_i - \underline{u}_i$. The obtained quadratic form $1/2. \bar{\varepsilon}.\bar{\bar{M}}.\bar{\varepsilon}$ is positive with a matrix $\bar{\bar{M}}$ the coefficients of which can be explicited as

$$M_{i,j} = \frac{\partial^2 \Phi}{\partial u_i . \partial u_j} = \delta_{i,j} - \sum_{\nu,\nu'} \sigma_\nu (1-\sigma_{\nu'}) \; \frac{\Psi_i^{\nu *} \Psi_i^{\nu'} \Psi_j^{\nu' *} \Psi_j^{\nu}}{E_{\nu'} - E_\nu} - CC \tag{13c}$$

where the eigen functions Ψ_i^ν and eigen energies E_ν are given by eq.11b for $u_i = \underline{u}_i$. CC means complex conjugate of the previous term. The eigen values r_μ^2 of this matrix (which are necessarily positive), yields the renormalized phonon frequencies $\omega_\mu = r_\mu \, \omega_0$ of the phonon spectrum. This quadratic Born Oppenheimer approximation renormalizes the bare phonon frequency ω_0 by the factor r_μ for the mode μ.

Since the BO approximation assumes that the electrons remain in adiabatic equilibrium with the lattice, this approximation cannot describe in essence a superconducting state where the electrons form their own coherent state. It is often argue that this BO approximation is valid because the mass of the atoms is much larger than the mass of the electrons which means in our model that $\hbar\omega_0 << T$. This simple argument might be not sufficient even when β is small (see section 5). In fact, the validity of this approximation must be checked when it is applied and surely becomes wrong when the ground-state is superconducting. In that case, the quantum lattice fluctuations cannot be treated simply as a correction within the BO approximation because the ground-state is changed qualitatively. As we will see in the next, this approximation is also found to be doubtful when the phonon softening r_μ becomes smaller than a certain critical value.

2.4. The anti-adiabatic approximation and the standard superconductivity

The standard theory of superconductivity starts from an opposite assumption. It assumes that the phonon fluctuations can be eliminated and replaced by an effective interaction between the electrons. In text books[5], this elimination is performed through a unitary transformation calculated at the lowest order. The interaction between the electrons results from the exchange between two electrons of a single phonon. The reliability of this expansion becomes doubtful as soon that this interaction starts to involve many phonons processes that is just when $g >> \hbar\omega_0$. Thus, we believe that the property that the electrons can be treated by an effective Hamiltonian which is strictly decoupled from the phonon, is restricted to a domain with large enough β and also only close enough to the ground-state that is at temperature not too large.

The limit $\beta >> 1$ is the anti-adiabatic limit. Note that it has a rather simple physical meaning. It means that the distortion which would be created by a static electron at a given site is much smaller than the zero point motion of the atom at this site. In other word, in that case the electron-phonon coupling could only create lattice distortions which have negligible amplitude compared to the spontaneous quantum lattice fluctuations.

The Lang-Firsov unitary transformation[6,1] is well suited for studying the neighbourhood of the ideal anti-adiabatic limit. This Lang-Firsov unitary transformation applied to Hamiltonian (1) is defined as

$$\hat{H} = \exp(iS_{LF})\, H \exp(-iS_{LF}) \tag{14a}$$

with

$$S_{LF} = \frac{g}{\hbar\omega_0} \sum_i \Big(\sum_\sigma c^+_{i,\sigma} c_{i,\sigma}\Big)\left(a^+_i - a_i\right) \tag{14b}$$

In simple words, it is the unitary transformation which diagonalizes the part of the Hamiltonian corresponding to $H_{AI} + \beta^2 H_Q$ (or the initial Hamiltonian with T=0)

$$\hat{H} = \Big(\upsilon - \frac{2g^2}{\hbar\omega_0}\Big) \sum_i n_{i,\uparrow} n_{i,\downarrow} - \frac{g^2}{\hbar\omega_0} \sum_i (n_{i,\uparrow} + n_{i,\downarrow}) - \sum_{<i,j>,\sigma} \hat{t}_{i,j}\, c^+_{i,\sigma} c_{j,\sigma} + \hbar\omega_0 \sum_i \Big(a^+_i a_i + \frac{1}{2}\Big) \tag{15a}$$

with

$$\hat{t}_{i,j} = T \exp\Big[\frac{g}{\hbar\omega_0}\big((a^+_i - a_i) - (a^+_j - a_j)\big)\Big] \tag{15b}$$

The transfer integral of electron kinetic energy part now contains the phonons operators. In the limit $\beta=+\infty$, operator $\hat{t}_{i,j}$ becomes a constant T. Then, the electron Hamiltonian becomes independent of the phonon Hamiltonian but contains a Hubbard term with $U=(\upsilon - 2g^2/\hbar\omega_0)$. This Hamiltonian can be also obtained by assuming that the atomic oscillators are in adiabatic equilibrium with the electrons which is just the assumption opposite to those of the BO approximation. (When β is not infinite, there remains an electron phonon coupling coming from the transfer terms $\hat{t}_{i,j}$. By a perturbation theory at lowest order, this term gives also an effective but non local electron-electron interaction. This contribution is the most important one when $\hbar\omega_0 << T$ which is generally physically true. It is essential in the BCS theory for interpreting the isotopic effect in standard superconductors).

Superconductivity is a phenomena which requires that the electrons form coherent quantum states with permanent currents which are eigenstates of the whole system (that is with infinite lifetime). In principle, a distribution of current fulfilling the standard conservative laws determines an eigenstate. This property readily implies the Meissner effect. For a current density $\mathbf{j}=\rho e v$ (ρ is the density of electrons, e the charge of the electron and v their velocity), we only assume that the extra energy per unit volume $1/2.\rho\, m^* v^2 = m^*/(2\rho e^2)\, \mathbf{j}^2$ of the superconductor is the kinetic energy of the system of electrons moving without dissipation with density ρ and effective mass m^*. Since we have $\mathbf{j}=\mathrm{curl}(\mathbf{H})$, there is also a magnetic energy $1/2.\mu\, \mathbf{H}^2$ (μ is the magnetic permeability) in the system. For the equilibrium states, the total of magnetic energy and of the current energy $1/2.\ \mu\, \mathbf{H}^2 + m^*/(2\rho e^2)\ \mathrm{curl}(\mathbf{H})^2$ has to be minimized. At the 2d interface between a superconducting material and a non superconducting material (or the vacuum) in which there is a constant magnetic field $\mathbf{H}_0$, the minimisation of this energy yields that the the magnetic field decays exponentially in the superconductor from the interface with the London penetration depth $\xi = \sqrt{\frac{m^*}{\mu\, \rho\, e^2}}$.

Since the transport of the electronic structure with respect to the lattice can be done without energy exchange between the electrons and the phonons, in our opinion it is the fundamental question concerning the foundation of the microscopic theory of superconductivity to ask:

Is it physically correct to describe the electrons by an effective purely electronic Hamiltonian H_{eff} without phonons?

For self consistency reasons, the answer to this question is probably yes if the groundstate is superconducting (but only for the ground-state and the low temperature states). Otherwise, the physical fact that there is energy dissipation through the phonons of the lattice for the electronic transport (because it is not superconducting) proves that the electrons cannot be well described by a purely electronic effective Hamiltonian. In other words, assuming the existence of this effective Hamiltonian becomes a consistent assumption if and only if at the end of the calculation, its ground-state is indeed superconducting.

Our intuition is guided with possible similarities with the theory of perturbation of integrable systems[7] (Kolmogorov Arnol'd Moser or KAM theory). Considering the electron-phonon interaction as a perturbation, this one could be eliminated through a unitary transformation providing 1- that the Hamiltonian be restricted to a certain invariant subspace of states which contains the ground state and close-by states and 2-that this perturbation be not too large.

We conjecture that an effective Hamiltonian only could exist for the electrons far enough from the adiabatic limit that is when the amplitude β of the term describing the quantum lattice fluctuations becomes sufficiently large. The corresponding critical value β_c should depend on the other parameters t and U of the model. Once it is assumed that there exists an effective Hamiltonian for the electrons, the problem is to find the ground state for this model which has to be superconducting. When the interaction between the electrons is attractive at least at the Fermi surface, the standard BCS theory assumes that the ground state is a superconducting state which can be described as a coherent quantum state of pairs of delocalized electrons (Cooper pairs) with opposite spins and opposite wave vectors q and -q. The BCS wave function has the form

$$|\Phi\rangle = \prod_q (u_q + v_q \, c^+_{q,\uparrow} \, c^+_{-q,\downarrow}) \;|\emptyset\rangle \qquad (16)$$

where $|\emptyset\rangle$ is the vacuum and u_q and v_q are determined by minimization of $\langle\emptyset| H |\emptyset\rangle$.

3. Polarons and Bipolarons.

This concept is well defined in the literature only when the density of electrons is low. We also discuss in this section only isolated polarons and bipolarons. Because of the translation invariance of the system and of the quantum lattice fluctuations, they (in principle) always form bands. There is what is called large polaron and small polaron. The large polaron is in fact a (delocalized) quantum electron the mass of which is renormalized by the interaction with the phonons by a small factor. Thus this polaron keeps a strong quantum character, which means that their band width remains physically rather important.

By contrast, the small polaron has (in general) a very large effective mass or equivalently a very small band width which implies a very low (or negligible) quantum mobility. It corresponds physically to a localized electron trapped by the potential well which is created self consistently by its interaction with the lattice. Its large effective mass implies that it behaves practically as a classical particle.

If the repulsion between the electrons is not too large, the bipolaron is more stable than two polarons far apart. A large bipolaron is a pair of electrons bounded by the electron phonon coupling and is delocalized as well as a Cooper pair. It physically corresponds to the same concept although the mathematical treatment could be different. A small bipolaron is formed by two electrons localized self consistently in the same electronic state with opposite spins. They have generally a very small band width (or large effective mass) and thus cannot give rise to superconductivity at least with a non negligible critical temperature.

3.1. Effective Band Width for Large Polarons

The essential physical difference between small and large polarons (or bipolarons), lies in their effective mass which involves the quantum lattice fluctuations but as we will suggest for "spin resonant bipolarons" is not necessarily only related to the size. When at non vanishing density, the effective mass of the bipolarons remains comparable to the bare electron mass (large bipolarons) and while their interaction potential is not too large, they tend to form a superconducting ground state. When the effective mass becomes large (small bipolarons), the potential interaction between these bipolarons generally dominates the kinetic energy term. The bipolarons order in real space and the ground state becomes an insulating bipolaronic CDW. Thus, large polarons (and bipolarons) are better described within the anti-adiabatic approximation while the small polarons (and bipolarons) are better described within the opposite adiabatic approximation.

The Lang-Firsov transformation can be used for calculating the effective electron mass (i.e. of the band width) due the electron phonon coupling (large polarons). The effective band width is obtained by replacing $\hat{t}_{i,j}$ by its mean value in eq.15a

$$\hat{t} = T \exp\left(-\left(\frac{g}{\hbar\omega_0}\right)^2\right) \tag{17}$$

with the ground state phonon wave function. When $g \ll \hbar\omega_0$ that is close to the antiadiabatic limit, there is only a small renormalization of the band width. This regime corresponds to the large polaron limit.

3.2. Small polarons in the adiabatic limit. Dimensionality effects.

When $g \gg \hbar\omega_0$ the renormalized band width given by eq.17 collapses exponentially to zero. It can be considered to be practically zero as soon as g is larger than $\hbar\omega_0$ by few factors. Then polarons and bipolarons can be well described within the adiabatic approximation. They can be considered as localized in the real space and their real shape can be explicited. For convenience, we studied numerically the adiabatic limit of hamiltonian (1) with different notations. Taking the transfer integral 2T now as the energy unit instead of $E_0 = 8\,g^2/(\hbar\omega_0)$, Hamiltonian (1) becomes

$$\frac{H}{2T} = -\frac{1}{2}\sum_{<i,j>,\sigma} c^{+}_{i,\sigma} c_{j,\sigma} + \frac{\mathbf{k}}{2}\sum_i \mathbf{n}_i u_i + \frac{1}{2}\sum_i \left(u_i^2 + \frac{\gamma^2}{4} p_i^2\right) + \sum_i U\, n_{i,\uparrow} n_{i,\downarrow} \tag{18a}$$

with the definition $u_i = \frac{\sqrt{\gamma}}{2}\left(a^{+}_i + a_i\right)$ $p_i = \frac{i}{\sqrt{\gamma}}\left(a^{+}_i - a_i\right)$ $([u_i, p_i] = i)$ and

$$\mathbf{k} = \frac{1}{\sqrt{t}} = \frac{2g}{\sqrt{T\,\hbar\omega_0}} \qquad \gamma = \frac{\hbar\omega_0}{T} \quad \text{and} \quad U = \frac{\upsilon}{2T} \tag{18b}$$

The adiabatic limit is equivalent to take $\gamma = 0$ (and the anti-integrable limit is obtained for $\mathbf{k} \to +\infty$). Within the adiabatic limit (and for U=0), the electronic state $\{\phi_i^0\}$which is occupied for a bipolaron or a polaron is the ground-state ($\nu=0$) of the Hamiltonian

$$-\Delta\phi_i^0 + \mathbf{k}\, u_i\, \phi_i^0 = E_0(\{u_i\})\, \phi_i^0 \tag{19a}$$

Eq.13b yields the atomic distortion for a bipolaron ($\sigma_0 = 1$) or a polaron ($\sigma_0 = 1/2$)

$$u_i = -\mathbf{k}\, \sigma_0\, |\phi_i^0(\{\mathbf{u_i}\})|^2 \tag{19b}$$

When $\sigma_0=1/2$, by rescaling u_n as $v_n = \sqrt{2}\, u_n$, one readily find that the energy function of a polaron versus **k**, $E_{pol}(\mathbf{k})$ is related to the energy function of a bipolaron $E_{bip}(\mathbf{k})$ versus **k**

$$E_{pol}(\mathbf{k}) = \frac{1}{2}\, E_{Bip}\left(\frac{\mathbf{k}}{\sqrt{2}}\right) \qquad (20)$$

Thus, it is sufficient to study only a single bipolaron ($\sigma_0=1$). Eq.19b, yields with eq.19a the non linear eigen equation

$$-\ \Delta\phi_i^0 \ - \mathbf{k}^2\, |\, \phi_i^0\,)\, |^2\, \phi_i^0 = E_0\phi_i^0 \qquad (21a)$$

which is solved with the condition $\sum_i |\phi_i^0|^2=1$. Let now assume that the continuous approximation can be done for k small that is ϕ_i^0 is supposed to be a smooth function of i=**r**.Then this equation becomes the non linear Schroedinger equation (Δ is the d-Laplacian now)

$$-\Delta\phi - (2d + \mathbf{k}^2\, |\, \phi\, |^2)\, \phi = E_0\, \phi \qquad (21b)$$

In one dimension (d=1), this equation has a well known solution

$$\varphi(x) = \frac{\mathbf{k}}{2\sqrt{2}}\ \frac{1}{\cosh\ \mathbf{k}^2/4\ (x-a)} \qquad (22a)$$

with the energy

$$E_0 = -2 - \mathbf{k}^4/16 \qquad (22b)$$

and the total energy of a bipolaron $E_{bip}(\mathbf{k}) = -2 - \mathbf{k}^4/48$ which is smaller than the energy of a pair of extended electrons. At the anti-integrable limit, the bipolaron is localized on a single site and its energy is readily found to be $E_{bip}(\mathbf{k}) = -\,\mathbf{k}^2/2$. Fig1.a shows the energy of a bipolaron which has been calculated numerically compared to the approximations obtained for k small and k large.

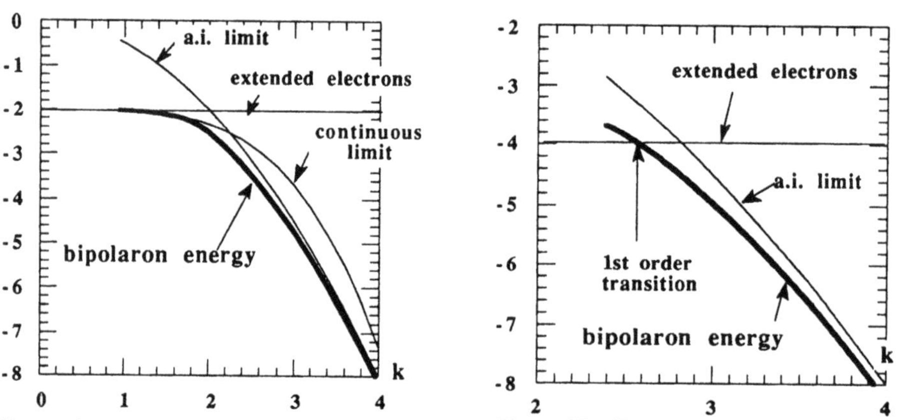

Figure 1a: Energy of a bipolaron versus k in the 1d adiabatic Holstein model (U=0)

Figure 1b: Energy of a bipolaron versus k in the 2d adiabatic Holstein model (U=0)

The situation is very different for models in two and more dimensions. A scaling argument due to D.Emin[8], shows that in that case, there is no solution to eq.21b. Indeed if $\phi(\mathbf{r})$ is the ground-state solution of (21b), it is straightforward to check that $\tilde{\phi}(\mathbf{r}) = \lambda^{-d/2}\,\phi(\lambda\mathbf{r})$ is also a ground-state of eq.21b for $\tilde{\mathbf{k}} = \lambda^{1-d/2}\,\mathbf{k}$ and $\tilde{E}_0 = \lambda^2(E_0 + 2d)$ -2d. The bipolaron energies are then related by $E_{bip}(\tilde{\mathbf{k}}) = \lambda^2 E_{bip}(\mathbf{k}) + 2d\,(\lambda^2 - 1)$. We then find a contradiction in this property for d>2 since for $\lambda \to \infty$, we have $\tilde{\mathbf{k}} \to 0$ but $E_{bip}(\tilde{\mathbf{k}}) \to -\infty$ (instead of the right limit which should be $E_{bip}(0) = -2d$) which cannot be true. Consequently, no solution can exist to eq.21b for d>2. This result also hold for d=2 because then we would have $\tilde{\mathbf{k}} = \mathbf{k}$ but $E_{bip}(\tilde{\mathbf{k}}) \neq E_{bip}(\mathbf{k})$. However, this solution does exist for the discrete model for **k** large enough (see next theorem). In fact, a numerical calculation shows that there is a first order transition at the critical value of $\mathbf{k}_{f.o} \cong 2.58$ for the 2d model. Below this critical value, the bipolaron energy becomes larger than the energy of two extended electrons. For lower value of **k**, it does not exist as predicted by the continuous model. This result points out the important role of the discrete lattice for allowing the existence of bipolarons and polarons in real systems in more than 1d models. In fact, the first order transition observed fig.1b persists for very anisotropic systems. Considering a kinetic energy for the electron with 1 as the transfer integral in the direction x and α in direction y, fig.2 shows the variation of $\mathbf{k}_{f.o}$ versus the anisotropy parameter α. Even for α as low as 10^{-2}, $\mathbf{k}_{f.o}$ is still 2/5 of its value for α=1.

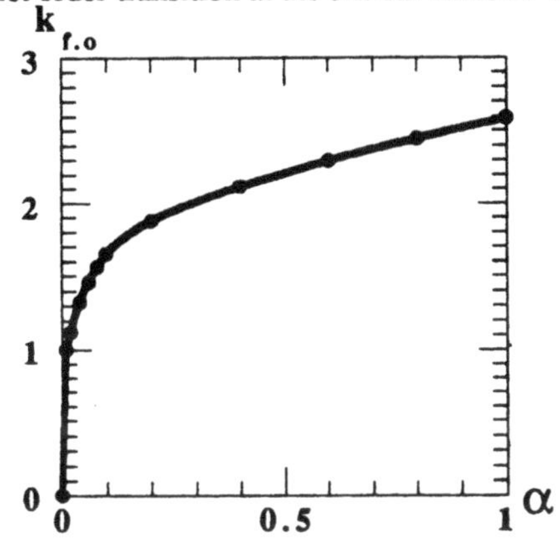

Figure 2: Variation of $k_{f.o}$ versus the anisotropy parameter for the 2 d adiabatic Holstein model.

4. Exact Results for the Adiabatic Holstein Model

4.1. The anti-integrable limit

Single polarons and bipolarons exist at large enough electron phonon coupling. Can they still be defined when their density is finite? We are presenting theorems which assert their existence for the adiabatic Holstein model at any density, providing the electron phonon coupling be large enough. At the anti-integrable limit, the existence of polarons and bipolarons is trivial. Since t=0 (and β=0) in eq.7, any *arbitrary* distribution of electrons $\{\mathbf{n}_i\}$ yields an eigenstate. The sites i are independent. When a site i is occupied by two electrons with opposite spins ($\mathbf{n}_i$=2), the local distortion obtained by minimization of (8a), is $\mathbf{u}_i = -\mathbf{n}_i/2$. Then, there is a bipolaron at site i . For the sites with only one electron ($\mathbf{n}_i$=1), we have $\mathbf{u}_i$=-1/2 (the spin of which has an arbitrary direction). These sites are occupied by polarons. Unoccupied sites are holes. If for a given configuration, there are bipolarons, polarons and holes, we say that the configuration is a mixed polaronic bipolaronic structure. If there is only bipolarons and holes, it is a bipolaronic structure.

Fig.3 schematizes such a configuration which is chosen at random but it could be also periodic or quasi periodic (or else).

4.2.Theorems for Bipolaronic and Polaronic Structures [3]

Up to now, the existence of polaronic and bipolaronic structures at arbitrary *non vanishing density* was not considered as obvious. The following theorem states the persistence of the bipolaronic and mixed polaronic bipolaronic states *at any non vanishing density* and at sufficiently large electron-phonon coupling which trivially exists at the anti-integrable limit (infinite electron-phonon coupling). The results are also valid a fortiori at zero density and for single polarons and bipolarons.

<u>1. Existence Theorem</u> (for the Holstein model (1) with $U=0$ and $\beta=0$)

There exists a non vanishing bound $t_m>0$, *such that for any arbitrary distribution of* $\{\mathbf{n}_i = 0,1 \text{ or } 2\}$ *on an (infinite) d dimensional lattice, the corresponding eigenstates (chaotic, periodic or else) defined at the anti-integrable limit, are preserved for* $t<t_m$. *In addition , the atomic distortion* $\{\mathbf{u}_i(t)\}$ *is a uniformly continuous function of* t *for* $t<t_m$ *and* $\mathbf{u}_i(0)= -\mathbf{n}_i/2$.

This theorem does not give any information concerning the properties of the single electronic eigen states. Indeed, these eigen states might be extended and form bands if the initial distribution of $\{\mathbf{n}_i\}$ is periodic. They also might be localized if this distribution is random. In fact, this single particle property is not essential. Since the electrons are indistinguishable particles, only the global subspace of occupied states is physically meaningful. As we will see in the next, theses polaronic and bipolaronic states are insulating.

This theorem holds for the adiabatic Holstein model defined on any lattice in finite dimensions, even non periodic. It also holds in the presence of a magnetic field. The orbital effects of the magnetic field can be easily taken into account in the kinetic part of the Hamiltonian H_k written as:

$$H_K = -\frac{1}{2} \sum_{<i,j>,\sigma} q_{i,j}\, c^{+}_{i,\sigma} c_{j,\sigma} \tag{23a}$$

where

$$q_{i,j} = |q_{i,j}| \exp(i\, \theta_{i,j}) \tag{23b}$$

with $|q_{i,j}|=1$ and $\theta_{i,j} = -\theta_{j,i}$. Then, the sum of the phases $\theta_{i,j}$ over a closed contour of the considered lattice divided by 2π, determines the number of quantum fluxes units hc/e (h=Planck cste, c=light velocity, e=electron charge) passing through this contour.

The magnetic field also interacts directly with the spin of the electron (Zeeman effect). There is an additional term in the Hamiltonian (1) which is ($g \cong 2.$ is the Lande factor)

$$H_M = -\frac{1}{2} g\, \mu_B H \sum_i (c^{+}_{i\uparrow} c_{i\uparrow} - c^{+}_{i\downarrow} c_{i\downarrow}) \tag{24}$$

In its present form, the theorem holds with an arbitrary non uniform magnetic field if only the orbital effect is considered. It holds only with a uniform magnetic field if the Zeeman effect is also considered. Disorder can be also included in the model. The theorem still holds when the moduli of the transfer integral $|q_{i,j}|$ is a random variable or is chosen with any more ordered distribution providing only that it is bounded (e.g. $|q_{i,j}| \leq 1$).

As a consequence of this theorem, these polaronic and bipolaronic states remains *all distinct* after perturbation for $t<t_m$. This bound t_m depends on the dimension of the lattice and on its geometry (square, triangular, honeycomb etc...), on the magnetic field... For a given lattice, the corresponding bound $t_m>0$ is valid for any mixed polaronic bipolaronic configuration, polaronic and bipolaronic structures. In fact for each choice of $\{\mathbf{n}_i\}$, the corresponding bipolaronic or polaronic structure exists for t up to $t_c(\{\mathbf{n}_i\})$. This best bound depends on $\{\mathbf{n}_i\}$ and then we have of course $t_m<t_c(\{\mathbf{n}_i\})$. If one restricts to the bipolaronic structures only, a larger value can be chosen for t_m. For example in the 1 d Holstein model, R. MacKay et C.Baesens[9] found the condition $t< 0.09$ for the existence bipolaronic structures.

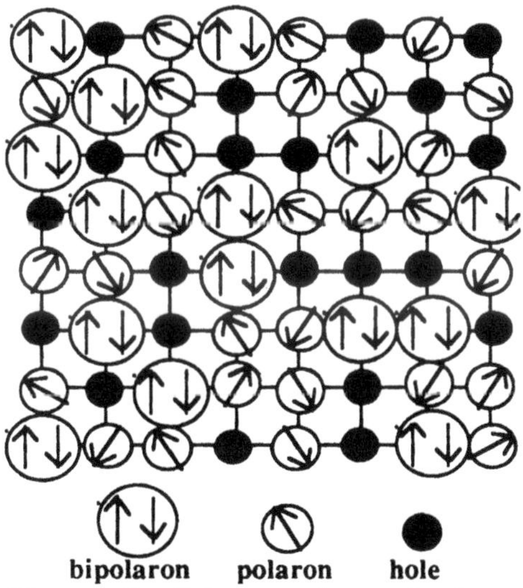

Figure 3: Scheme of a mixed polaronic bipolaronic structure on a 2d square lattice at the anti-integrable limit.

The proof of this theorem is based on the concept of anti-integrability initially developed for dynamical systems (see the appendix). It consists in proving that each extrema of the variational form (12a) which is trivial when t=0, still exists providing that t be not too large. The technical details of the proof turns out to be rather complex and described in ref.3 This method is extendible to much larger classes of models where a non singular anti-integrable limit can be defined. The proof of this theorem has been recently improved significantly by R.MacKay and C.Baesens[9] for the 1d adiabatic Holstein model and for purely bipolaronic structures.

Clearly, for physical applications, this theorem needs to be completed by both quantitative analytical calculation (which turns out to be very tough) or by numerical calculations which are easily affordable with modern computers. Indeed, the bounds $t_c(\{\mathbf{n}_i\})$ which we observed numerically, were found to depend on the configuration $\{\mathbf{n}_i\}$ and to be always larger than the bound predicted analytically. We are now going to describe some properties of these bipolaronic and polaronic states which can be also proven rigourously.

4.3.Exact Properties of the Bipolaronic and Polaronic Structures

a: electronic spectrum: The single particle spectrum of the electrons in the potential generated by the lattice distortion has the structure which is schematized by fig.4a for the bipolaronic structures and fig.4b for the mixed polaronic bipolaronic structures. In the first case there is a non vanishing gap in energy which separates the doubly occupied states from the empty states. However, the structure of the lower and of the upper band is not necessarily smooth. In the second case, the polarons correspond to midgap states. Thus there is a gap which separates the doubly occupied electronic states from those who are singly occupied, and another gap between the singly occupied states and the empty states.

b. Metastability-Phonon spectrum, Peierls-Nabarro Energy Barrier:

If the electronic population factors are fixed, the bipolaronic and mixed polaronic bipolaronic configurations are local minima of the adiabatic energy (11c). As shown by fig.4b for the mixed polaronic bipolaronic states, the global electronic state is not in its ground state relatively to the lattice distortion. However, they become ground states if an appropriate (strong) uniformly magnetic field is applied. Then, the degeneracy of the free spins of the polarons is raised and the electronic ground state is obtained when they are alined in the direction of this magnetic field. Then, both the bipolaronic and the mixed polaronic bipolaronic states become physically metastable states which means that they local minima of the total energy both with respect to the lattice and the electron relaxations.

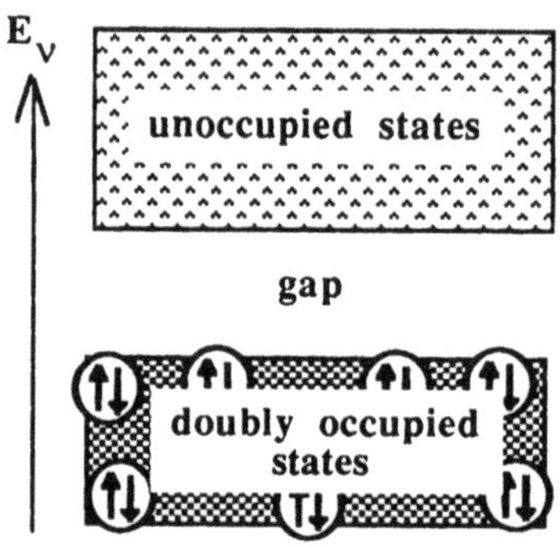

Figure 4a Distribution of the electronic eigen energies for Bipolaronic Structures

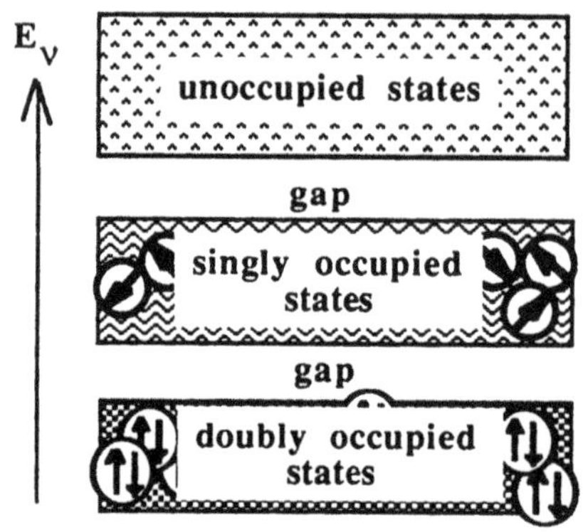

Figure 4b Distribution of the electronic eigen energies for mixed Polaronic Bipolaronic Structures

The lattice vibration around this local minima renormalizes the bare phonon frequencies ω_0. The new frequencies are obtained by diagonalization of matrix $\bar{\bar{M}}$ given by (13b). It is proven for all bipolaronic and polaronic structures that 1- the renormalized frequencies ω_η are strictly smaller than the bare phonon frequency ω_0 except for one phonon mode corresponding to the zero wave vector. 2- the lowest frequency ω_η (phonon gap) is strictly larger than zero. Thus it cannot vanish and cannot give a zero frequency mode such as the sliding mode of a soliton or the gapless phason mode of a Peierls Frohlich incommensurate CDW.

Another important result which comes out from our theory is the existence of a non vanishing Peierls Nabarro Energy Barrier (PNEB). Given two bipolaronic or mixed polaronic bipolaronic configurations C_0 and C_1 which differ by the location (at the anti-integrable limit) of one bipolaron (or one polaron), there exists continous paths $C(\tau)$ such that $C(0)=C_0$ and $C(1)=C_1$ which connects these two configurations. Along this path, the energy variation of the configuration (which remains finite) reaches a maximum $E(C(\tau))$. The minimum of $E(C(\tau))$ over all possible paths $C(\tau)$ connecting C_0 and C_1 is the Peierls Nabarro energy barrier (PNB) between C_0 and C_1. For all bipolaronic or mixed polaronic bipolaronic configurations as predicted by the theorem, the PNB is not zero and bounded from below.

c.Coherence Length, Defectibility.. It is also proven that a local perturbation of these polaronic and bipolaronic structures, at a given site which is produced for example by an impurity decays exponentially as a function of the distance to this impurity.

Let us now consider two mixed polaronic-bipolaronic structure such that at the anti-integrable limit we have $\mathbf{n}_i = \mathbf{n}_i'$ except at a finite number of sites i_0. Then these two configurations $\{\mathbf{u}_i\}$ and $\{\mathbf{u}_i'\}$ are identical at infinite distance of the perturbed site and the difference between these two configurations decays exponentially far away from from these sites i_0. In other word, the injection to or the extraction of electrons from a mixed polaronic-bipolaronic structure only creates a local distortion of the structure. This is nothing else than the characteristic property of an insulator (at 0 K). We say by definition that the system is charge defectible.

This is a global property of the electron-phonon system which is not due to the localization of the single wave function of the electrons. Indeed, this property holds for random as well as for periodic mixed polaronic-bipolaronic structure where the single electron waves functions are extended. Thus, charge defectibility is a concept which extends the concept of localization and applies for non independent particles.

The energy $\mathbf{H}$ of the bipolaronic and mixed polaronic bipolaronic structures depends on the configuration $\{\mathbf{n}_i\}$ of the bipolarons and polarons at the anti-integrable limit which is determined by the electron distribution $\mathbf{n}_i=0,1$ or 2 (which can be viewed as pseudo spins). Thus, our theorem implies that at low temperature, the thermodynamics of this system is described by a pseudo-spin Hamiltonian $\mathbf{H}(\{\mathbf{n}_i\})$ or in other words a lattice gas model.The exact form of the Hamiltonian is not known, but the lowest order terms can be obtained[7] by an expansion in $1/\mathbf{k}$. There is also multi spins interactions appearing.

d.Effect of a Magnetic field We noted above that the existence theorem is unchanged because of the presence of a magnetic field. In fact, we even get better bounds for t_m. Recent numerical investigations on the 2d adiabatic Holstein model, shows that indeed the formation of polaronic and bipolaronic structures is favoured by a magnetic field.

In addition, we have recently proven analytically that single polarons and bipolarons do exist in 2 and 3 dimensions for the continuous model in a magnetic field while in the absence of magnetic field, eq. (21a) have no solution. The size of these polarons is the cyclotron radius divided by a factor depending on the electron phonon coupling.

e. Ground States We prove the following essential exact results valid for t is small enough (that is for an electron phonon coupling large enough):

In the absence of magnetic field or with a uniform magnetic field smaller than some non zero H_m, the ground-state of the adiabatic Holstein model is a bipolaronic structure.

For a uniform magnetic field larger than some $H'_m > H_m$, the ground state is surely a mixed polaronic bipolaronic structure (note that then the spin degeneracy of the polarons is raised, and the free spins aline in the magnetic field). This result is due to the Zeeman term (24) which has to be added to the Hamiltonian (1).

Of course, for these ground-states, the distribution of the bipolarons (and polarons) is (likely) not chaotic but well ordered. It corresponds to the ground-state of the pseudo spin Hamiltonian $\mathbf{H}(\{\mathbf{n}_i\})$. It might be a periodic superstructure but then the concentration of electrons has to be a rational number. It might also be more generally an incommensurate structure. We have no general exact results about the exact structures of these ground states. Nevertheless, we prove at large enough electron phonon coupling that 1-Bipolaronic Charge Density Waves are necessarily bipolaronic 2- Field Induced Spin Density Waves (FISDW) must exists for large enough magnetic field. (For having a FISDW, the Zeeman energy $2\mu_B H$ for this magnetic field H can estimated to be 1/8 of the gap of the electronic spectrum. This estimation is made at the anti-integrable limit by calculating the energy for breaking a single bipolaron into two polarons using eq.20. In 1d, the same result is obtained within the continuous limit).

4.4.Numerical observation of Bipolarons in 1d

In agreement with the predictions of Peierls, the ground-state of the 1d adiabatic Holstein model is found to be a CDW which is incommensurate for an irrational band filling ζ and commensurate otherwise. It has been noted[10,7] some years ago, that the behaviour of this CDW as a function of the reduced electron phonon coupling $\mathbf{k}$, is similar to those of the incommensurate ground-state of the Frenkel-Kontorowa model as a function of its coupling constant k and at the same incommensurability ratio ζ. Both models exhibit a "transition by breaking of analyticity" (TBA) with the same critical exponents (within our numerical accuracy). (However, for the FK model, the existence of this transition has been proven analytically while it is only observed numerically for the Holstein model). In both models, when the coupling constant increases above a certain critical value, the hull function which describes the incommensurate modulation of ground-state is not smooth anymore but exhibits infinitely many discontinuities and takes its values in Cantor set. For the 1d Holstein model, this 2π periodic hull function g(x), yields the ground state modulation by

$$\mathbf{u}_i = g(2\pi\zeta\, i + \alpha) \qquad (25)$$

where α is an arbitrary phase. There exists $\mathbf{k}_c(\zeta)$ such that for $\mathbf{k} < \mathbf{k}_c(\zeta)$, g(x) is a smooth function (see fig.5). This ground-state is in fact the expected Peierls-Frohlich CDW with gapless phason modes. For $\mathbf{k} > \mathbf{k}_c(\zeta)$, g(x) becomes a discontinuous function with infinitely many discontinuities. The CDW then becomes pinned to the lattice and the phason modes exhibit a non vanishing gap. This TBA is fact a second order transition between a Peierls-Frohlich CDW and a bipolaronic CDW.

According to a "decomposition theorem"[11] (proven only for the FK model), we apply the same technique to the non-analytic CDW for finding the effective bipolarons which form this CDW. The lattice distortion $\{u_i\}$ which is proportional to the electronic density, is searched with the (non trivial) form

$$u_i = -\sum_n \sigma_n \, b_{n+i} \tag{26a}$$

where σ_n is a pseudo spin which is 1 if a bipolaron is present at site n and 0 if not. $\{b_n\}$ is a *summable* vector which describes the electron density associated with a single effective bipolaron. For the ground state $\sigma_n = \chi(2\pi n\zeta + \alpha)$ (26b) has a 2π-periodic hull function $\chi(x)$ which is defined as $\chi(x) = 1$ for $0 \le x < 2\pi\zeta$ and $\chi(x) = 0$ for $2\pi\zeta \le x < 2\pi$. In addition, form (26a) remains an "asymptotically exact" approximation for describing the low energy excitations of the ground-state which correspond to sequences $\{\sigma_n\}$ which are "weakly disordered". The numerical calculations yield a well defined $\{b_n\}$ which physically is the "effective bipolaron" distortion. This bipolaron is different from the single bipolarons calculated figs.1. because its shape takes into account its interactions with the other bipolarons.

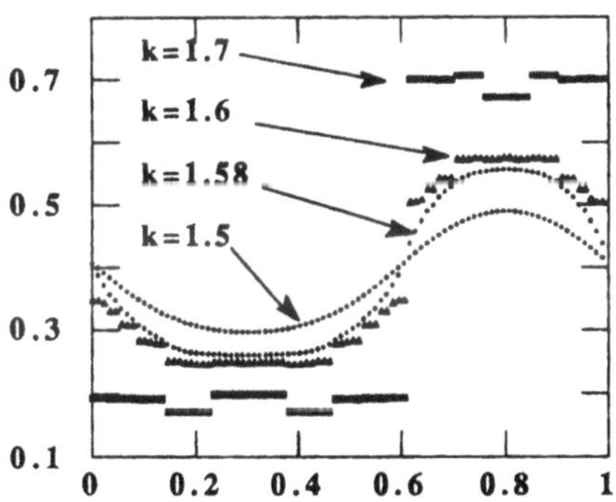

Figure 5: Hull function g(x) for the adiabatic Holstein model for several values of k and $\zeta = 34/89$

Fig.6a shows that the size of effective bipolaron diverges at $\mathbf{k}=\mathbf{k}_c(\zeta)\cong 1.58$ for $\zeta = (3-\sqrt{5})/2$ (with a critical exponent calculated elsewhere). Another physically very important feature is that the size of the bipolaron depends not only on **k** but on the band filling. Fig.6b, shows the effective shape of bipolarons at constant **k** but for different band filling.

For $\mathbf{k} < \max \mathbf{k}_c(\zeta) \cong 1.58$, we observe that the TBA occurs also as a function of the band filling (physically the doping). At small band filling, the ground-state of the CDW is an incommensurate array of pinned small bipolarons.

By doping this size increases and diverges in one or several intervals in the neighbourhood of "good irrationals" ζ such as $\zeta = (3-\sqrt{5})/2$ and other noble numbers and the system then becomes conducting. (In fact, these intervals are not "full" intervals but are "fat Cantor sets" which have finite measure. Nevertheless, these intervals can be considered physically as full because in the small intervals of this Cantor set, the size of the bipolarons although not strictly infinite, remains extremely large). In the vicinity of low order rational band filling such as 1/2 or 1/3, the size of the bipolaron again becomes finite.

Thus when **k**<1.58, the adiabatic Holstein model exhibits a transition from an insulator to a metal by doping from the 0 band filling (or from other simple rational such as 1/2). It can return to an insulator for larger doping when approaching other simple rational band filling. Similar phenomena are also observed in high Tc superconductors. In the FK model, this feature is well understood on the basis of exact theories (KAM) describing the structure of the invariant tori in the standard map. For **k**>1.58, the ground state of the Holstein model always remains a BCDW, although the size of the bipolaron depends on the band filling or doping. The numerical analysis of the 1d adiabatic Holstein model gives a qualitative phenomenology which we found to be completely identical to those of the FK model.

When introducing a magnetic field in this 1 d model, it is rather easy to observe a FISDW but then the phenomenology appears to be in general rather different of those of the FK and not yet understood. For arbitrary band filling (except in the half filled band case), we easily find mixed polaronic bipolaronic states but the ground state is difficult to select and could have an unusual ordering (weak periodicity?). In the half filled band case, we can find the ground-state which corresponds to a bipolaron and polaron ordering which can be either commensurate or incommensurate. In addition, we recover the phenomenology of the FK model, with for example a devil's staircase magnetization response to the magnetic field which can be either incomplete with TBA (at low **k**) or complete (at large **k**).

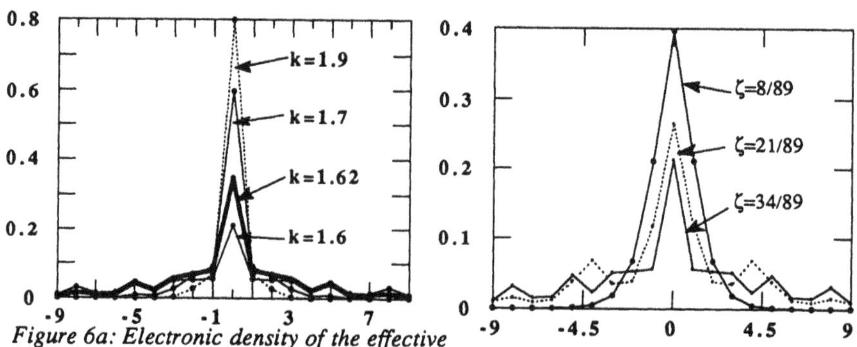

Figure 6a: Electronic density of the effective bipolaron in the 1d adiabatic Holstein model for several values of k and ζ=34/89

Figure 6b: Same as figure 6a but for k=1.6 and several values of ζ

The phenomenology of the 2d adiabatic Holstein model, looks to be rather different from those of the 1d model. Up to now, although we can again easily find bipolaronic and polaronic structures in the 2d model, we could not find the ground-state with certitude and did not observe yet a TBA but only first order transition between bipolaronic structures and unmodulated metallic states. Are the transitions between bipolaronic structures and metallic states always first order in models with more than 1d? Does the size of the bipolarons also depends on their density? These are open questions on which numerical analysis can give useful informations.

5. Quantum Lattice Fluctuations Close to the Adiabatic Limit.

5.1. The Mean Field Correction

The theorems above described, have been established for the adiabatic Holstein model(β=0). Thus, since although small β is never strictly zero, it is physically essential to understand the perturbation of the bipolaronic structures by the quantum lattice fluctuations. We pointed out above that the adiabatic approximation is equivalent to the mean field approximation on the electron phonon coupling. Given a bipolaronic of mixed polaronic bipolaronic structure $\{u_i\}$ and the corresponding set of occupied electronic wave functions $\Psi_i^\nu = |\nu\uparrow(\{u_i\})\rangle$ with their spins, the corresponding mean field wave function of Hamiltonian (7) (with U =0)

$$|\{u_i\}\rangle = \prod_{\nu\ \mathrm{occ}\ ,\nu'\ \mathrm{occ}} c^+_{\nu,\uparrow}\, c^+_{\nu',\downarrow} \exp \sum_n \frac{2g}{\hbar\omega_0} u_n\,(a^+_n - a_n)\ |\emptyset\rangle \qquad (25)$$

is obtained from the vacuum by creating the electrons in their occupied states and applying at each site i, the distortion operator of the lattice by u_i. The overlap between the wave functions of two bipolaronic or polaronic states $|\{u_i\}\rangle$ and $|\{v_i\}\rangle$ with the same number of electrons $\uparrow$ and $\downarrow$ are

$$\langle\{u_i\}|\{v_i\}\rangle = \det(A^\uparrow).\det(A^\downarrow).\exp(-2\,(\frac{g}{\hbar\omega_0})^2 \sum_i (u_i-v_i)^2)) \qquad (26a)$$

det $(A^\uparrow)$ is the determinant of the matrix of Hermitian products $A^\uparrow_{\nu,\nu'} = \sum_i \Psi_i^{\nu *}\,\Psi_i^{\nu'}$ $=\langle\nu\uparrow(\{u_i\})|\ \nu'\uparrow(\{v_i\})\rangle$ between the occupied electronic states with spin $\uparrow$ of the wave function $\langle\{u_i\}|$ and $|\{v_i\}\rangle$. Its modulus is smaller than 1. $A^\downarrow_{\nu,\nu'}$ is defined similarly for the occupied electronic states with spin $\downarrow$. The projection of Hamiltonian (7) in the space of bipolaronic and polaronic states defined by (25) yields:

$$\langle\{ui\}|H|\{v_i\}\rangle = \frac{1}{2}\Big(N\beta + \Phi(\{u_i\}) + \Phi(\{v_i\}) - \frac{1}{2}\sum_i (u_i-v_i)^2\Big)\langle\{u_i\}|\{v_i\}\rangle \qquad (26b)$$

$\Phi(\{u_i\})$ is the adiabatic energy (11c) of the bipolaronic (or polaronic) structures $\{u_i\}$. These overlaps are in fact underestimated because the softening of the renormalized phonons softening is not taken into account in the approximate wave function (25). However, we do not expect that this correction is qualitatively important for small t or large **k.**

The adiabatic approximation is acceptable when the overlaps (26) between two distinct arbitrary states is small. For t small or **k** large, u_i and v_i are close to -1, -1/2 or 0 which implies that the sum $\sum_i (u_i-v_i)^2$ in (26a) is at least larger than 1/4 for polaronic structures and 1 for bipolaronic structures whatever are $\{u_i\}$and $\{v_i\}$. Thus, we have $|\langle\{u_i\}|\{v_i\}\rangle| < \exp(-4\beta)$ (or $\exp(-16\beta)$ for bipolaronic structures). Since $\beta = (\hbar\omega_0)^2/(8g^2)$ is supposed to be much smaller than 1, we find that close to the anti-integrable limit the adiabatic approximation becomes very accurate. This result appears to be in qualitative agreement with the effective band width calculation (17).

The overlap (26) is the product of an electronic term and of a phonon term. At the anti-integrable limit, the main contribution to this overlap is given by the phonon term. Physically, it is an approximation for describing the tunnelling of the bipolarons or polarons through the Peierls Nabarro barrier. Consequently, the bipolaronic structures predicted by the above theorem exists with the two conditions $g \gg \sqrt{T \hbar \omega_0}$ and $g \gg \hbar\omega_0$. Since generally the electronic band width is much larger than the phonon energy, we have $T \gg \hbar \omega_0$. The physical conclusion is that the bipolaronic and polaronic structures predicted by our theorem, physically exist with negligible quantum lattice fluctuations effects for $g \gg \sqrt{T \hbar \omega_0}$. The validity of the adiabatic approximation is questionable when the phonon overlap starts to be important. This situation occurs at least in 1d models when the size of a bipolaron diverges. Then, for configurations $\{\mathbf{u}_i\}$ and $\{\mathbf{v}_i\}$ which differs only by the location of a polaron or of a bipolaron by one lattice spacing $\sum_i(\mathbf{u}_i-\mathbf{v}_i)^2$ goes to zero so that the main contribution to the overlap become the electronic part. This one turns out to be also close to 1 because the electronic wave functions tend to extend. Consequently the overlap $<\{\mathbf{u}_i\}|\{\mathbf{v}_i\}>$ becomes practically 1, when the bipolarons (or polarons) extend. This situation is associated with phonon softening and the vanishing of the PNB. The adiabatic approximation then becomes unreliable.

5.2. The Wagner Unitary Transformation

The analysis presented now can be done for any adiabatic solution for which the electrons are in their ground state which means $\sigma_\nu = 1$ for $E_\nu < E_F$ and $\sigma_\nu = 0$ else. The question is now: Is the BO approximation acceptable when there is phonon softening? There exists a transformation analogous to the Lang Firsov transformation. Instead of considering the unitary transformation $\exp(iS_{LF})$ which diagonalizes the Hamiltonian $H_{AI}+\beta^2 H_Q$, we consider the unitary transformation $\exp(iS_{BO})$ which diagonalizes $H_{AI}+t\, H_K$ (with U=0) $\mathbf{H}_{BO} = \exp(iS_{BO})\, \mathbf{H} \exp(-iS_{BO})$. This transformation can be carried out explicitly[2]. We get

$$\mathbf{H}_{BO}=\frac{1}{2}\sum_{\nu,\sigma} E_\nu(\{\mathbf{u}_i\})\, c^+_{\nu,\sigma} c_{\nu,\sigma} + \frac{1}{2}\sum_n \mathbf{u}_n^2 + \frac{\beta^2}{2}\sum_n \tilde{\mathbf{p}}_n^2 \tag{27a}$$

where (11b) yields $$c^+_{\nu,\sigma} = \sum_i \Psi^\nu_i\, c^+_{i,\sigma} \tag{27b}$$

This Hamiltonian has the form of the Born Oppenheimer Hamiltonian (13a) except that the momentum $\mathbf{p}_n$ has to be replaced by $\tilde{\mathbf{p}}_n = \mathbf{p}_n + \mathbf{f}_n$ where

$$\mathbf{f}_n = i \sum_{\nu\neq\nu',\sigma} \frac{\Psi^{\nu'}_n \Psi^{\nu *}_n}{E_{\nu'} - E_\nu}\, c^+_{\nu,\sigma} c_{\nu',\sigma} \tag{28}$$

These extra terms $\mathbf{f}_n$ (28) in the BO Hamiltonian involve electron transfers between different states. The BO approximation will be consistent if these terms are indeed negligible. We check this property by tracing out the electronic degrees of freedom in (27a) and expanding the potential $\Phi(\{\mathbf{u}_i\})$ at second order around an adiabatic solution $\underline{\mathbf{u}}_i$ as in (13b) Then , the BO Hamiltonian becomes

$$\hat{H}_{BO} = \Phi(\{\mathbf{u}_i\}) + \frac{1}{2}\varepsilon_i M_{ij}\varepsilon_j + \frac{\beta^2}{2}\sum_i (\mathbf{p}_i + \mathbf{f}_i)^2 \tag{29a}$$

$\bar{\bar{M}}$ given by (13c) is diagonalized with the eigen modes α_i^μ and the eigen value r_μ^2

$$\hat{H}_{BO} = \Phi(\{\mathbf{u}_i\}) + \frac{1}{2}\sum_\mu r_\mu^2 \varepsilon_\mu^* \varepsilon_\mu + \frac{\beta^2}{2}\sum_\mu (\mathbf{p}_\mu + \mathbf{f}_\mu)^2 \tag{29b}$$

with $\quad \mathbf{p}_\mu = \sum_i \alpha_i^\mu \mathbf{p}_i \quad$ and $\quad \mathbf{f}_\mu = \sum_i \alpha_i^\mu \mathbf{f}_i \qquad$ (29c)

In the standard BO approximation, the "anti-adiabatic" term $\mathbf{f}_\mu$ is neglected. Considering $\mathbf{f}_\mu$ as a perturbation, we find $<\mathbf{f}_\mu>=0$ and

$$<\mathbf{f}_\mu^* \mathbf{f}_\mu> = \sum_{n,\nu,\nu',\sigma} \sigma_\nu(1-\sigma_{\nu'})\left(\frac{\alpha_n^\mu \Psi_n^{\nu'} \Psi_n^{\nu *}}{E_{\nu'} - E_\nu}\right)^2 \tag{31a}$$

Since we have the exact bounds $2t\,d+1 > E_{\nu'} - E_\nu > 2\Delta > 0$ for ν occupied and ν unoccupied, where 2Δ is the electronic gap in energy units (7a), we find

$$2\Delta <\mathbf{f}_\mu^* \mathbf{f}_\mu> < \sum_{i,j} \alpha_i^\mu (\delta_{ij} - M_{ij})\alpha_j^\mu = 1 - r_\mu^2 < (2t\,d+1) <\mathbf{f}_\mu^* \mathbf{f}_\mu> \tag{31b}$$

The validity of the BO approximation requires $<\mathbf{f}_\mu^* \mathbf{f}_\mu> \ll <\mathbf{p}_\mu^* \mathbf{p}_\mu> = r_\mu/(2\beta)$. If there is a phonon softening in the considered structure $\{\mathbf{u}_i\}$, r_μ can be chosen close enough to zero in order that this condition be not fulfilled. The maximum phonon softening allowed for accepting the consistency of the BO approximation is obtained when $<\mathbf{f}_\mu^* \mathbf{f}_\mu> \ll <\mathbf{p}_\mu^* \mathbf{p}_\mu>$ which gives the condition

$$\frac{r_\mu}{1 - r_\mu^2} \gg \frac{\beta}{\Delta} \tag{32}$$

(choosing the left member of (31b) for the estimation of $<\mathbf{f}_\mu^* \mathbf{f}_\mu>$). For real CDW, β is generally very small and Δ is a non negligible fraction of the band width so that the right member of (32) can range around 10^{-2}. Then, the BO approximation can be still valid with r_μ very small and thus the phason modes almost gapless. (We should analyse with more details the consequence of the strong coupling of the soft oscillator μ (which is physically a phase mode) with the electronic states). For the bipolaronic structures at large $\mathbf{k}$, the phonon softening is small and the gap 2Δ is large. Thus, (32) is fulfilled which confirms that the BO approximation is valid without ambiguity. When there is a strong phonon softening for the bipolaronic structure as in the 1d case, the BO approximation should not be valid when approaching very close to $\mathbf{k}_c(\zeta)$ from above, and below $\mathbf{k}_c(\zeta)$ which suggests that the Peierls Frohlich CDW are not well described. It should not be valid also for describing sliding solitons. We conjecture that the bipolaronic structures disappear in that limit because there exists paths with almost constant energy along which the bipolarons can move almost freely without significant PNEB. A superconducting state could then appear.

6.Effects of an electron-electron repulsion. A new conjecture for the interpretation of high Tc superconductors.

In the physical situation where β is small, increasing the electron phonon coupling necessarily leads to an insulating bipolaronic structure. Although the pairing energy of the electrons in the bipolaron increases, the effective band width of this bipolaron drops sharply to zero. Thus, the interaction potential terms between these bipolarons becomes essential and order the structure in real space as a BCDW. This situation can be changed by the presence of a Hubbard term.

6.1.Stability of the Bipolaronic and Polaronic structures of the adiabatic Holstein Hubbard model close to the anti-integrable limit

We consider only the physical situation $U > 0$ of repulsive electrons. With a Hubbard term, the electronic state is not described by a Slater determinant and it is not possible to determine the electronic excited states by population factors. For the metastable states, we have to require that the electron are in their ground state relatively to the lattice distortion potential.Then, it is intuitive to believe that when U becomes too large, the bipolarons become unstable. This can be easily proven at the anti-integrable limit (t=0, β=0) where the bipolarons become unstable for U >1/4. It is sufficient to consider an empty site 0 and a doubly occupied site 1 with two electrons with opposite sites. We choose a variational normalized wave function for the electrons which depends continuously on parameter a.

$$|\Psi> = \left(\sqrt{1-2a^2}\, c_{1\uparrow}^{+} c_{1\downarrow}^{+} + a\, c_{0\uparrow}^{+} c_{1\downarrow}^{+} + a\, c_{1\uparrow}^{+} c_{0\downarrow}^{+}\right) |\emptyset> \qquad (33a)$$

For a=0, it is the 2-electrons wave function of a bipolaron on site 1, and for $a=1/\sqrt{2}$ it is the resonant 2-electrons wave function of two polarons with opposite spins at sites 0 and 1.Then the energy of H_{AI} given by (8a) is

$$<\Psi|H_{AI}|\Psi> = U\,(1-2a^2)^2 + 1/2\,(\mathbf{u}_0^2+2a^2\,\mathbf{u}_0) + 1/2\,(\mathbf{u}_1^2+2(1-a^2)\mathbf{u}_1) \qquad (33b)$$

which after minimization over $\mathbf{u}_0$ and $\mathbf{u}_1$ yields the energy

$$\phi(a) = -1/2.[\, a^4 + (1-a^2)^2\,] + U\,(1-2a^2)^2 = (4U-1)\,(a^4-a^2) + U - 1/2 \qquad (33c)$$

When U <1/4, the minimum of (33c) is obtained for a=0 which shows that the solution with two polarons on different sites is unstable. Otherwise when U >1/4, the bipolaron is unstable. For this value U =1/4, the energy of a bipolaron is just equal to the energy of two polarons. (The same instability for a pair of polarons also exists for the mixed polaronic bipolaronic structures if the electronic spins are allowed to relax. As already noted, the stability can be restored by application of a uniform magnetic field. Consequently, if our theorem obtained for U=0 could be extended). We conjecture:

For U <1/4, there should exists t(U) such that for t<t(U) the bipolaronic configurations only are preserved by the electron kinetic energy perturbation.

For U >1/4, the situation becomes less intuitive. The polaronic structures only could be perturbed but at t=0, the polarons have degenerate free spins. For t≠0, the spin degeneracy should be raised and yields some unknown magnetic structure superimposed to the polaronic structure. This structure has to be likely a spin glass structure is the underlying polaronic structure has been chosen random.

However, the energy of this structure should remain invariant by the global rotation of the spins. According to the Goldstone theorem, there must exist gapless spin excitations (magnons). In 1d systems, we know that long range order for the spins is impossible due to these excitations although in 3d systems long range order becomes possible. It is not clear wether these magnetic fluctuations preserve the stability of the underlying polaronic structure or not or in other words can be decoupled of the polaronic structure or not. Thus, the extension of our theorem for the existence of polaronic structures is still questionable and might depend for example on the dimensionality of the model and on the ordering of the polaronic structure. However, it is known that there exists in many physical systems local crystal fields which breaks this rotational symmetry, then global orientation of the spins can be fixed to a finite number of equivalent discrete directions. (For obtaining this effect, we should introduce extra electronic terms in the Holstein Hubbard Hamiltonian for breaking its rotational spin symmetry. We will not discuss here neither their form nor their origin). If we assume for example, that the spins are forced to be in the two directions +z and -z only by a strong crystal field, then the magnetic excitations get a non zero gap and we expect that our theorems could be extended again for predicting the stability of polaronic structures under perturbation by the electronic kinetic energy. This is what is implicitly assumed when doing the standard mean field approximation

$$H^{mf}_{AI} \equiv \sum_i \left[\frac{1}{2}(u_i^2 + \mathbf{k}\, n_i u_i) + U\, (n_{i,\uparrow} <n_{i,\downarrow}> + <n_{i,\uparrow}> n_{i,\downarrow} - <n_{i,\uparrow}><n_{i,\downarrow}>\right] \quad (34)$$

for Hamiltonian (18a). Then, we get a variational form for which we expect that our theorem can be extended without major problems.

6.2. Numerical observation of a single bipolaron in the 1 d Holstein-Hubbard model

Thus it appears rather interesting to perform a numerical study of the bipolaron of the adiabatic Holstein Hubbard model where the Hubbard term is taken into account exactly. For a single pair of electrons with opposite spins, the global wave function has the form

$$|\Psi> = \sum_{i,j} \Psi_{i,j}\ c^+_{i\uparrow} c^+_{j\downarrow}\ |\varnothing> \quad (35a)$$

with $$-\frac{1}{2}\Delta\ \Psi_{i,j} + \frac{\mathbf{k}}{2}\ (u_i + u_j)\, \Psi_{i,j} + U\ \delta_{i,j}\, \Psi_{i,j} = E_{electr.}\ (\{\mathbf{u}_i\})\Psi_{i,j} \quad (35b)$$

The numerical problem has a form similar to those of the standard bipolaron problem except that the dimensionality of the eigen problem has been doubled. It is instructive to compare our results with those of mean field calculation (34) in order to evidence an interesting effect which is the binding of two polarons by spin resonance.

For the mean-field calculation of a bipolaron, we have $u_i = -\mathbf{k}/2.(<n_{i,\uparrow}> + <n_{i,\downarrow}>)$ and for symmetry reasons $<n_{i,\uparrow}> = <n_{i,\downarrow}> = -u_i/\mathbf{k}$, which yields

$$H^{mf}_{AI} = \sum_i \frac{1}{2}\left[(1-2U/\mathbf{k}^2)\, u_i^2 + (\mathbf{k} - 2U/\mathbf{k})\, n_i u_i\right] \quad (36a)$$

By a change of variable $v_i = \sqrt{(1-2U/\mathbf{k}^2)}\, u_i$, the mean field bipolaron energy $E^{mf}_{\mathbf{bip}}$ is found to be the same as the bipolaron with U=0 at

$$E^{mf}_{\mathbf{bip}}(\mathbf{k},U) = E_{\mathbf{bip}}(k',0) \qquad \text{with} \qquad \mathbf{k}' = \sqrt{\mathbf{k}^2 - 2U} \tag{36b}$$

This mean field calculation yields the stability domain of this bipolaron when the energy of a bipolaron is larger than the energy of two polarons

$$E_{\mathbf{bip}}(\mathbf{k}',0) = E^{mf}_{\mathbf{bip}}(\mathbf{k},U) > 2E_{\mathbf{pol}}(\mathbf{k}) = E_{Bip}(\frac{\mathbf{k}}{\sqrt{2}}) \tag{37a}$$

which yields $\mathbf{k}' > \mathbf{k}/\sqrt{2}$ or $U < \mathbf{k}^2/4$ (37b)

(let us recall that these calculations do not depend on the dimensionality of the model). When (37b) is not fulfilled, the bipolaron should break up into two polarons. The numerical calculations shows that it does not break up in two polarons far apart although the shape of the bipolaron becomes usual, it does not spread as much as a bipolaron at parameter $\mathbf{k}'$. In fact, instead of a bipolaron we have a bounded pair of polarons. We can interpret this binding by the following scheme. Considering two polarons with opposite spins, there is a spin degeneracy ↑ ↓ and ↓ ↑, the system has to gain energy by forming the symmetric resonant state $1/\sqrt{2}$ (| ↑ ↓ > + | ↓ ↑ >. Such a pair of bounded polarons is called "spin resonant bipolaron". Fig.7 suggests the physical mechanism of this binding.

This binding is in fact reminiscent of those proposed by P.Anderson for binding electrons in the positive U Hubbard model which was called Resonant Valence Bond (RVB). However, in the pure Hubbard model, this binding which undoubtly exists for two electrons on a single pair of sites, has not been proven to be effective for the infinite system. By adding also a relatively strong electron phonon coupling which is not treated as an effective interaction between the electrons, such a binding becomes efficient between polarons. This effect can be considered as maximum when $U \cong \mathbf{k}^2/4$. However, it is negligible for t small or very large $\mathbf{k}$ (and then the mean filed calculation tends to be exact) but becomes very significant for $\mathbf{k}$ moderately large. By varying U, the bipolaron continuously evolves from the standard bipolaron (U=0), to the spin resonant bipolaron $U \cong \mathbf{k}^2/4$. For larger values of U, the binding still exists but becomes much weaker. The calculated bipolarons in the 1d adiabatic Holstein Hubbard model were always found with a centre of symmetry either at a lattice site or at half distance between two lattice sites. Thus for calculating the Peierls Nabarro energy barrier of this bipolaron, we calculated this one with two symmetry conditions. Figure 8a shows the variation of the energy of this spin resonant bipolaron for $\mathbf{k}$=2, with a centre of symmetry at site 0 (open circles) (o-bipolarons) and with

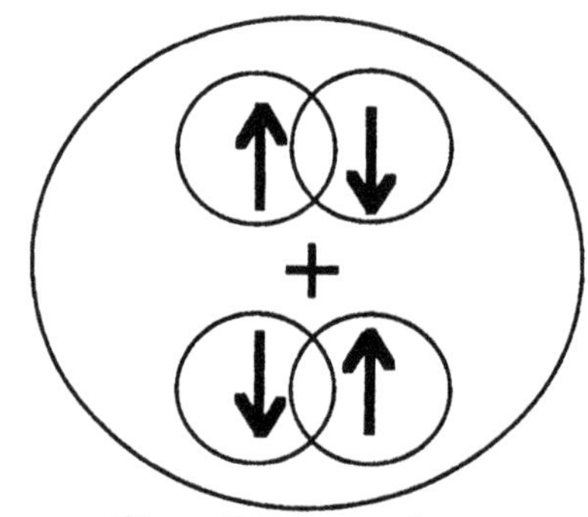

Figure 7: Scheme of a "spin resonant bipolaron"

a centre of symmetry at half distance between sites 0 and 1 (crosses) (a-bipolarons). For U=0, the ground-state of the bipolaron is the o-bipolaron. The difference between the energies of the a-bipolaron and the o bipolaron yields the Peierls Nabarro Energy Barrier (PNEB). The mean field energy of this bipolaron has been also calculated. This approximation starts to be inaccurate for U≅.5 then showing the effect of the spin resonance. At U=1, where the bipolaron should break up, it is still bounded and there is still a significant binding energy between the two polarons. But the most important fact is that the Peierls Nabarro energy barrier almost vanishes as soon as the spin resonance effect becomes effective (see fig.8b). Then the energies of the o-bipolaron and the a-bipolaron becomes almost the same. Since in that regime the energy of a bipolaron is close to the energy of 2 polarons, the bipolaron which is now better interpreted as a bounded pair of polarons is rather soft and it requires very little energy to move through the lattice.

Fig.9a shows the electronic densities of a bipolaron for several values of U. It tends to spread when increasing U. For very large values of U, the bipolaron appears as two polarons far apart (fig. not shown). Figure 9b, shows a spin resonant o-bipolaron with a centre of symmetry at the origin and the corresponding a-bipolaron with a centre of symmetry at 1/2.

As we suggested by the discussion of section 5, the vanishing of the Peierls Nabarro barrier (which also implies a gapless sliding mode) is associated with a sharp enhancement of the effect of the quantum lattice fluctuations. Although, we have not calculated yet this band width of the spin resonant bipolaron, but taking into account the discussion of the effect of the quantum lattice fluctuations of section 5, it is reasonable to expect that this band width should take significant values.

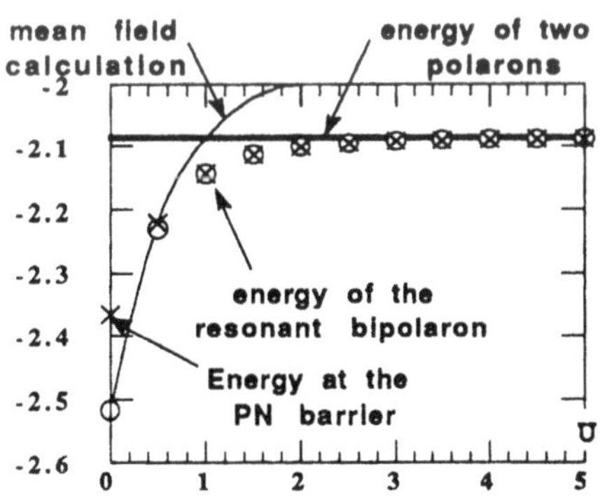

Figure 8a: Energy versus U of a spin resonant Bipolaron for the Holstein Hubbard Model at k=2

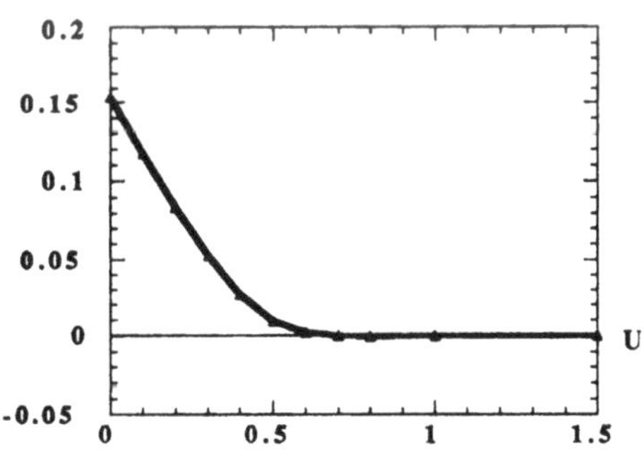

Figure 8b: Peierls-Nabarro energy barrier versus U for the bipolaron of fig.8a

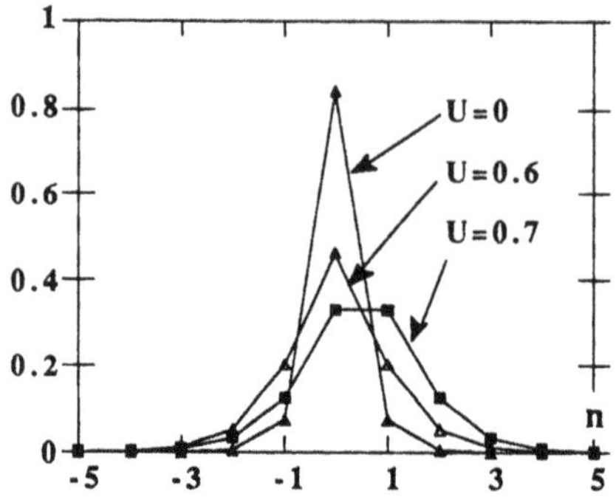

Figure 9a: Electronic density of the bipolarons at k=2 and several values of the Hubbard term U

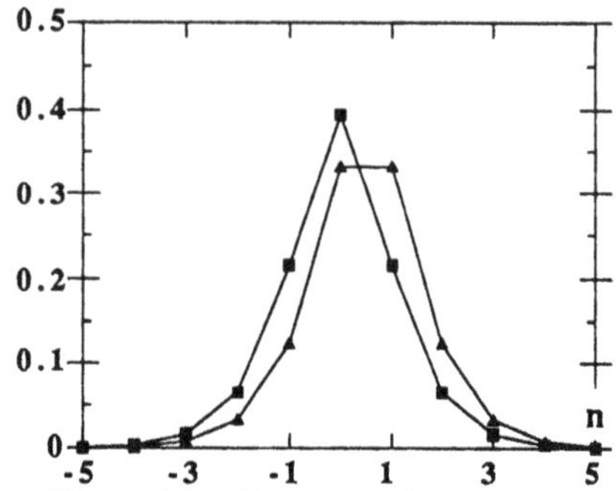

Figure 9b: Electronic densities of two bipolarons with almost identical energy at k=2 and U=0.7

6.3. A scenario for real high Tc superconductors

We propose to explain the microscopic origin of high Tc superconductors by a superfluid state of spin resonant bipolarons. Superconductivity requires the formation of electron pairs. It seems reasonable to think that the electron phonon coupling only (directly or indirectly) can create an attractive force between the electrons since the direct Coulomb force between the electrons is strictly repulsive. We can roughly distinguish three energies for these bounded pairs: 2Δ is the binding energy of the pairs, B is the band width energy of the pairs and V is the potential energy of interaction between the pairs.

For having a superconductor, it is necessary that the kinetic energy of the pairs B be more important than their potential energy V in order to prevent the ordering of these pairs in real space and the formation of a bipolaronic CDW. If the binding energy of these pairs 2Δ is smaller than B, we have a "displacive case" (by analogy with the structural phase transitions), that is the pairs disappear just above the critical superconducting temperature Tc. There is almost no critical fluctuations and this transition is well described in a mean field theory such as the standard BCS theory. The size of the Cooper pairs (coherence length) is very large compared to the lattice spacing.

When this binding energy 2Δ becomes much larger than B, the electronic pairs still exist above Tc. This "order-disorder case" is usually called bipolaronic superconductivity and has a coherence length comparable to the lattice spacing. This kind of superconductivity has been predicted to exist but never has been really observed. There exists a continuum of situations between the (displacive) BCS superconductivity and the (order-disorder) bipolaronic superconductivity. The reason is that when the pairing energy increases too much, generally the potential V also increases while the band width B collapses to zero. The result is that the system ends up into an insulating bipolaronic structure before that this superconductivity becomes effective.

However, in the case of spin resonant bipolarons, the situation becomes more favourable to bipolaronic superconductivity. Because of this Hubbard term, the pairing energy 2Δ of the electrons although it is reduced, keeps a significant value but the bipolaronic band width is drastically enhanced. A superconducting state thus is possible if the band energy B becomes larger than the potential energy V.

However, this superconductivity has unusual property. Since the electron pairs are in fact polaron pairs, the quantum spin fluctuations should remain present in the superconducting state (at sufficiently large energy). The typical energy of these fluctuations corresponds to the energy for exciting the spin state $1/\sqrt{2}\,(|\uparrow\downarrow\rangle + |\downarrow\uparrow\rangle$ into the anti-binding state $1/\sqrt{2}\,(|\uparrow\downarrow\rangle - |\downarrow\uparrow\rangle$ that is approximately half the energy for breaking the spin resonant bipolaron into two polarons far apart.

If this superconducting phase exists, it should be very sensitive not only to the model parameters but also the doping. We have pointed out in section 4 that the size and the shape of the bipolarons depend on their concentration and that commensurate structures of low order favour the formation of polaronic and bipolaronic structures.

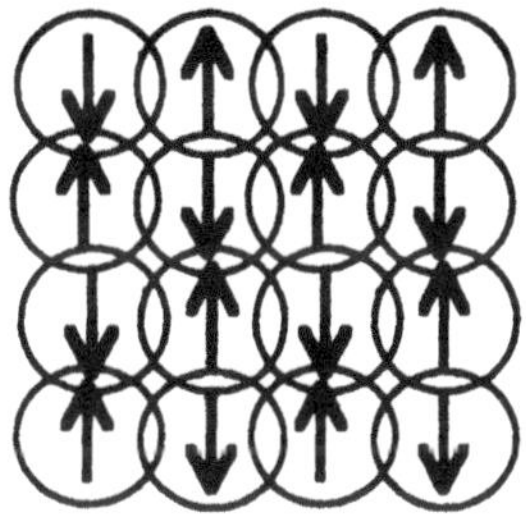

Figure 10a: Scheme of a polaronic AF or SDW ordering favored at and close to 1/2 filled band

Figure 10b: Scheme for the formation of delocalized resonant bipolarons and Superconductivity by doping

Therefore, we conjecture that in the cuprate superconductors the antiferro magnetic structure at half band filling is a polaronic (commensurate) SDW. First by doping the system (by holes), this antiferro magnetic structure becomes a polaronic incommensurate SDW while the polarons extend but still remains too small to gain enough energy for forming spin resonant bipolarons. At some critical doping, this extension becomes sufficient so that the superconducting state with spin resonant bipolarons becomes more stable. This transition could be of first order because as suggested by the polaron calculation in 2d. (fig.1a). However, *this transition does not imply any charge separation.* If the doping can be increased more, we may arrive at a point where the spin resonant bipolaron returns to the state of a normal bipolaron (that is the PNEB is not zero anymore which raise up its effective mass, and localize these particles). Then, the system should become a bipolaronic CDW but could also undergo structural instabilities.

In the superconducting cuprates, it is known now that 1-the band of electrons which becomes superconducting hybridizes the copper and the oxygen wave functions. 2-the oxygene bounds O_2 and O_2^- are highly polarizable which means a strong coupling between the electrons and the oxygen lattice 3- the electron electron interaction on the copper is also strong and known to be at the origin of the anti-ferromagnetism for the undoped samples. Thus, we think that the Holstein Hubbard model catches some the essential characters which a more realistic model for the cuprates should also have, that is the competition between an electron phonon coupling and a Hubbard term both strong.

Although not giving precise predictions, our scenario for the Holstein Hubbard model looks possible. It is roughly qualitatively in agreement with many experimental features and for example with one of the most puzzling which is the coexistence of magnetic features and superconductivity in these materials.

Summary: We have discussed in that paper the effect of the electron-phonon coupling and of the competition with the electron-electron interaction in the adiabatic limit. We also discussed the role of the quantum lattice fluctuations

When the electron phonon coupling is large enough, the existence of bipolaronic CDWs as ground state of the system is definitely proven for the adiabatic model and the role of the quantum fluctuations is proven to be negligible in that limit. We show by numerical simulations in 1d that the size of the bipolarons depends on its density and that it becomes smaller for commensurate configurations of low order.

When a strong Hubbard term is also present, we suggest (without a full mathematical proof) that at large enough electron phonon coupling, the ground state could be a polaronic SDW or AF.

We have also shown that the role of the quantum lattice fluctuations becomes essential if there is a phonon softening or equivalently a depressing of the Peierls Nabarro Energy Barrier.

We have predicted and found numerically in a 1d model that when the energy of a bipolaron is approximately the energy of two polarons, a new state appears which is a pair of polarons bounded by a spin resonance and thus called spin resonant bipolaron. For this new kind of bipolaron, the Peierls Nabarro barrier is sharply depressed, which should produce a sharp enhancement of the role of quantum lattice fluctuations.

Finally , on the basis of the discussion concerning the quantum lattice fluctuations, we suggest that high Tc cuprate superconductors, could be interpreted as spin resonant bipolaron superconductors.

Although we have already obtained much results with certainty, many conjectures remain to be confirmed and more quantitative calculations to be developed.

Appendix: What is anti-integrability?

The existence of chaotic trajectories in dynamical systems may look rather mysterious for those who are not expert in this field. The proof for the existence of chaotic trajectories is most cases rather technical and based on the construction of a "horseshoe". In fact, it is very easy to prove and to predict in some condition, the existence of chaos at least for dynamical systems with a discrete time using the concept of anti-integrability. Instead of looking to the system in a limit where it looks understandable and predictable, that is the integrable limits, and to try perturbations of these limits, t, we look to it at another limit where it is perfectly chaotic. Let us explain what these limits are:

A dynamical system on a manifold $\mathcal{M}$ is a map of this manifold onto itself $X_{n+1} = F(X_n) \in \mathcal{M}$ which generate recursively a trajectory $\{X_n\}$ from any arbitrary initial point $X_0 \in \mathcal{M}$. $\mathcal{M}$ could be for example the usual Euclidian space but could be also any space even at infinite dimension providing it has a "good topology".

Let us consider a one parameter family of C_1 (that is differentiable with continuous derivatives) of dynamical system $F: \mathcal{M} \to \mathcal{M}$ on the arbitrary smooth manifold $\mathcal{M}$ (with a finite or an infinite dimension) with the general form

$$X_{n+1} = F(X_n) = G(X_n) + \lambda\, H(X_n) \qquad \text{(A1)}$$

$G(X)$ and $H(X)$ are smooth C_1 functions and λ is a real parameter. The anti-integrable limit is obtained for $\lambda \to \infty$. Apparently, the dynamical system has no well defined limit in that case. In fact, we can define formally this limit: Since a trajectory $\{X_n\}$ is a set of X_n which fulfils (A1), this equation can be rewritten

$$H(X_n) = \frac{1}{\lambda}(X_{n+1} - G(X_n)) \qquad \text{(A2)}$$

When taking the limit $\lambda \to \infty$, we define the limit dynamical system by dropping the second member in (A2) so that at $\lambda = \infty$, a trajectory is now by definition a sequence $\{X_n\}$ which fulfils

$$H(X_n) = 0 \qquad \text{(A3)}$$

If the equation $H(X)$ has several zeros (or infinitely many zeros) Y_α, the trajectories defined by (A3) are obviously chaotic since at time n, X_n is equal to one of the zeros Y_{α_n} of $H(X)$ chosen at random. This limit $\lambda = \infty$ is called anti-integrable limit. The chaos is "perfect" since the coordinate of the trajectory at time n is totally independent of the coordinate of the trajectory at time n-1. (More precisely, this dynamical system is is a Bernoulli shift). Then it becomes easy to prove under some assumptions which will not detailed here that many of these chaotic trajectories still exist at large but finite λ.

The technique of the proof consists first to define an operator S_λ in a neighbourhood of each trajectory Y_{α_n} at $\lambda = \infty$ which from any arbitrary sequence $\{X_n\}$ in that neighbourhood yields another sequence $\{X'_n\} = S_\lambda(\{X_n\})$ through the implicit equations

$$H(X'_n) = \frac{1}{\lambda}(X_{n+1} - G(X_n)) \qquad \text{(A3)}$$

and second to prove that when λ is large enough 1- that it can be indeed defined for a neighbourhood chosen in appropriate manner 2- that it applies this neighbourhood into itself and 3- that it is contracting. Then this operator has a fixed point $Y_n(\lambda)$ which continuously depends on λ. The fixed point of S_λ is a trajectory of the initial dynamical such that $Y_n(\infty) = Y_{\alpha_n}$. Thus if the initial trajectory is chaotic, $\{Y_n(\lambda)\}$ is also a chaotic trajectory. The theorem proving the existence of bipolaronic and polaronic configurations follows in its principle the same idea.

At the anti-integrable limit (no kinetic energy for the electrons) the distribution of these electrons is chosen randomly as above for the anti-integrable trajectories.

For a given population factor $\{\sigma_v\}$, we define the operator $\{u'_i\} = S_t(\{u_i\})$ by

$$u'_i = -\sum_{\nu} \sigma_\nu |\Psi_i^\nu(\{u_i\})|^2$$

where Ψ_i^ν are given by (11b). Thus a fixed point of this operator is a solution of eq.13b, that is an extrema of the variational energy (11c). In a neighbourhood of each anti-integrable configuration of polarons and bipolarons and for t not too large, S_t will be proven to have a unique fixed point. This is the main part of the proof of our theorem section 4.

Acknowledgements. I thank A.Bishop, D.Campbell, G.Doolen and H.Trugman for useful discussions and their interest in this work while I was visiting the Los Alamos Lab. I acknowledge the hospitality of CNLS supported by contract W705/ENG-36. I also thank Robert MacKay and Claude Baesens for their interesting e-mail comments on the resonant spin bipolaronic state.

References:

[1] S.Aubry, G. Abramovici, D.Feinberg, P. Quemerais and J.L. Raimbault in *Non-Linear Coherent Structures in Physics, Mechanics and Biological Systems* (Lect. Notes in Physics, Springer vol **353** p.103-116 (1989)

[2] S.Aubry in *Microscopic Aspects of Non-Linearity in Condensed Matter Physics* ed. A.R.Bishop, V.L.Pokrovsky and V.Tognetti NATO ASI Series, Series B vol **264** Plenum p.105-114 (1991)

[3] S.Aubry, G.Abramovici and J.L. Raimbault *J. of Stat.Phys.* **67** 675-780 (1992)

[4] S.Aubry in *Twist Mappings and their Applications* Proc. of the IMA, Minneapolis, Springer (1992) to appear

[5] C.Kittel *Quantum Theory of Solids* J.Wiley &Sons Inc. New York.(1963)

[6] A.S.Alexandrov, J. Ranninger and S. Robaszkiewicz *Phys.Rev.***B33** p.4526 (1986)

[7] S.Aubry and P. Quemerais in "*Low Dimensional Electronic Properties of Molybdenum Bronzes and Oxides*" p.295-405 Ed.Claire Schlenker (1989) Kluwer Acad. Publ. Group

[8] D.Emin and T.Holstein *Phys.Rev.Letts* **36**, 323 (1976),D.Emin *Physics Today* June 1982 p.34

[9] R.MacKay and C.Baesens *A New Paradigm in Quantum Chaos:...*to appear in *Proceedings of The International Summer School on Quantum Chaos*, Enrico Fermi Institute, Varenna, Italy (22 July-4 August 1991)

[10] P.Y. Le Daeron and S.Aubry *J. Phys.* **C16** 4827-4838 (1983), *J.Physique* (Paris) **C3** 1573-1577 (1983)

[11] S.Aubry, J.P. Gosso, G. Abramovici, J.L. Raimbault, P. Quemerais *Physica* **D47** p.461-497 (1991)

[12] P.Anderson *Science* **235** 1196 (1987) P.Anderson, G.Baskaran, Z.Zou and T.Hsu *Phys.Rev.Letts.* **58** 2790 (1987)

PHASE SEPARATION AS RESULT OF A THERMODYNAMICAL VARIATIONAL METHOD FOR THE EMERY MODEL

K. DICHTEL

and

J. CARSTENSEN

Institut für Theoretische Physik
der Universität Kiel
D 2300 Kiel, Germany

ABSTRACT

The variational principle of Bogoljubov is used to derive approximations for the free energy and for the temperature dependent magnetic phase diagram of hole doped HTSC cuprates. The coexistence regions of the phase separation of carriers occurring up to some critical doping are given.

1. Introduction

One of the concepts of the last years towards an understanding of the unconventional properties of the HTSC cuprates [1] was the idea of nonuniform densities of the carriers within the copper oxygen planes and possible separation of the carriers into phases of different concentration. Experimental evidences as e.g the strong dependence of the HTSC properties on the previous thermal treatment [2], the chemical phase separation in $LaCuO_{4+\delta}$ [3] and the production of metallic regions from photogenerated carriers [4] renewed the interest in the phase separated solutions suggested previously for theoretical models with strong on site electron correlations. As formerly proposed [5] for the Hubbard model, the different kind of dependence of the energy on the number of holes in a ferromagnetic band and on the number of bonds in the antiferromagnetic phase favours spatial separation of carriers into neutral AF and highly concentrated ferromagnetic domains. For the *t–J* model according to [6] similar arguments hold: For small values of the hopping constant *t* compared to the AF exchange energy *J*, the energy to break AF bonds would be too high compared to the gain in kinetic energy of delocalized carriers as to prevent the destruction of neutral AF regions. Despite of the still unresolved question whether these solutions for models with strong on site correlation would also exist in models including the long range Coulomb interactions, a variety of papers in the last years discusses phase separation resulting from different approximations e.g. [7–15]. Whereas most calculations on this topic are done on the t–J model or on the one band Hubbard model, we perform a variational calculation directly on the three band Emery model [16] which is more realistic. Thus we avoid the mapping on simpler models which is restricted to the strong coupling case and in its validity for high carrier concentration still questioned. The variational method has the advantage to give upper bounds of thermodynamical quantities even in temperature and doping regimes where Monte Carlo evaluations of path integrals still run into difficulties and to provide a decision between different types of solutions. Furthermore because of the finite extension of phase separated regions, calculations in the thermodynamic limit and in the grand canonical ensemble seem to be more suitable for the discussion of phase separation than the exact diagonalization studies still restricted to few lattice units.

2. The Model and the Method

We start with the standard three band model [16] for the $Cu\ O_2$ planes with hole concentration $n_h = 1$ for the neutral case $Cu\ 3d^9$

$$\mathbf{H} = U_d \mathbf{H}_{dd} - \epsilon_0 \mathbf{N}_d + \gamma \mathbf{H}_1 + U_{pd} \mathbf{H}_{pd} \tag{1}$$

with

$$\mathbf{H}_{dd} = \frac{1}{2} \sum_{n,\sigma} \mathrm{x}_{n,\sigma}^{1+} \mathrm{x}_{n,\sigma}^{1} \mathrm{x}_{n,-\sigma}^{1+} \mathrm{x}_{n,-\sigma}^{1}, \tag{2}$$

$$\mathbf{N}_d = \sum_{n,\sigma} \mathrm{x}_{n,\sigma}^{1+} \mathrm{x}_{n,\sigma}^{1}, \tag{3}$$

$$\mathbf{H}_1 = \sum_{\substack{\langle n,(m,j)\rangle \\ \sigma}} t_{n,(m,j)} \mathrm{x}_{n,\sigma}^{1+} \mathrm{x}_{m,\sigma}^{j} + \text{h.c.} \quad , \tag{4}$$

$$\mathbf{H}_{pd} = \sum_{\substack{\sigma,\sigma' \\ \langle n,(m,j)\rangle}} \mathrm{x}_{n,\sigma}^{1+} \mathrm{x}_{n,\sigma}^{1} \mathrm{x}_{m,\sigma'}^{j+} \mathrm{x}_{m,\sigma'}^{j} \tag{5}$$

where $\mathrm{x}_{n,\sigma}^{j+}$ creates holes on copper sites n for $j = 1$ and holes on adjacent oxygen sites m for $j = 2, 3$. The constant U_d denotes the on site Coulomb correlation on copper places which we take as energy unit in the following. The absolute value of the transfer integral is γ and ϵ_0 denotes the charge transfer energy. In the first calculations we set the next neighbour Coulomb repulsion constant $U_{pd} = 0$. We use the thermodynamic variational principle of Bogoljubov [17,18] for the grand canonical ensemble

$$\Omega \le \Omega_0(\mathbf{U}) = \Omega_T + \left\langle \mathbf{U}^+ \mathbf{H} \mathbf{U} \right\rangle_T - \left\langle \mathbf{H}_T + \mu \mathbf{N} \right\rangle_T \tag{6}$$

where the test–Hamiltonian $\mathbf{H}_T$ and the canonical transformation $\mathbf{U}$ can be chosen arbitrarily in a manner that analytic calculation of the traces is possible. Ω_T means the grand canonical potental of $\mathbf{H}_T$ and $\left\langle ... \right\rangle_T$ the thermodynamic trace average with $\mathbf{H}_T$.
Choosing a one particle Hamiltonian in k-space as test–Hamiltonian one recovers standard weak coupling Hartree Fock solutions as given e.g. in [19]. In the strong coupling case however a local test–Hamiltonian diagonal in real space like the atomic limit $\mathbf{H}_{dd} - \epsilon_0 \mathbf{N}_d$ is appropriate, improved by a canonical transformation

$$\mathbf{U} = \exp(-i\gamma \mathbf{S}_1) + o(\gamma^2). \tag{7}$$

With

$$\left\langle \mathbf{U}^{+}\mathbf{H}\mathbf{U} \right\rangle_a = \left\langle \mathbf{H} \right\rangle_a + \gamma^2 M_s + o(\gamma^2) \tag{8}$$

$\mathbf{U}$ is optimized to second order of γ taking $\delta M_s = 0$. This procedure in operator algebra is strictly analogous to the one used in [20,21] for the one–band–Hubbard–model. A simple solution for $\epsilon_0 = \frac{1}{2}$ is obtained:

$$\mathbf{S}_1 \; = \; -2\exp(i\pi \mathbf{H}_{dd})[\mathbf{N}_d, \mathbf{H}_1]\exp(-i\pi \mathbf{H}_{dd})),$$

where $[\ldots,\ldots]$ denotes the commutator and yields

$$\begin{aligned} M_s \; &= \; 2 \sum_{\substack{(n,(m,j))\\ \sigma}} \{2\left\langle \mathrm{x}_{n,-\sigma}^{\;1+} \mathrm{x}_{n,-\sigma}^{\;1} (\mathrm{x}_{n,\sigma}^{\;1+} \mathrm{x}_{n,\sigma}^{\;1} - \mathrm{x}_{m,\sigma}^{\;j+} \mathrm{x}_{m,\sigma}^{\;j}) \right\rangle_a \\ &\quad - \left\langle \mathrm{x}_{n,\sigma}^{\;1+} \mathrm{x}_{n,\sigma}^{\;1} - \mathrm{x}_{m,\sigma}^{\;j+} \mathrm{x}_{m,\sigma}^{\;j} \right\rangle_a \}. \end{aligned} \tag{9}$$

One advantage of analytical methods is the possibility to compare the numerical contributions from the different occuring terms. Thus the computations show, that for small doping the main contribution in the expectation values of the quantity M_s comes from the last expression, the simple difference of occupation numbers on copper and oxygen sites (not present in single orbital Hubbard models). The strong coupling method allows for a larger occupancy of copper sites and thus produces a lower energy than the Hartree Fock method which hybridizes the orbitals. If the carrier concentration increases, the correlation terms in M_s instead of the density terms dominate.
The following calculations are done with the test–Hamiltonian

$$\begin{aligned} \mathbf{H}_T := \; & \sum_{c_1,c_3} u_{c_1 c_3} \frac{1}{2} \sum_{\sigma,n} \mathrm{x}_{n,\sigma}^{\;c_1+} \mathrm{x}_{n,\sigma}^{\;c_1} \mathrm{x}_{n,-\sigma}^{\;c_3+} \mathrm{x}_{n,-\sigma}^{\;c_3} \\ + \; & \sum_{c_1,c_3} (1 - \delta_{c_1,c_3}) v_{c_1 c_3} \frac{1}{2} \sum_{\sigma,n} \mathrm{x}_{n,\sigma}^{\;c_1+} \mathrm{x}_{n,\sigma}^{\;c_1} \mathrm{x}_{n,\sigma}^{\;c_3+} \mathrm{x}_{n,\sigma}^{\;c_3} \\ + \; & \sum_{n,c_1} h_n^{c_1} \frac{1}{2} (\mathrm{x}_{n,\uparrow}^{\;c_1+} \mathrm{x}_{n,\uparrow}^{\;c_1} - \mathrm{x}_{n,\downarrow}^{\;c_1+} \mathrm{x}_{n,\downarrow}^{\;c_1}) \\ - \; & \tilde{\epsilon}_0 \mathbf{N}_d - \lambda \mathbf{N}, \end{aligned} \tag{10}$$

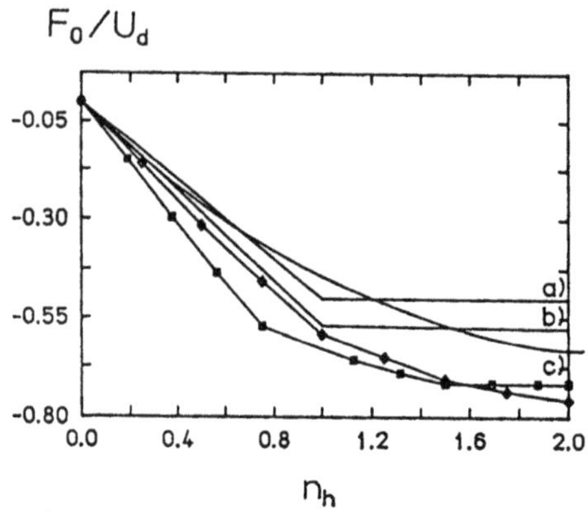

Fig. 1: Free energy versus hole concentration n_h for $\gamma/U_d = 0.1$ and $\epsilon_0/U_d = 0.5$: a) Atomic limit; b) strong coupling solution; c) Hartree Fock. Curves with squares and diamonds: lower bounds according to [23].

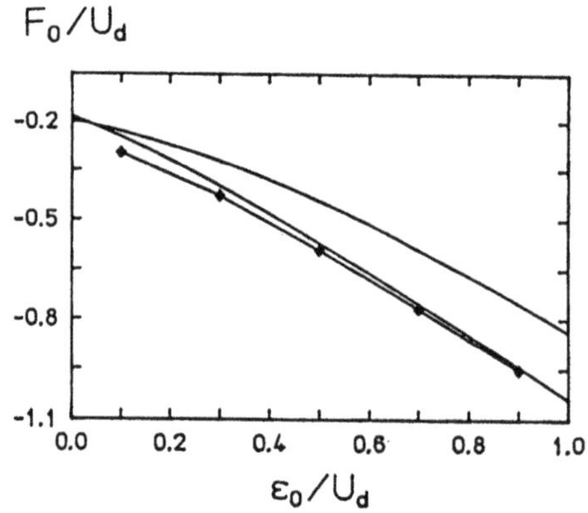

Fig. 2: Free energy for $n_h = 1$ and $\gamma/U_d = 0.1$ versus charge transfer constant ϵ_0/U_d. Upper curve: Hartree Fock; lower curve: strong coupling solution; curve marked with diamonds: lower bound according to [23].

containing as variational parameters the effective correlations $u_{c_1c_3}$ and $v_{c_1c_3}$, the effective charge transfer energy $\tilde{\epsilon}_0$, the antiferro– (ferro–) magnetic effective fields $h_n^{c_1}$ and the variational chemical potential λ. The latter one is necessary to ensure thermodynamic consistency [18].

In order to adapt our transformation to finite values of γ and to values $\epsilon_0 \neq \frac{1}{2}$ we also introduced a variational parameter φ replacing γ in Eq. 7. The straightforward but lengthy procedure of performing the transformation, calculating the expectation values of the grand canonical potential, minimalizing with respect to all parameters, solving the variational equations and passing to the free energy as a function of the hole concentration n_h, are to be published elsewhere [22].

3. Numerical Results for the Free Energy

Fig. 1 shows the curves of the free energy F_0 for $\gamma = 1$ versus hole concentration n_h for a) the atomic limit, b) the strong coupling variational approximation as described above and c) the standard paramagnetic Hartree Fock approximation. The curves a) and b) have a very similar shape: the descending part for $n_h < 1$ corresponds to the filling of the copper places (note we chose $\epsilon_p = 0$) with maximum one hole per site, whereas the nearly constant parts for $n_h > 1$ correspond to the filling of the oxygen sites. For low hole concentrations $n_h \leq n_h^*$ the strong coupling solution yields a lower free energy than the HF solution which is in turn the better one for large hole concentrations. Thus the critical concentration n_h^* marks an abrupt nearly tempera-

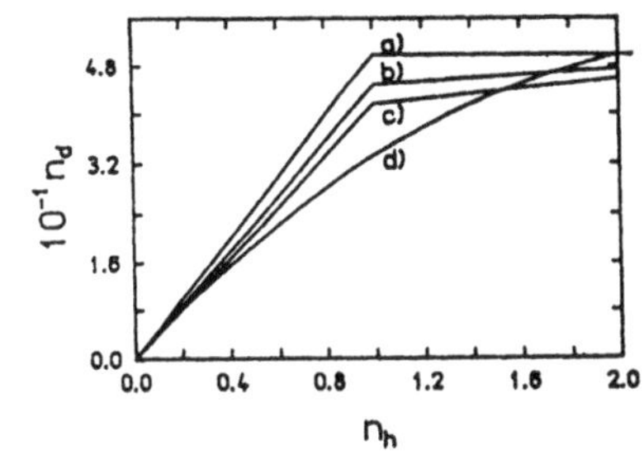

Fig. 3: Occupancy n_d of copper sites versus total hole concentration n_d: a) atomic limit; b) for $\gamma/U_d = 0.05$; c) for $\gamma/U_d = 0.1$; d) Hartree Fock.

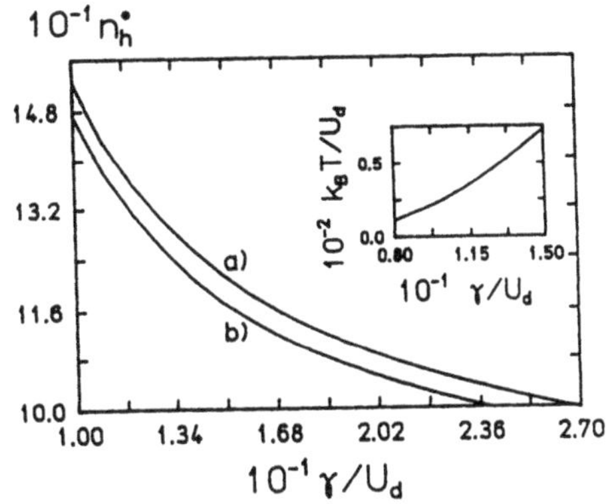

Fig. 4: Critical concentration n_h^* as a function of γ/U_d: a) for $\epsilon_0/U_d = 0.5$; b) for $\epsilon_0/U_d = 0.4$. Only in the region below this curve phase separation occurs. Inset: Maximum Phase Separation Temperature as function of γ/U_d.

ture independent crossover from highly correlated to Fermi liquid behaviour possibly related to the change in experimental properties, when passing from underdoped to overdoped materials.

In addition to our upper bounds some numerically lower bounds resulting from an exact diagonalization of small lattice clusters are displayed [23]. The method previously was applied to the Hubbard model [24–26]. As seen in Fig. 1 and 2, our result is close to the lower bounds in the neutral case whereas especially the region of critical doping shows the largest discrepancy and requires more effort in the future.

Expectation values of physical quantities are in this method to be calculated from the operators transformed with **U** [20]. Thus e.g. the total copper occupancies per spin n_d are given as

$$n_d := \frac{1}{2N}\left\langle \mathbf{U}^{+}\mathbf{N}_d\mathbf{U}\right\rangle_T . \tag{11}$$

The oxygen occupancies n_p can be determined from $n_p + n_d = n_h/2$.

Fig. 3 shows the copper occupancy as a function of the hole doping n_h for different values of γ. The discontinuities in slope at $n_h = 1$ reflects the fact that upon doping the holes change from their preferance of the copper to those of the oxygen sites. The smooth curve of the HF approximation on the contrary indicates the hybridization of the copper and oxygen orbitals in mean field theory.

The parts of the curves with $n_h > 1$ are calculated with respect to the phase separated situation described below.

Fig. 4 shows the critical doping as a function of γ. As expected, there exists some maximum value of γ where n_h^* passes $n_h = 1.0$ and HF is the better solution.

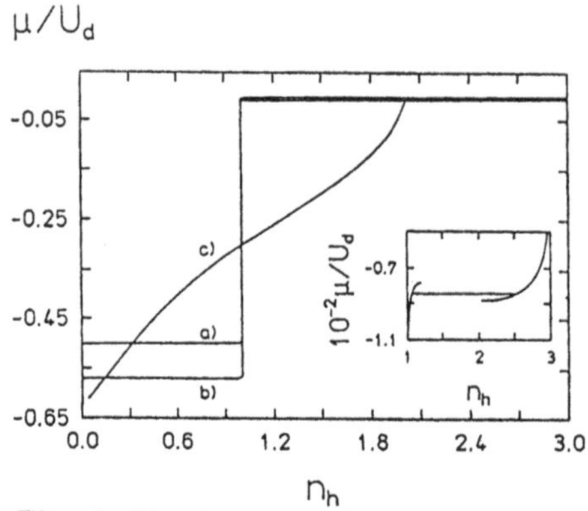

Fig. 5: Chemical potential μ/U_d versus n_h for $\epsilon_0/U_d = 0.5$ and $\gamma/U_d = 0.1$ shows the "charge transfer" discontinuity at $n_h = 1.0$. Curves: a) Atomic limit; b) strong coupling variational method; c) Hartree-Fock solution as weak coupling limit. Inset: Enlarged part of the constant line for $n_h > 1$ and Maxwell construction.

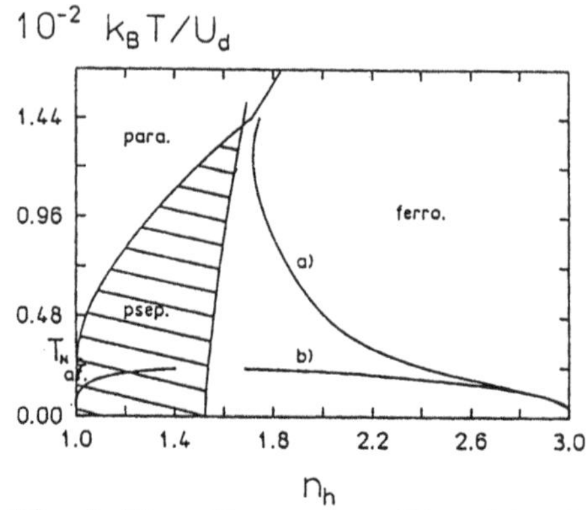

Fig. 6: Phase diagram for $\epsilon_0/U_d = 0.5$ and $\gamma/U_d = 0.1$: a) of separated states between antiferro– (para-for high temperatures) and ferromagnetic states. The hatched coexistence region is again reduced by the "Hartree–Fock limiting curve"; b) for comparison phase separation line as in Fig. 7 between paramagnetic states.

Fig. 5 shows the chemical potential μ versus doping n_h with the "charge transfer gap" as discontinuity at $n_h = 1.0$ in the strong coupling solution compared to the HF curve. The latter goes smoothly up to the value zero at $n_h = 2.0$ where the nonbonding oxygen band in this weak coupling picture is reached.
The most important feature is seen from the inset: an enlarged portion of the nearly constant value for $n_h > 1$ reveals a negative slope which indicates a region of thermodynamic instability and phase separation. The usual Maxwell construction in the free energy gives the value of the chemical potential and the concentration values of the separated phases.

4. Phase Diagrams

At low temperatures the variational equations have nonvanishing magnetic molecular field solutions h_n^c. As Fig. 6 shows, pure antiferromagnetism prevails on the axis $n_h = 1$ up to the Néel temperature T_N and disappears already at concentrations $n_h = 1.01$. The Néel temperature in the neutral case is $T_N/U_d \approx 29°K/eV$ for $\epsilon_0/U_d = 0.5$ and $\gamma/U_d = 0.1$. This is in accordance with the experimental phase diagram of $La_{2-x}Sr_xCuO_4$ for $U_d = 10eV$. Ferromagnetism is for large n_h the phase with lowest energy in this case where we only allow for these collinear magnetic symmetries.

In the temperature versus doping phase diagramm as shown for the magnetic phases in Fig. 6 and for the paramagnetic case in Fig. 7 phase separation occurs in both cases within the hatched areas of coexistence bounded from the right by the HF limit at the critical concentration n_h^*. For a given temperature there exists a nearly neutral and a hole rich phase, into which separation occurs. Note that phase separation can provide small regions of much higher hole concentration up to $n_h = 3$ than the mean hole concentration of $n_h < n_h^*$ of the material. First calculations on spiral magnetic phases show that phase separation still persist, but if the system can choose spiral symmetries varying continously from AF to ferromagnetic behaviour the coexistence regions are reduced considerably compared to those in Fig. 6. In the magnetic case phase separation also provides a mechanism to continue antiferromagnetism in small regions up to much higher carrier concentrations. The inset in Fig. 4 shows the maximum temperature T_{max} of paramagnetic phase separation as function of γ. Fig. 8 gives the paramagnetic phases with nonvanishing next neighbour repulsion U_{pd}. We see that (starting with the upper curve for $U_{pd} = 0$) even small values of U_{pd} reduce drastically the maximum phase separation temperatures and brings T_{max} down to a temperature range near T_c. The maximun concentration of the hole rich regions decreases from $n_h = 3$ to $n_h = 2$.

5 Approximate Density of States

Despite of the fact that according to the simple test–Hamiltonian the density of states in the strong coupling approximation consists only of delta peaks, some characteristic features in location and spectral weight result from the adjustment of the variational parameters during the minimalization procedure. Figs. 9a and 9b show the resulting DOS in the neutral and highly doped case respectively. In the large delta peaks (artificially broadened) of the upper and lower Hubbard band of holes on copper sites and of the band of oxygen states situated in between small delta peaks indicate the part of the corresponding other type of carriers. The chemical potential λ (marked as a vertical line) lies in the charge transfer gap for the neutral case and within the oxygen states for the highly doped case. Consequently a spectroscopic measurement probing the DOS of such a phase separated system should see a combination of both pictures put together at the same scale with a relative weight according to the relative part of the neutral or hole rich phase respectively. Thus phase separation would manifest itself by the appearence of oxygen states from the highly doped regions directly at the chemical potential or Fermi level of the whole system. Even in this oversimplified picture another important feature can be seen: a splitting of the oxygen states in Fig. 9b. The oxygen part with spin direction antiparallel to the spin direction of the lower Hubbard band (at the left side in this hole picture) splits upon doping from the total of oxygen states to lower energies and thus reaches the chemical potential. This splitting is proportional to the parameter $u_{1,2}$ in the test–Hamiltonian Eq. 10

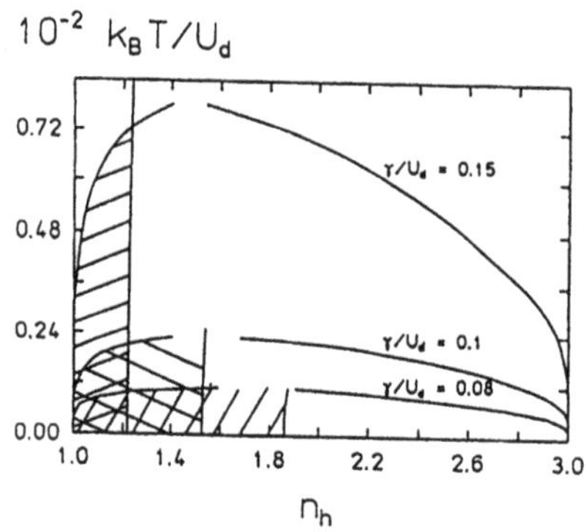

Fig. 7: Phase diagram of the carrier-separated paramagnetic states for $\epsilon_0/U_d = 0.5$ and different values of γ/U_d. The hatched coexistence region in the phase diagram is bounded to the right by the "Hartree-Fock limiting curve" at n_h^*.

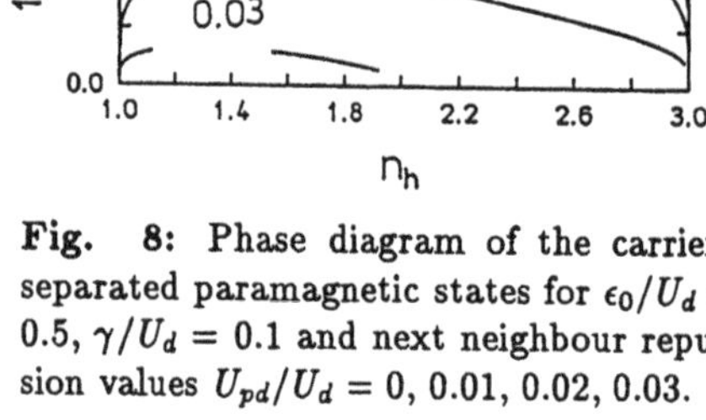

Fig. 8: Phase diagram of the carrier-separated paramagnetic states for $\epsilon_0/U_d = 0.5$, $\gamma/U_d = 0.1$ and next neighbour repulsion values $U_{pd}/U_d = 0, 0.01, 0.02, 0.03$.

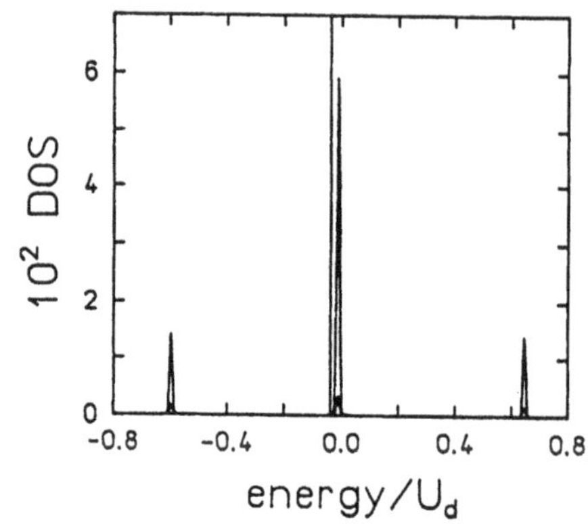

Fig. 9a: Density of States of H_T for $\gamma/U_d = 0.1$, $\epsilon_0/U_d = 0.5$ for $n_h = 1$.

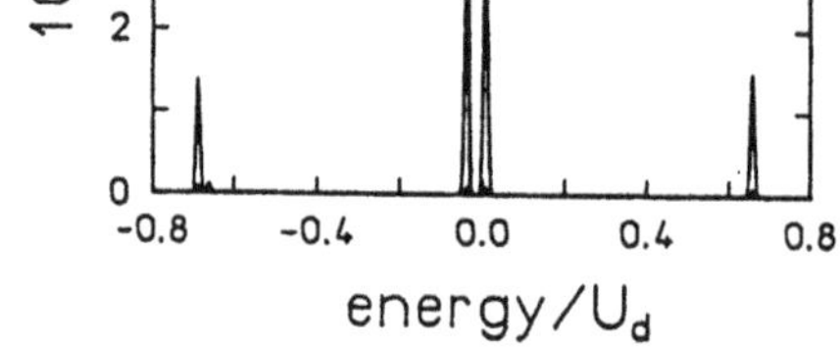

Fig. 9b: Density of states for the same parameter values and chemical potential as in Fig. 9a, but for the highly doped region of phase separation $n_h = 2.92$.

and (see the expansions in the next chapter) proportional to γ^2. Thus it corresponds to the appearance of the Zhang Rice singlet band [27] or to the appearance of the quasiparticle subband below the oxygen band in Fig. 4 of [28].

6. Expansion for small Ratio Hopping Constant to Correlation Value

An analytical insight into the type of solutions can be obtained also by an expansion of all calculated quantities with respect to powers of the parameter φ for small values of the ratio γ/U_d. The occuring combinations of φ are abbreviated

$$\begin{aligned} \Phi_2 &= \varphi^2 + 2\gamma\varphi \approx -(\frac{\gamma}{\epsilon_0})^2(\epsilon_0 - \frac{1}{4}) \\ \Phi_4 &= \varphi^4 + 4\gamma\varphi^3 \approx -(\frac{\gamma}{\epsilon_0})^4(\frac{\epsilon_0}{2} - \frac{1}{16}). \end{aligned} \tag{12}$$

where the second parts of Eqs. 12 follow from the first approximation of the optimized value

$$\varphi = -\frac{\gamma}{2\epsilon_0}. \tag{13}$$

This means that small values of the charge transfer energy ϵ_0 produce the same tendencies in the model than large values of the transfer integral γ. In most cases we used a special value $\epsilon_0 = U_d/2$ which gives

$$\Phi_2 = -\gamma^2 \quad \text{and} \quad \Phi_4 = -3\gamma^4. \tag{14}$$

Approximate values of the effective correlations in the test–Hamiltonian Eq. 10 up to order γ^2 are

$$u_{1,1} = 1 - 32\,\Phi_2 + o(\varphi^2) \quad \text{and} \quad u_{1,2} = 4\,\Phi_2 + o(\varphi^2). \tag{15}$$

The result for $u_{1,1}$ means an enhancement of the on-site correlation on copper sites. The possibility for the holes to move to oxygen places for $\gamma > 0$ reduces the tendency for double occupancies of copper places and indeed increases the effective correlation on copper sites. The negative value for $u_{1,2}$ means an effective attraction between holes on copper and oxygen places, the way in which hybridization can take place in this local picture.

Now we proceed with an expansion of the variational equations. Regarding the magnetic variational equation for the field h^1 we see [22] that an expansion for small order parameter near the second order antiferromagnetic phase transition in the neutral case for Eq. 14 gives a transition temperature

$$k_B T_N = -16\Phi_4 + o(\varphi^4) \approx 48\gamma^4. \tag{16}$$

In a mean field approximation [29] this corresponds to an effective exchange constant $J_{eff} = 24\gamma^4$ of order γ^4 as expected in superexchange.

The expectation values with the test–Hamiltonian appearing in the following equations are denoted

$$\mathrm{n}^c := \frac{1}{2}(\left\langle \mathrm{x}_\sigma^{c+}\mathrm{x}_\sigma^c \right\rangle_T + \left\langle \mathrm{x}_{-\sigma}^{c+}\mathrm{x}_{-\sigma}^c \right\rangle_T). \qquad (c = 1,2,3) \tag{17}$$

For $n_h \geq 1$ we use the approximations

$$n^1 = \frac{1}{2} \quad \text{and} \quad n^2 = n^3 = \frac{n_h - 1}{4}. \tag{18}$$

The renormalized charge transfer constant $\tilde{\epsilon}_0$ appearing as slope for $n_h < 1$ of Fig. 1 writes

$$\tilde{\epsilon}_0 - \epsilon_0 \;=\; -4\Phi_2(n^2 + n^3 - n^1) + \frac{6\gamma^2}{\epsilon_0} + o(\varphi^2) \quad , \tag{19}$$

which shows that the value of variational charge transfer is enhanced for all doping values in our parameter range and independent of doping for $\epsilon_0 = U_d/4$ (charge transfer instability). Note that in our variational picture of dipersionless bands quantities like $\tilde{\epsilon}_0$ only mark differences between the center of bands and cannot be compaired with real band gaps.

The variational equation which determines the variational chemical potential λ and produces formally the phase separation is in an expansion up to γ^4

$$\begin{aligned} \lambda - \mu \;=\; & \{-4\,\Phi_2 + \frac{272}{3}\,\Phi_4\}\,n^1 - 40\,\Phi_4(n^2 + n^3) \\ & + 8\gamma(\varphi - \frac{40}{3}\varphi^3) + 8\epsilon_0(\varphi^2 - \frac{20}{3}\varphi^4) + o(\varphi^4). \end{aligned} \tag{20}$$

The DOS of the test–Hamiltonian consists of spikes as shown in the Figs. 9a and 9b. When passing a spike for $n_h > 1$ large variations in n^2 and n^3 for only small variations in λ arise. With a prefactor $-\Phi_4 > 0$ in our standard region of ϵ_0 the consequence of this increase in the r.h. side of Eq. 20 is a decrease in μ of the l.h. side for compensation and hence phase separation is a phenomenon of order γ^4 in our approximation. Note that it occurs because of the narrow bands in the highly correlated limit quite independent of the variational equations for the molecular h_n^c which determine the magnetic symmetry. Thus we expect the phenomenon to persist even in calculations including more magnetic phases of different symmetry. From these equations follow, that the lower the temperature, the larger γ and in our parameter range the lower ϵ_0, the more is the tendency of carriers to separate.

Finally we compare these results for the special case $\epsilon_0 = 1/2$ (in our unit U_d) to the standard high correlation expansions in the Hamiltonians mapping Hubbard like models on models of the t-J type [27,28,30]. Following [27] the second order terms are given by the effective hopping parameter $t = 2\gamma^2$. According to [28] the parameter choice above means no O-O hopping and the second order term in γ with a coefficent

proportional $J_K = 4\gamma^2$ represents the "Kondo" scattering of the oxygen holes on the Cu spins. The parameter J_K appears in our effective attraction $u_{1,2}$ in Eq. 15. Relating our expansions of thermodynamic quantities to these expansions in Hamiltonians, one recognizes, that due to the local ansatz necessary for the explicit variational calculations certain expectation values of the Hamiltonians vanish. Thus instead of the complete spin–carrier interaction terms including spin flip in the Kondo scattering Hamiltonian [28], our corresponding part of Cu–O interaction in Eq. 9 contains only a site diagonal part of the latter one.
The fourth order term of all expansions essentially means a Heisenberg model with an superexchange constant $J = 24\gamma^4$ (see [28]) in agreement with our exchange constant J_{eff} extracted from Eq. 16.
The curve critical concentration n_h^* versus γ in Fig. 4 gives furthermore within our approximation a rough estimate of the range where the strong coupling solutions and their expansions to t–J like models are valid.

7. Conclusions

In a strong coupling variational treatment of the Emery model phase separation into nearly neutral and hole rich domains arises within a large coexistence region of the phase diagram up to some maximum temperature and some critical doping. This coexistence region is enlarged when regarding phases of magnetic order and drastically reduced introducing a next neighbour Coulomb interaction. The maximum temperature of phase separation increases with increasing γ and decreasing ϵ_0. The rudimentary DOS shows that for the hole rich phase an oxygen band near the chemical potential splits of from the total oxygen states and is occupied by holes. The critical concentration marks a nearly temperature independent crossover from strong coupling to HF Fermi liquid behaviour and as a function of the transfer integral a limiting range of the mapping on t–J like models.

Acknowledgements

The authors are indebted to Prof. Stolze and cand. phys. Wittman in Dortmund for their first results on the lower bounds shown in Fig. 1 and 2 . They also profited from discussions with Prof. R.J. Jelitto, Frankfurt.

References

[1] J.G. Bednorz and K.A. Müller. *Z.Phys.* **B 64** (1986) 189.

[2] R.K. Kremer, E. Sigmund, V. Hizhnyakov, F. Hentsch, A. Simon, K.A. Müller, and M. Mehring. *Z.Phys.* **B 86** (1992) 319.

[3] M.F. Hundley, J.D. Thompson, S.W. Cheong, Z. Fisk, and J.E. Schirber. *Phys. Rev.* **B 41** (1990) 4062.

[4] G. Yu, C.H. Lee, A.J. Heeger, N. Herron, E.M. McCarron, L. Cong, C.G. Spalding, C.A. Nordman, and A.M. Goldman. *Phys.Rev.* **B 45** (1992) 4964.

[5] P.B. Vischer. *Phys. Rev.* **B 10** (1974) 943.

[6] V.J. Emery, S.A. Kivelson, and H.Q. Lin. *Phys. Rev. Lett.* **64** (1990) 475.

[7] N. Cancrini, S. Caprara, C. Castellani, C. Di Castro, M. Grilli, and R. Raimondi. *Europhys. Lett.* **14** (6) (1991) 597.

[8] W.O. Putikka, M.U. Luchini, and T.M. Rice. *Phys. Rev. Lett.* **68** (1992) 538.

[9] E. Arrigoni and G.C. Strinati. *Phy. Rev.* **B 44** (1991) 7455.

[10] M. Grilli, R. Raimondi, C. Castellani, C. DiCastro, and G. Kotliar. *Phys. Rev. Lett.* **67** (1991) 259.

[11] G. Castellani, M. Grilli, and G. Kotliar. *Phys. Rev.* **B 43** (1991) 8000.

[12] A. Auerbach and B.E. Larson. *Phys. Rev.* **B 43** (1991) 7800.

[13] M. Marder, N. Papanicolaou, and G.C. Psaltakis in G. Reiter, P. Horsch and G.C. Psaltakis (eds.). *Dynamics of Magnetic Fluctuations in HTSC.* Plenum Press, New York, 1991.

[14] H. Fehske, V. Waas, H. Röder, and H. Büttner. *Phys. Rev.* **B 44** (1991) 8473.

[15] H. Röder, H. Fehske, V. Waas, and H. Büttner. *Phys.Rev.* **B 45** (1992) 13092.

[16] V.J. Emery. *Phys. Rev. Lett.* **58** (1987) 2794.

[17] H. Koppe in. *Werner Heisenberg und die Physik unserer Zeit, F. Bopp (ed.).* Vieweg, 1961.

[18] A. Huber. Self-consistency and stability in the variational calculation of thermodynamic quantities. In *Boulder Lectures in Theoretical Physics, K.T. Mahanthappa and W.E. Brittin (eds.)*, volume XI D, page 355. Gordon and Breach, 1969.

[19] X.Y. Chen, W.P. Su, C.S. Ting, and D.Y. Xing. *Solid State Comm.* **67** (1988) 349.

[20] M. Heise and R.J. Jelitto. *Z. Phys.* **B 25** (1976) 381.

[21] M. Heise and R.J. Jelitto. *Z. Phys.* **B 26** (1977) 177.

[22] J. Carstensen and K. Dichtel to appear in. *Phys.Rev.* **B** (1992).

[23] J. Stolze and Wittman. private communication (1992).

[24] R. Valenti, J. Stolze, and P.J. Hirschfeld. *Phys. Rev.* **B 43** (1991) 13743.

[25] R. Valenti, P.J. Hirschfeld, and J.C. Angles d'Auriac. *Phys. Rev.* **B 44** (1991) 3995.

[26] R. Valenti, C. Gros, P.J. Hirschfeld, and W. Stephan. *Phys. Rev.* **B 44** (1991) 13203.

[27] F.C. Zhang and T.M. Rice. *Phys. Rev.* **B 37** (1988) 3759.

[28] J. Zaanen and A.M. Oles. *Phys. Rev.* **B 37** (1988) 9423.

[29] W Nolting. *Quantentheorie des Magnetismus, Vol.2.* Teubner, 1986.

[30] V.J. Emery and G. Reiter. *Phys. Rev.* **B 38** (1988) 4547.

GENERAL DISCUSSION

During the workshop ample time had been scheduled to discuss each paper delivered. Advantage of this possibility was indeed taken. However, as the workshop drew to its close, it was felt that the results should be reviewed in perspective of each other and, if necessary, the important points reemphasized. Therefore, a two-hour long general discussion was arranged in the afternoon of the last day. The session opened with the structural aspects, which naturally led over to the electronic properties related to them and their unique behaviour per se. From this we moved on to the theoretical considerations and reviewed those achievements. The entire session was recorded on tape. Each contributor has either performed a direct transcription of his contribution, including all subsequent comments, or revised the text produced by the editors(°). This material, which represents and important complement to the workshop, is hereafter reproduced.

Giorgio Benedek and K. Alex Müller

H. Lütgemeier

Two aspects of phase separations have been considered in this workshop. Here we consider the first, the structural one, most evident in the 123 compounds of the type $YBa_2Cu_3O_x$ where at increasing x from 6 to 7 at least one tetragonal and two orthorhombic structures appear with different superconducting and magnetic properties.

The information on the oxygen order in these compounds obtained by nuclear resonance of Cu is based on principles quite different from those in the diffraction methods. In these methods the coherence of the lattice structure is studied whereas in our case only the relative amount of sites with a special symmetry and charge distribution is depicted in the spectra. Thus the three different lines for twofold coordinated Cu(1) at an oxygen concentration near y = 6.5 prove the existence of the ortho–II structure where only each other chain is filled by oxygen, but the strong intensity of threefold coordinated Cu(1) sites reveals a short average length of about 10 oxygen ions per chain. The systematic developement of the spectra with the oxygen concentration and the correlation with T_c proves that the condition for the onset of superconductivity is the existence of extended domains of the ortho–II structure containing chains of at least 6 oxygen ions. This is not in contradiction with the results of neutron diffraction where much larger coherent domains of ortho–II are found , but the number of defects within the domains cannot be detected. Thus the informations gained by both methods are complementary and help to improve the picture of the separation into a superconducting ortho–II phase and an oxygen–poor, probably tetragonal phase.

(°) Dr. Tagaki's comments could not be revised by the author.

More difficult is the question of the boundary between the ortho–II and the oxygen rich ortho–I phase near x = 7.0. Here the NQR spectra of Cu(2) give evidence that in a wide range of x the randomly diluted ortho–I structure exists and only in a narrow region of the 60 K plateau ortho–I and –II coexist. The pure ortho–II structure is characterized by two sites of different EFG but equal probability which must be ascribed to Cu(2) sites connected via to an apex oxygen tó a full or empty chain, respectively.

P. Burlet

In the cuprate superconductors the important role of the apical oxygen and of its distance to the CuO_2 plane has been recognized long time ago. During this workshop very interesting experimental results, obtained by various techniques (Moessbauer Spectroscopy, NMR, photoemission, X ray and neutron diffraction) on several superconducting systems (La_2CuO_4, YBCO, Bi:2212, etc.) have been presented. In any case it appears that is exists two different Cu sites in the CuO_2 plane which are characterized by short and long Cu-O (apical) distances. This differentiation is static and there is no fluctuation of a two level system.

We have presented the results of a neutron diffraction study of a single crystalof $YBa_2Cu_3O_{6+x}$ with x = 0.45. In the 123 system the insertion of the oxygen (O4) from which results the hole doping occurs in the Cu! planes and leads, for large enough concentration, to orthorombic phases as a consequence of the formation of Cu_1-O_4 chains. In the vicinity of x = 0.5 it has been evidenced, mainly by electron and X ray diffraction, that a superstructure consisting in a doublingof the unit cell along the a a xis occurs. This corresponds to an ordering of full and empty chains. In our neutron experiment we have measured the scattering due to this ordering and all the paramenters of the so called ortho-2 phase have deen derived. The atomic displacements of the apical oxygen as well as those of O_2, O_3, Ba, Cu_2 and Y were obtained. In a xy plane containing a full chain, the apical oxygens move away from Cu_1, which leads to a short Cu_2-O(apical) distance (2.31 A), while antiparallel displacements occur in the planes containing empty chains leading to a long Cu_2-O(apical) distance (2.42 A), as shown in the figure. These values are in excellent agreement with those obtained by other experimental techniques and they appear to be a general feature in the cuprate superconductors.

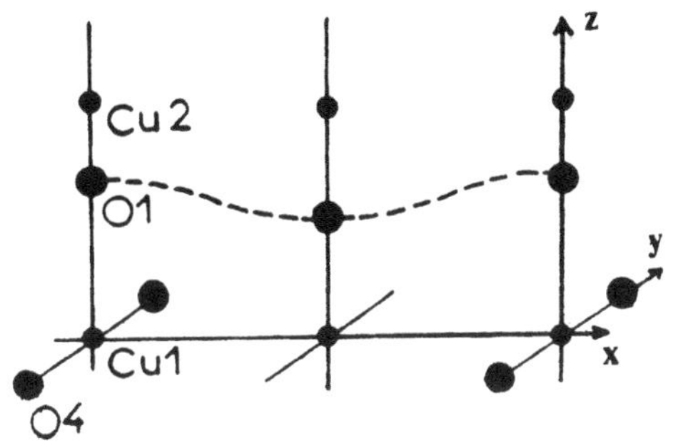

An important fact is that this ordering is only of short range and, moreover, it is very anisotropic. Indeed the chains are rather well formed and extend over more that 300 A, but the correlation length in the y direction is shorter (90 A). In the z direction the correlation length is very short since it covers only three lattice units, reflecting a strongly 2-dimensional character of this ordering. A quite important question is to know whether this short range character is intrinsic and reflects physical aspects, or whether it s extrinsic and results from non-stoichiometry and defects. It is clear that both aspects have to be considered. The thermal treatment by which the oxygen content is obtained is very important. Long treatment in controlled oxygen atmosphere at relatively low temperature and slow cooling are necessary to ensure good oxygen diffusion and sample homogeneity with room-temperature equilibrium properties. Samples with the same oxygen content can alternatively be obtained by quenching from equilibrium temperature in air and then one can obtain a different final state.

More experiments are obviously needed to determine the dependence of this oxygen ordering as a function of the concentration, temperature and thermal treatment in order to check the different theoretical models proposed to interpret the formation of superstructures due to the oxygen ordering.

A. Bianconi

At this meeting finally the disagreement between two classes of structural probes on the Cu local structure was solved for two classes of high Tc superconductors. A long standing controversy has been going on for about five years between X-ray absorption spectroscopy (EXAFS and XANES), pulsed neutron diffraction, experimental methods that probe the local structure, and crystallography investigations (neutron and x-ray diffraction), which probe the long range order, on the presence of two positions for the apical oxygen O(4) in $YBa_2Cu_3O_{7-\delta}$ and second, on the anomalies of the local structure at the normal to superconductor phase transition, which have been found by XANES, EXAFS and pulsed neutron scattering but they have not been found by crystallography.

At this meeting we have presented the evidence that the Cu sites in $Bi_2Sr_2CaCuO_{8+\delta}$ are not homogeneous. We can distinguish two main types of Cu sites:

type a), where the Cu ion is displaced by about 0.2 Å above the distorted square plane of the planar oxygen, and the Cu-O (apical) distance is shorter 2.3 Å and

type b), where the Cu ion is close to the plane of planar oxygens and the Cu-O (apical) distance is longer, about 2.5 A.

It is relevant that the two types of Cu site con-figurations are similar in $YBa_2Cu_3O_{6.5}$ and $Bi_2SrCaCuO_{8+\delta}$. $Bi_2Sr_2CaCu_2O_{8+\delta}$ provides the ideal system to compare the results of experimental investigations probing the local order or the long range order. In fact this system exhibits a well defined 4.75 x 1 superstructure and therefore the structure of different Cu sites can be obtained also by diffraction methods if they solve the superstructure. The EXAFS investigation of the local structure by our group has shown the presence of two Cu-O(apical) distances and of two different Cu sites. These results are in agreement with the diffraction works that solved the superstructure. The EXAFS investigation of the local structure by our group has shown the presence of two Cu-O(apical) distances and of two different Cu sites. These results are in agreement with the diffraction works that solved the superstructures, but cannot be compared with the other works that report the average structure. The superstructure of $YBa_2Cu_3O_{6.5}$ and $Bi_2Sr_2CaCuO_{8+\delta}$ have been presented where the experimental methods probing the long and short order are in agreement.

I think that also the controversy on the Cu-O(apical) distance in $YBa_2Cu_3O_{7-\delta}$ can be interpreted as due to superstructure in small domains or with the dynamic character of the superstructure that cannot be seen by diffraction. The EXAFS experiments probe both the static and dynamic disorder because of the very short measuring time of 10^{-16} sec. Therefore the anomalies at the superconducting phase transitions can be assigned to the different dynamics of the two types of Cu sites. Egami et al. by pulsed neutron diffraction have observed the increasing probability for Cu sites of type b below the temperature. I think that we have to look at the superstructure as having both static and dynamic character.

How common is the presence of a superstructure with linear arrays of Cu sites of types a and b in the Tc superconductors? At this meeting Grenier et all. have reported the presence of a superstructure in La_2CuO_{4+x} $(0.6<x<0.8)$ similar to that of $Bi_2Sr_2CaCuO_{8+\delta}$. On the other side no superstructure has been solved up to now in several high T_c superconductors like in $La_{2-x}Sr_xCuO_4$ and $YBa_2Cu_3O_7$. However it is possible that the size of the ordered domains can be as small as the superconducting coherence length 20 Å, so that they escape the detection by diffraction methods. I would like to stress that the aspect that seems to me more relevant for the physics of high T_c superconductivity is the fact that alternate lines of Cu sites are formed in the CuO_2 plane. The new scenario that appears now for the CuO_2 plane is similar to the shape of a corrugated iron foil. Linear arrays of type a Cu sites and of type b Cu sites are present in the plane. Therefore the plane is no more isotropic and the carriers move more freely in one direction that in the other.

C. Di Castro

I would like to make a general comment. First of all, the involvement of the apex oxygen changes the symmetry of the holes in the plane. This means that there is a change from the simplified picture we had for many years from the Zhang and Rice singlets, in the sense that there is a relevant part of holes with a_1 symmetry. Concerning the number of bands there is consensus that only one band crosses the Fermi energy. However this band is made by holes with different local symmetry. This means that they respond differently to local distortions and to the effective potential which comes out from different techniques and mechanisms. I have a question for experimentalists. As Bianconi said the super-structure where the apex oxygen is involved has a complicated behavior where several crystallographic parameters are modulated involving also tilting of the square pyramids and so on. Now you have to be good with theorists. We cannot introduce in the theory many types of distortions. To get information on the holes on the plane one has to identify the parameter modulated by the superstructure that is relevant for local distortion and symmetry. This may be helpful for the selection of the effective pairing.

A. Bianconi

With the present experimental information, the main aspects of the modulation of the Cu site configurations that I think are relevant for the pairing mechanisms are the following:

1) The Cu sites can be identified by the long or short Cu-O(apical) distance. This modulation induces that the local symmetry a_1 of the holes in the plane is large for type a sites and smaller for type b sites.

2) The different Cu sites form linear arrays. I think that the superstructure can be different in different systems, but the formation of linear arrays seems to be a common aspect of an instability driven by the electron lattice interaction in the planes.

3) The third point is that the period of the modulation λ satisfies the condition $k_F.\lambda = 2\pi$ where k_F is the wavevector of the itinerant holes at the Fermi level in the direction of higher dispersion. Moreover it has been found that the super-conducting coherence length x is close to that modulation period. Therefore the condition $k_F.x = 2\pi$ is satisfied.

I would like to suggest a pairing mechanism that is inferred by these new data. We have the presence of two components in the carrier spectrum in the planes. The first component is formed by charges that moves freely mainly in one direction like the water running on the grooves of a corrugated iron foil. The first type of charge carriers have mostly b_1 symmetry and associated with Cu type b sites move freely in that direction. The second type of charge carriers, with larger components of states with a_1 local symmetry, associated with the Cu sites of type a, that have a strong electron lattice interaction and form probably linear arrays of condensed polarons separated by the modulation period λ. The pairing of the charges in the first set of states is mediated by virtual excitations with a wavevector $2\pi/\lambda$.

H. Lütgemeier

The similarity mentioned before between the modulated structure of the Bi-2212 superconductor with a modulation length of *5a* and the ortho-II structure of YBCO with a modulation period of *2a* is also seen in the NQR spectra of the planar Cu ions in both compounds (Fig. 1). The main difference is that in the case of *5a* a nearly homogeneous distribution of the EFG appears, in the case of *2a* only two discrete values. Different reasons for the modulation of the EFG can be considered, e.g., a modulation of the charge density in the planes or, as suggested by Prof. Eschrig in the next comment, of the occupation of the Cu d_{z^2} orbital due to the modulation of the distance to the apex oxygen ion. In this case the larger width of the spectrum in the case of Bi-2212 should be connected with a larger modulation of this distance, which might be checked in the near future.

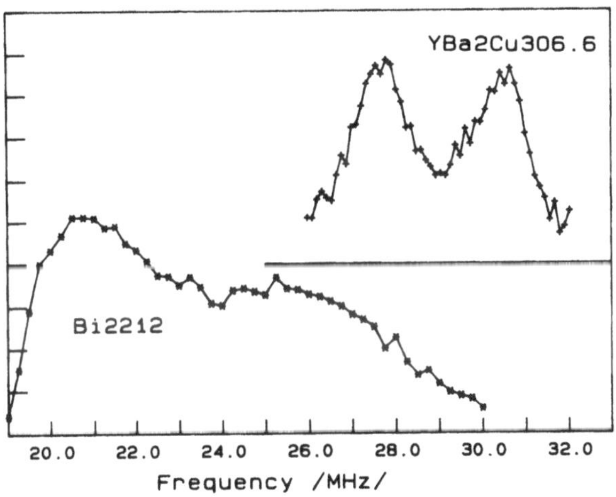

Fig. 1: ^{63}Cu NQR spectra of the Cu(2) sites in $YBa_2Cu_3O_{6.6}$ (top, calculated from the NMR spectrum in [1], Fig. 6) and $Bi_2Sr_2Ca_1Cu_2O_{8+x}$ [2]. This figure shows the influence of the modulation on the ^{63}Cu NQR spectrum of the Cu(2) sites in YBCO (top) and Bi-2212 (bottom) with modulation periods of *2a* and *5a*, respectively. The stronger splitting in the Bi compound indicates a larger amplitude of the EFG modulation in this compound; the wavelength of the modulation has no influence.

1. H. Lütgemeier and I. Heinmaa, these proceedings.
2. R. Michalak et al., Proc. Colloque Ampère Zurich, Switzerland, 1991, p. 114.

A. J. Heeger

A very nice experiment has been done in Santa Barbara on 123-O_7 crystals showing that they are ferroelectric with polarization in the z direction (c-axis); see D. Mihailovich et al in Solid State Commun. 75, 319 (1990). This seems to have been forgotten (or, alternatively, maybe it is not important). The question is as follows: If there is a distortion in the structure - I'm thinking now along the lines of the apex oxygen being displaced from the symmetry position as it would be if there were a double well potential - cannot it lead to a dipole moment along c and to ferroelectricity? It would not be the case if the apex oxygen experienced a symmetric potential, but a displacement in the position of the apex oxygen (which would locally distort the hole concentration) could give a dipole moment.

K.-A. Müller

If you shift the corrugated apex oxygen lines by half a lattice distance below or above the yttrium plane then you may get it.

C. Hucho (Freie Universität, Berlin, Germany)

Ferroelectricity may have been forgotten in these 123 compounds, but there exist other experiments which show a sort of ferroelectric transition in 123 compounds as for example in the group of Volker Müller at FU Berlin, where we see a ferroelectric transition in the region between 106 and 98 K in the 123-O_7 compounds from ultrasonic experiments. I like to mention that there is another very interesting systems which indeed has been forgotten: the solid solution BaPbBi oxide. This can be obtained by doping from the BaBi oxide, which is ferroelectric, to a $BaPb_{0.25}$ oxide, which is low T_c superconductor, and there is a ferroelectric transition which can also be seen via ultrasonic experiments just above this T_c.

H. Takagi

Is there any evidence of ferroelectricity in BaBi oxide? As far as I remember there is a kind of breathing mode distortion.

C. Hucho

We are just doing these experiments now. It's quite a difficult task because they are not just simple ultrasonic experiments. Here you have to detect sound-induced electromagnetic fields in presence of a high magnetic field.

H. Eschrig

I'd like to go back to the splitting of the NQR frequency for the Cu II site shown by Lütgemeier. I asked two days ago why is the splitting so large compared to that for the Cu I site. It seems that the field gradient should change much more because there is a change in the surrounding 1st neighbors, whereas there is a change only in the 2nd neighbors of this Cu site. Now I realize that, if this reflects a modulation in the Cu II - apex oxygen distance, then this modulation should be reflected in the change of the $3z^2$ occupation number, which (we know from calculation by Schwartz), very strongly affects the quadrupole resonance. This same explanation should also apply to the Bi system.

C. Di Castro

Yes, definitely it does!

A. J. Heeger

I quickly summarize the experimental evidence for and against phase separation that we heard during the last few days.

Some evidence for phase separation:

The photodoping experiments that we have done in Santa Barbara provide clear evidence of electronic phase separation. The remarkable set of experiments on magnetic susceptibility by Kremer et al at the MPI Stuttgart certainly show phase separation. There are also the structural data on LaCuO system (Jorgensen et al) and I think there are still other data.

Where are the data presented against phase separation? The only data that I can recall were the extensive NMR experiments on $La_{2-x}Sr_xCuO_4$. In this case, the Sr ions would be least mobile (the electrons in our photodoping experiments would be the most mobile species). Perhaps the entropy at high temperature (where the crystal is grown) prevents phase separation. Then at lower temperature, where phase separation might give a lower free energy, the phase separation would be inhibited by the poor mobility of the heavy ions.

There was also some discussion on the neutron data. It seems that neutron data are typically interpreted as not being due to phase separation. However, if I understand Vic Emery's comment, what is happening is that when the data show the change from a peak at (π,π) to an incommensurate structure, there is not really a shift. Rather the intensity at (π,π) decreases while the other feature grows - as might be interpreted in terms of phase separation. The experimental situation is perhaps still not completely clear. Some comment by the neutron scattering people might be useful.

Two obvious questions:

1. Very little is known about the length scale. Certainly nothing is known on the length scale in the case of our photodoping experiments. Clearly the length scale must be greater than the coherence length for superconductivity. However one doesn't really have information on the length scale.

2) Another fundamental question - assuming that there is phase separation, what is the relationship of phase separation to the high T_c problem? I do not have a strong opinion on this important issue. The existence of phase separation has given us some doubts about some models (dilute holes, eT_c.). That is a kind of negative progress. Vic Emery and others have proposed that maybe there is a pairing mechanism coming from a frustrated phase separation. That is certainly a really interesting question, maybe we should return to that later in our discussion.

C. Hammel

There is really a nice experiment showing that there's no chemical phase separation in LaSr cuprates. I think it is made more interesting by the fact that in copper NQR in LaSr cuprate there are two clearly different Cu sites observed (actually they are observed by two different groups at ISSP and Northwestern). It is a very misterious observation. Something is distinguishing the two copper sites and these relative intensities of the two Cu sites are doping dependent, not T dependent: as function of x one peak grows up ant the other goes down. There is some local effect.

H. Takagi

Actually I am not sure about my memory. There are two peaks and if you measure the relaxation rate, both peaks show superconductivity at the same T. Could you comment on that?

P. C. Hammel

I think this inhomogeneity is very microscopic. I mean on the scale of Sr disorder or something like that. Definitely both peaks are at the same T_c, definitely they are in the same phase. So I don't know whether this should be under phase separation or structure or what, but one way to think about it is two different phases - one for one site, one for the other.

K.-A. Müller

I have a rather experimental question to Alan with regard to his nice experiments. This has to do with this large photon energy you shine on your samples. You have a pulse and then a change in resistivity. You get this continuing increase in conductivity. Can you definitely rule out that this increase has to do with some transfer of energy to the lattice, i.e., heating?

A. J. Heeger

Heating is of course a very serious: we have though a great deal about the heating question. I summarized most of the evidence in the notes which I distributed. The principal points are as follows: To obtain the major change in conductivity which we observe from heating, you would have to raise the sample surface temperature to essentially the melting point. With thousands of "shots" we see no surface degradation. The non-monotonic behavior (with temperature) is inconsistent with heating. And finally - perhaps the most convincing evidence - if you change the doping concentration by a trivial amount, the optical constants are essentially unchanged and yet the photoinduced conductivity data change dramatically. Recall that at $x = 6.3$ you do not see the long-lived metastable states, whereas at 6.4 you do. So I feel pretty confortable about the heating issue; we are confident that heating is not an important contributor to the data we have obtained in the transient photo-excitation experiments.

V. J. Emery

I want to show one viewgraph, which I have already shown in my talk, addressing the question that Alan raised. Is neutron evidence consistent with phase separation and how big are the domains? I would remind you of the curve from Keimer et al [Phys. Rev. Lett. 67, 1930 (1991): fig. 1] who looked at 2% and 4% Sr-doped materials. They were able to write the widths of their peaks as function of x and T in the form quoted in Eq. (5) of my paper. It consists of two parts, one $\kappa_0(x)$ which is independent of T, the other independent of x. Even more significantly, the part which depends on T follows the form expected for the undoped antiferro-magnet. You could imagine this to arise from domains of the size given by $\kappa_0(x)^{-1}$, which is 200 Å at 2% doping and 40 Å at 4% doping. In this sense there is a little evidence on both points.

P. Imbert

Local probe techniques such as Mössbauer spectroscopy and muon spin relaxation (μSR) experiments are particularly relevant to study phase separations between magnetically ordered regions and non magnetic regions. I will outline and compare the main results obtained by these techniques in the $La_{2-x}Sr_xCu0_4$ and $YBa_2Cu_3O_x$ series.

1. Mössbauer experiments

In the Mössbauer experiments, we used dilute $^{57}Fe^{3+}$ probes substituted for Cu in $La_{2-x}Sr_xCuO_4$ and dilute $^{170}Yb^{3+}$ probes substituted for Y in $YBa_2Cu_3O_x$.

In a certain sense, the $^{57}Fe^{3+}$ probe in the former series is "less innocent" than the $^{170}Yb^{3+}$ probe in the latter series (contrary to Yb, the Fe impurity lowers T_C and can induce localized magnetic moments on Cu neighbours in the superconducting phase). Another difference concerns the paramagnetic fluctuation rate of the two probes and therefore the type and the analysis procedure of the low temperature Mössbauer spectra (effective field spectra are observed on the ^{57}Fe nucleus, whereas slow relaxation paramagnetic hyperfine structures are observed on the $^{170}Yb^{3+}$ ion).

Despite these differences on the probes, the methodology used was essentially the same in the $La_{2-x}Sr_xCuO_4$ and $YBa_2Cu_3O_x$ samples:

- In a preliminary step, we have clearly identified in both series the response of the Mössbauer probe in the two following limiting situations: first, for small x values, when a Cu magnetic ordering is present throughout the sample (antiferromagnetic or spin-glass like semiconducting phase); and second, for large x values, when no intrinsic Cu magnetic ordering is present (superconducting or metallic phase).

- In a further step, we have observed the presence of mixed subspectra of both types for some intermediate x values, so evidencing the presence of a "phase separation" into magnetically ordered regions and regions without spontaneous Cu magnetic ordering. This two-fraction behaviour has been observed in superconducting samples of both series for moderate x doping levels. The relative volume P_{SG} of the (presumably semiconducting) magnetic fraction is given versus x in Figure 1.

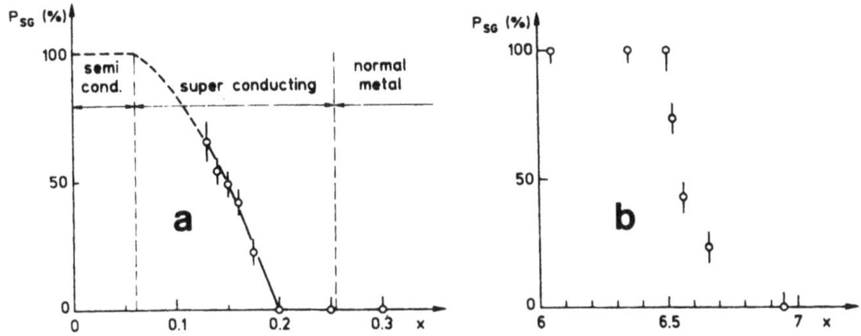

Figure 1: Relative volume $P_{SG}(\%)$ of the magnetic fraction (Cu spin-glass), respectively in (a): $La_{2-x}Sr_x(Cu_{0.995}Fe_{0.005})O_4$ and (b): $(Y_{0.97}Yb_{0.03})Ba_2Cu_3O_x$, from the Mössbauer data.

With the Mössbauer technique, we have no direct information on the spatial size of the magnetic and non magnetic regions in the samples. Concerning the dynamical Cu spin behaviour at the lowest temperature ($T = 1.5$K), we can assess the following limits to the fluctuation rates from our experiments:

– In the magnetically ordered regions, the local Cu antiferromagnetic arrangement is either static or slowly fluctuating (Cu spin fluctuation rate $W < 10^8 s^{-1}$).

– In the non magnetic regions, the Cu magnetic moments do not exist or they are rapidly fluctuating ($W > 10^{12} s^{-1}$ from ^{170}Yb measurements in $YBa_2Cu_3O_x$ and $W > 10^{10} s^{-1}$ from ^{57}Fe measurements in $La_{2-x}Sr_xCuO_4$).

2. μSR measurements.

The left part of Figure 2 shows zero field μSR spectra at $T < 1$K in $YBa_2Cu_3O_{6+x}$ samples, from the measurements of Weidinger *et al.* The muon asymmetry signal for $x = 0.57$ ($T_C = 29$K) and $x = 0.67$ ($T_C = 42$K) is fitted to 2 components. The exponentially decaying component comes from regions in the sample where internal magnetic fields of electronic origin are present. The relative volume of this magnetic fraction is plotted versus T_C in the right part of Figure 2. This magnetic fraction tends to vanish for compositions corresponding to $T_C \simeq 55$K. These results present strong analogies with our Mössbauer data in the same series.

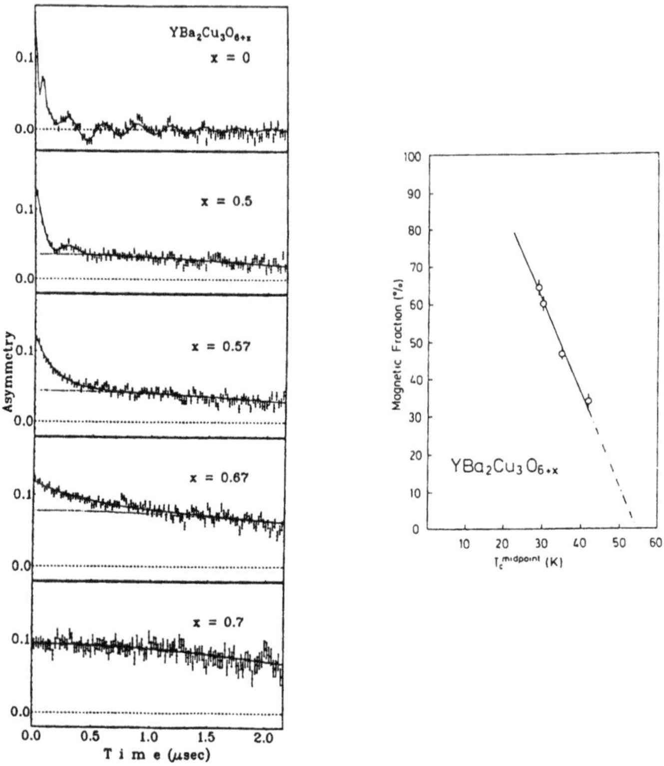

Figure 2: μSR results in $YBa_2Cu_3O_{6+x}$, from Weidinger *et al.* (*Hyperfine Interactions* **63** (1990) 147). Left part: zero field μSR spectra at $T < 1K$. Right part: magnetic fraction (%) versus T_C.

From low temperature μSR experiments in $La_{2-x}Sr_xCuO_4$, Weidinger *et al.* (*Phys. Rev. Lett.* **62** (1989) 102) have also observed the persistence of internal fields of electronic origin in superconducting samples for x ranging up to at least 0.15, as in our ^{57}Fe Mössbauer measurements in the same series. Let us mention that the presence of a spin-glass like fraction predominant for $x < 0.15$, has also been detected from nuclear spin lattice relaxation measurements on ^{139}La in superconducting samples of $La_{2-x}Sr_xCuO_4$ by Kobayashi *et al.* (*J. Phys. Soc. Jpn. (Japan)* **58** (1989) p. 779 and 2662). The presence of a phase separation in these compounds is thus evidenced by different local probe experiments.

In addition, from the decay time of the exponential component in the μSR signal, Weidinger *et al.* have extracted the average internal field $\langle|B|\rangle$ in the magnetic fraction. The value of $\langle|B|\rangle$ is plotted versus T_C in the Figure 3 for both $YBa_2Cu_3O_x$ and $La_{2-x}Sr_xCuO_4$. From the observed correlation between $\langle|B|\rangle$ and T_C when x is varied, these authors argue that magnetic regions and superconducting regions in the same crystal can influence each other and that they should be not separated by more than a few lattice constants. This argument could be however questioned, since T_C reflects percolation phenomena rather than the intrinsic occurence of superconducting properties in small parts of the crystal.

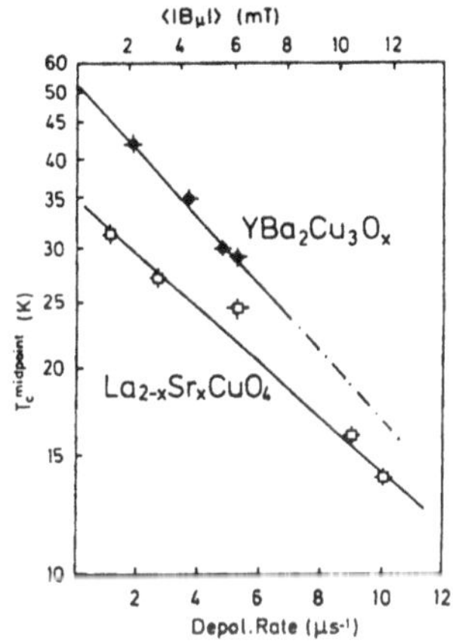

Figure 3: Correlation between T_C (left scale) and the internal magnetic field $\langle|B|\rangle$ (upper scale) from μSR data in $YBa_2Cu_3O_x$ and $La_{2-x}Sr_xCuO_4$ (from Weidinger *et al. Hyperfine Interact.* **63** (1990) 147).

3. Concluding remarks and open questions.

– As a conclusion, local probe experiments such as Mössbauer spectroscopy and μSR strongly suggest the coexistence of mixed magnetic and non magnetic fractions in superconducting samples of $La_{2-x}Sr_xCuO_4$ and $YBa_2Cu_3O_x$ for moderate doping levels x.

– When comparing these results with those obtained in NQR-NMR experiments, one should not forget that the NQR-NMR signal can be lost in the magnetically ordered regions: for example, the Cu NQR signal cannot be detected in the semiconducting spin-glass composition range in $La_{2-x}Sr_xCuO_4$, probably because the spin-echo decay time T_2 is much too short (Kitaoka *et al.*, *Physica C* **170** (1990) 189).

– The consistency of the local probe experiments with neutron experiments is not straighforward and should be examined in more details. From neutron scattering experiments in superconducting $La_{1.89}Sr_{0.11}CuO_4$, Birgeneau *et al.* (*Phys. Rev. B* **39** (1989) 2868) have reported the occurence of a gradual Cu spin freezing between 350K and 5K, which apparently suggests the presence of a large Cu spin-glass fraction. In contrast, Dr. Burlet has not observed a Cu spin-glass like fraction in superconducting samples of $YBa_2Cu_3O_x$ near $x = 6.5$.

– Is the observed "phase separation" in superconducting samples of $La_{2-x}Sr_xCuO_4$ and $YBa_2Cu_3O_x$ driven by composition fluctuations (such as oxygen occupancy modulation in the chains in $YBa_2Cu_3O_x$ or local modulation of the Sr concentration in $La_{2-x}Sr_xCuO_4$)? Or alternatively by electronic segregation into hole rich and hole poor regions? Or by a combination of the two effects?

H. Takagi

We have to figure out whether this kind of inhomogeneous picture is due to composition distributions or to electronic phase separation, this is an important point. At the moment, we do not have any evidence that it is due to an electronic phase separation, I think that it is likely to be due to composition fluctuations.

C. Di Castro

I have a comment which refers both to Drs. Imbert and Heeger. Both show that a phase separation appears for moderate doping. All have been questioning on the right miscibility gap from Jorgenson results on La-doped Sr cuprate between superconducting and metal phase. If the phase separation has an electronic origin, the absence of phase separation at high doping may be due to the fact that the metallic phase + attraction is always unstable to superconductivity. So if phase separation would occur in some region, it would be just washed out by the presence of superconductivity. Of course the point raised by Takagi — whether this is due to phase separation of electronic origin — is hard to answer. However if it is of electronic origin, then it is very relevant, otherwise we have to look for some other mechanism, e.g., as Bianconi suggests, the involvement of the lattice structure and superstructures in the pairing mechanism.

P. Burlet

About the consistency between Mössbauer and neutron results, we should not forget that in neutron experiments we have to extract the magnetic signal on large backgrounds. So we cannot completely exclude that a small part of the sample behaves like spin-glass, although we have never seen it.

P. Emery

Can you, from the interpretation of your data, tell us the difference between cluster spin-glass and other kinds of spin-glass? (...) Do you need to have small regions correlated antiferromagnetically in the immediate neighbourhood in order to explain your experiments, so that we would say we have cluster spin-glass?

P. Imbert

In the magnetic fraction, the low temperature Mössbauer spectra show that the electronic spin of the Mössbauer ion (Fe^{3+} or Yb^{3+}) experiences a static or slowly fluctuating molecular (exchange) field induced by the Cu neighbours. The presence of this quasi-static molecular field implies that the Cu spins are correlated in the vicinity of the probe, over distances of at least a few lattice constants. But we cannot set an upper limit to the correlation length from a local technique such as Mössbauer spectroscopy, and in that sense we cannot distinguish between cluster spin-glass and other kinds of spin-glass.

H. Takagi

Could you have iron clustering in your samples?

P. Imbert

I do not think that iron clustering is at the origin of the magnetic fraction, because we have also observed the persistence of a magnetic fraction in Mössbauer emission measurements where the $La_{2-x}Sr_xCuO_4$ samples are not doped with ^{57}Fe, but with the radioactive parent ^{57}Co at extremely low concentrations.

P. Emery

Dr. Takagi has raised the question of whether effects seen in Sr-doped materials should be attributed to chemical inhomogeneity rather than fluctuating electronic phase separation. The difference is that chemical inhomogeneity is static. In private discussion with Dr. Heeger and Dr. Hammel, we asked if it would be possible to use NQR to see "charge-glass" effects to go along with the spin-glass behaviour that has already been observed.

C. Hammel

The answer is that one needs to find a system in which quadrupolar relaxation dominates magnetic relaxation, and that is something which can be determined. Candidates are Ba in YBCO and La in LSCO. We shall be looking for that shortly.

P. Emery

I have two questions for the NMR experimentalists.

First, Dr. Mehring fitted T_1^{-1} on Cu and the Knight shift (K) to similar spin gap formulae. This suggests that KT_1 for Cu might have a simple form, for example

$$(KT_1)^{-1} = a + bT\,.$$

This looks roughly right for reduced-oxygen samples of YBCO, where $b \sim 0$. For YBCO with oxygen content $y = 6.92$, K is roughly constant and T_1^{-1} on Cu is of the form $a + bT$. Dr. Mehring has shown us the fits for 248 materials. So the questions are:

1) Is it generally true that one can fit K to the kind of formula used by Dr. Mehring?
2) Does $(KT_1)^{-1}$ for Cu have a simple form, just as $(KT_1T)^{-1}$ for oxygen has?

The second question I wish to ask of the experimentalists is how does the spin gap vary as a function of oxygen content y in YBCO? A combination of neutron scattering data (from Brookhaven and Grenoble) and NMR data suggests that the spin gap increases from about 60 K to 200 K in crossing the 60 K plateau region, with increasing y. If one goes to $y = 6.9$ ($T_c = 90$ K), we can put an upper limit on the spin gap by saying that we do not see it above T_c which implies that it must be less than 200 K. However it should be noted that the spin gap was fitted below T_c by neutrons and above T_c by NMR.

C. Hammel

The questions are difficult to answer carefully without analyzing the data. If y is reduced starting from O_7, the spin gap increases but then levels off.

P. Emery

How much data do you have across the plateau region in YBCO? I referred to YBCO because there is a lot of neutron data, but one should generalize and combine information from different materials.

Y. Berthier

In the Tl system, you can go to $T_c = 0$ in the overdoped regime and in this case NMR experiments have shown that the AF fluctuations are completely suppressed. This means that superconductivity is always accompanied by antiferromagnetic fluctuations.

P. Emery

OK but what kind of variation of the spin gap do you have there.

Y. Berthier

If I understand, you say that $1/T_1$ on Cu would be reflected on O to explain $1/KT_1 \sim T$. This means that you do not have a perfect filtering from Cu to O.

P. Emery

It may mean different things — it was simply that there was an empirical observation that a spin gap formula seems to work for K and T_1^{-1} on Cu. If the same gap appears in both, then dividing one by the other removes the fast variation. There are two examples, 90 K and 60 K materials.

Y. Berthier

I would say that this gap fit is only true for the 6.63 concentration, not for $y = 7$, where there is no variation.

P. Emery

That's right, it is consistent with zero gap for both K and T_1 on Cu. So there are three cases where it works, 123 on the 60 K plateau, 248 and 123 with $y = 7$, where it is trivial. So one might analyze any material in this way to see if it rationalizes the data and gives us something to think about.

H. Takagi

An obvious comment: a lot of discussions on electric phase separation and chemical phase separation. I think that the possibility of chemical phase separation promoted by electric phase separation is pretty interesting. We saw a nice example on oxygen-doped lanthanum cuprate. Actually in this phase there seems to be phase separation into antiferromagnetic insulator and hole high T_c superconductor. This will be very important in electric phase separation. However, in order to establish chemical phase separation we have to be very careful in examining the phases. This has already been addressed by Professor Heeger.

Furthermore the argument of inhomogeneity depends on the length scale. Actually if you look at the X-ray data you are probing on the length scale of thousand Å. If you measure transport the characteristic length is the free mean path of electrons, and here this would be of order 102 Å. The coherence length is pretty short in HT_C (about 20 Å). On this scale you can count Sr ions and the solution cannot be homogeneous. But if you look by X rays the material looks homogeneous. So you may look at completely different properties of the material, depending on the length scale.

Finally, the composition distribution is inherent of solid solution systems ans also we can have a kind of composition distribution due just to pure processing. I'd like to stress that composition distribution and phase separation are different comcepts. When you have chemical inhomogeneity you should always keep asking whether it is phase separation or composition distribution and we have to be careful about the terminology of phase separation.

P. C. Hammel

Neutrons and X rays are on the same length scale?

H. Takagi

The X-ray resolution is vastly better.

V. J. Emery

The width of the neutron peaks is giving you some information on the coherence length and domain sizes, then you are down onto a few lattice spacings, but it is not to answer questions about global inhomogeneities because neutron is a bulk probe, so it's going to average over the sample. It's not going to give variations across the sample, but it can see quite short length scales - averaged!

K.-A. Müller

Probably the new machines with high luminosity like the synchrotron to be built in Grenoble will be able to give also magnetic information, because magnetic scattering can be studied at the energies they have.

V. J. Emery

Magnetic scattering is already done in Brookhaven, of course, but you can do better with the bigger machines. But the other thing you may be able to do is to focus on very small parts of your sample, and then you will get more information on inhomegeneities.

C. Di Castro

Following the question that Takagi has posed to us, I ask Dr. Imbert: with Mössbauer probes are you really seing phase separation on the doping-induced destruction of the underlying antiferromagnetic clusters which are left from the original 3D antiferromagnetic order This would be very important for theoretical models. You remember that I presented a model (which I called Kondo-lattice model) where mobile holes of p character are moving in an antiferromagnetically correlated substrate. The pairing effect is due to this underlying correlation which is frustrated by the motion of the mobile holes. With increasing doping there is a reduction of effective pairing due to the substrate because you spoil the local residual antiferromagnetic correlation.

Thus your experiment would suggest that, by varying the doping, the number of sites where this correlation does not hold anymore is increased continuously, and then you spoil superconductivity. The T_c would first increase by doping, because you increase the carriers, then it decreases because you are spoiling the effective interaction. Thus the question is how much is phase separation and how much is indication that we are locally disturbing by doping the residual antiferromagnetic correlation. This connects to previous Vic Emery's question on how much is spin glass (cluster spin glass) and how much of this remains.

J.-C. Grenier

Following Dr. Takagi's comment, we could agree on one point: in lightly doped systems we have phase separation between antiferromagnetic and metallic phases; in heavily doped systems we have usually a metallic phase and only fluctuations of composition which are fixed, e.g., in 123 compounds, by annealing at high T, and, amazingly, in the case of $La_2CuO_{4+\delta}$ it seems to be related to mobile oxygen.

H. Takagi

I agree that there is the final evidence for phase separation on oxygen-doped La cuprate and in this case you see a separation into antiferromagnet and hole-superconductor. The coincidence of antiferromagnetism and superconductivity is accidental or not? This is in my opinion still an open question. In the Sr-doped case I think we don't have yet the final evidence for phase separation

K.-A. Müller

This is precisely what Dr. Kremer is going to address in his presentation.

R. F. Kremer (Max-Planck-Institut für Festkörperforschung,
Heisenbergstr. 1, D-7 Stuttgart 80, Germany)

I want to address a little bit this question which was just raised by Dr. Grenier and discussed by Dr. Tagaki. Let me briefly display the transparency again which you have already seen some days ago in the lectures of Sigmund and Mehring.

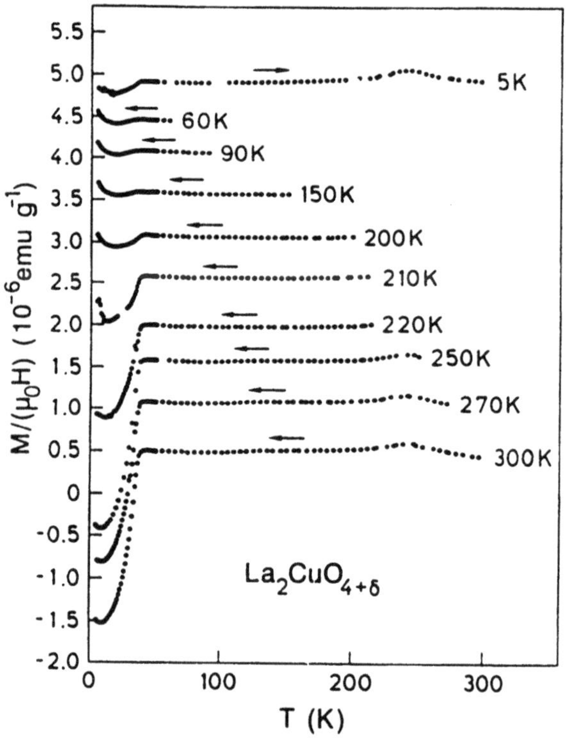

Fig. 1 Magnetization of an 'as-prepared' $La_2CuO_{4+\delta}$ sample as a function of the temperature. The starting temperatures T_s to which the sample was quenched from room temperature are indicated. The uppermost curve was obtained after zf quenching the sample to 5 K and slowly heating to room temperature. Beginning from the lowest data set each curve was shifted upwards by a value of 5×10^{-7} emu g^{-1} compared to the preceding one. The measuring field was 90 G, arrows indicate the direction of temperature change during the measurement.

From the results shown here we think we can draw the conclusion that we have a phase separation in oxygenated lanthanum cuprate. We have delt with as prepared lanthanum cuprate and measured the Meissner effect in very low fields of about 10 mT. The lowest curve is obtained by starting from room temperature and slowly cooling down the sample. We observe quite an appreciable Meissner signal from

which we estimate the amount of superconducting phase to about 200 to 300 ppm. Then we take the same sample and quench it down from room temperature to several 'starting' temperatures, T_s, from which on we slowly cool down and measure the Meissner signal. The quenching process was performed with a cooling rate of the order of 10 K sec^{-1}. We successively lower T_s and we observe that the Meissner signal decreases sharply in the temperature range between 250 K and 200 K. From this sudden drop we conclude that a diffusive process is suppressed when the sample is quenched below this temperature range.
The uppermost curve was obtained after the sample had been quenched from room temperature to 5 K in zero-field. A field was applied and the shielding signal was measured. You see that that there is hardly a diamagnetic signal left. It is about one and a half orders of magnitude lower than the Meissner signal obtained after slow cooling the sample.
The next transparancy demonstrates that the Meissner signal not only can be affected by a thermal treatment but also by applying a polarizing magnetic field.

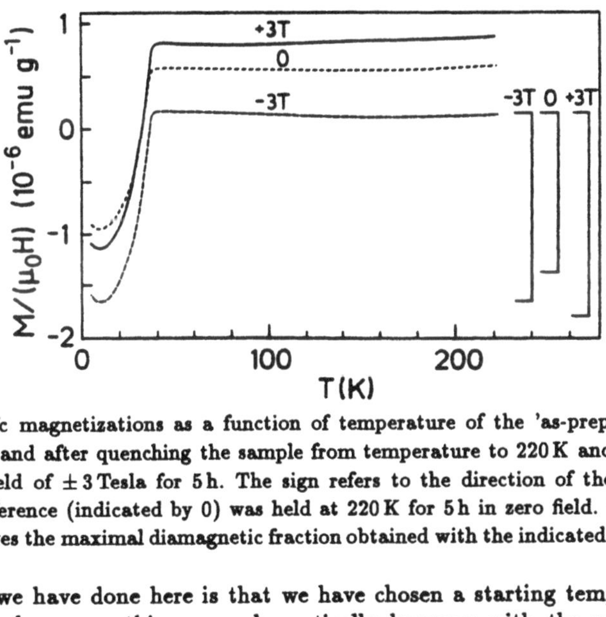

Fig. 2 fc magnetizations as a function of temperature of the 'as-prepared' $La_2CuO_{4+\delta}$ sample and after quenching the sample from temperature to 220 K and applying a magnetic field of $\pm$ 3 Tesla for 5 h. The sign refers to the direction of the measuring field. The reference (indicated by 0) was held at 220 K for 5 h in zero field. The length of the bars gives the maximal diamagnetic fraction obtained with the indicated polarization field.

What we have done here is that we have chosen a starting temperature around 220 K where something very dramatically happens with the sample. We have quenched the sample down from room temperature to 220 K and held it there for 5 h. We quenched down in zero field and then apply a field which again is about 10 mT and obtain the dashed curve marked by '0'. This curve serves as reference. Then we do the same thermal procedure but apply at 220 K a polarizing field of ± 3 T for 5 hours, the sign refers to the direction of the measuring field. After 5 h we switch back to the measuring field of 10 mT. We obtain some polarization in the paramagnetic regime and also a significant enhancement of the Meissner signal as indicated by the bars on the right hand sight of the figure. The next two transparencies show you how the onset of superconductivity shifts if one quenches or slowly cools a sample of 'as-prepared' $La_2CuO_{4+\delta}$.

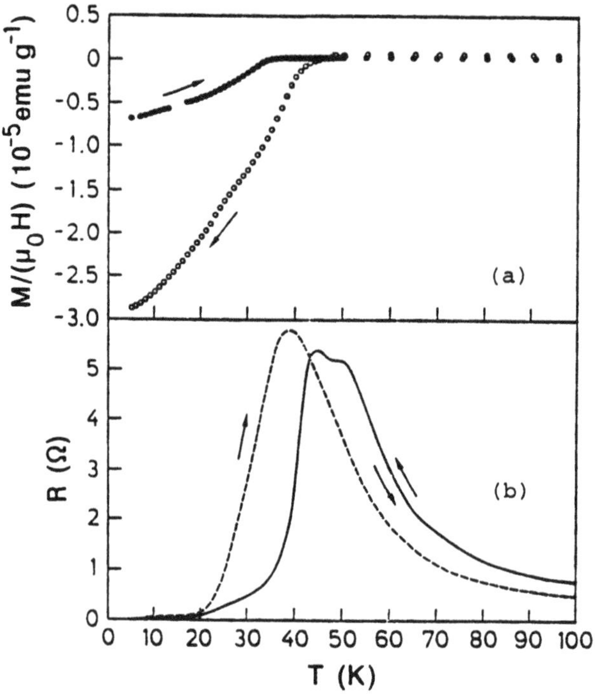

Fig. 3a Magnetization of 'as-prepared' $La_2CuO_{4+\delta}$ obtained at 10 mT. The curve marked by full symbols was obtained after the sample had been zero - field quenched from room - temperature to 5 K, the curve marked by open symbols represents the slow field - cooled result.

3b Resistance of the same sample measured during very slow heating and cooling cycles (full line) and after a very rapid quench from room - temperature to 5 K during a heating cycle (dashed line). Arrows indicate the direction of temperature change during the measurement.

Fig. 3a displays the susceptibility results. The lower curve is obtained from the slow cooled sample the upper curve represents the quenched sample. One observes a difference of the onset temperature of about 10 degrees. The quenched sample has the lower onset temperature. Similar observations have been made before by Wilson and coworkers[1] in resistivity measurements. The results of our resistivity experiments are displayed in Fig. 3b. If one puts the two transparencies on top of the other we see that the onset temperature obtained from susceptibility and resistivity experiments match perfectly.

In summary, we have demonstrated that by a proper thermal treatment and also by a suitable treatment with polarizing magnetic fields the phase separation in lightly oxygen doped lanthanum cuprate very effectively can be manipulated. One not only can suppress the diagmagnetic signal by more than an order of magnitude but also shift the onset temperature by about 10 degrees. Thank you.

1. J. Ryder, P.A. Midgley, R. Exley, R.J. Beynon, D.L. Yates, L. Afalfiz, J.A. Wilson, *Physica* C **173** (1991) 9.

J.-C. Grenier

Yes, I can make a comment. These are exactly the same results that we obtained. There are two different temperatures in the system, 32 K and 44 K, and the question is does it correspond to a peculiar ordering of the oxygen in the system? I think this is now the basic question which we will try to answer in a few months.

K.-A. Müller

So, if you cool slowly the excess oxygens order. It seems that if you have these donors ordered, the system likes it better.

J.-C. Grenier

If I can interpret these results. You have a phase separation now when you make a slow cooling between La_2CuO_4 and a very rich or very well ordered phase. This is what I can say today.

R. F. Kremer

The question remains what is really going on. Is it chemical phase separation, is some oxygen migration involved or is it just electronical or is it both or one pulling the other? From our experience a phase separation is confirmed for oxygen doped lanthanum cuprate whatever the details finally will be and whatever will go together finally. Magnetism is certainly involved as demonstrated by the magnetic field polarization experiment.

Oxygen diffusion can be ruled out for Sr doped samples. Presently, we are therefore investigating a whole series of lightly Sr doped samples. The only statement I want to give about these results so far is that everything seems to be much more complicated. If there is a phase separation the effects are by far not as mighty as we see it for the oxygen doped samples.

K.-A. Müller

May I guess, the times involved here are so much longer than the optical excitation that I feel the optical excitation gives you an intrinsic property.

R. F. Kremer

You have seen from Tagaki's presentation that even at 800 °C ages are involved.

V. Hizhnyakov

A short remark about whether it is electronical, chemical or joint phase separation. I think here we have really one experimental fact which he obtained it is just magnetic field influenced phase separation. So, if it is pure chemical the oxygen diamagnetic ion cannot feel magnetic fields. So, I think that for this concrete situation one should come to the conclusion if these experiments are correct it is a joint phase separation.

K.-A. Müller

A year ago I voiced such a possibility, if you remember.

P. C. Hammel

I think that from our data in the lanthanum cuprate it is really hard to doubt that magnetism plays some role in the phase separation but that does not allow you to conclude that it is either electronic or chemical but I think there is no doubt that magnetism plays some role.

E. Sigmund

I want to mark a few points concerning our model of the percolative phase separation.

A hole introduced into the antiferromagnetically ordered CuO_2 plane changes the surrounding Cu spins forming a spin-polarized cluster (magnetic polaron, ferron). From our calculations as well as from the measurements performed by Dr. Kremer we estimated the size of the averaged cluster to contain around 10 Cu spins. Such a cluster has a large effective mass and is practically immobile. Therefore and in order to reach a conducting or superconducting state of the system we have to allow the spin-polarized single-hole clusters to percolate building up conducting or superconducting percolation paths.

In order to check the percolation idea we thought about how to influence the formation of the percolation net. This led us to the experiments of Dr. Kremer, where the percolation is influenced by thermal treatments or by magnetic fields. It turned out that this influence is different in $La_2CuO_{4+\delta}$ and in $La_{2-x}Sr_xCuO_4$ samples. It seems that in the latter sample the mobility of the spin-polarized single-hole clusters is smaller than in the former one suggesting that the holes are stronger bound to Sr than to O.

At low hole concentrations each cluster is formed by one hole, which induces one or two Cu-spin flips. However, at higher hole concentrations the clusters contain two or more holes and the percolation net starts to become metallic. Within this picture the concentration thresholds for destruction of long-range antiferromagnetic order and beginning of conductivity can be calculated easily. When the in-plane percolation starts 30 to 40% of the plane is covered by metallic or spin-polarized clusters. When conductivity starts already 80% of the plane is occupied by the percolation net and by single isolated clusters. From these considerations we can calculate the amount of remaining antiferromagnetically correlated domains. This gives us an estimation of the coherence length of the remaining antiferromagnetic fluctuations, which seems to be in agreement with experiments.

C. Castellani

Dipartimento di Fisica, Università "La Sapienza"
2, Piazzale Aldo Moro, 00185 Roma, Italy

I will comment on the most relevant topics concerning phase separation and superconductivity as they emerged in this Workshop

Soon after the work by Emery, Kivelson and Lin [1], it became evident that phase separation appears as a "generic" feature in most of the short range models for high Tc materials [2]. This makes urgent and theoretically relevant the question whether phase separation and superconductivity are or are not related. Instead of supporting one of these views, let me discuss the various scenarios that would result from each of them in the light of the theoretical work presented in this Workshop.

1st Scenario: superconductivity at the boundary of phase separation from short range forces. I can envisage two ways in which the connection between phase separation and superconductivity can be realized in the cuprates. The understanding of the simplest one starts from the observation that both phenomena are caused by the presence of effective attractive forces in the system. Specifically, phase separation is related to the presence of a (unstable) region in doping with negative compressibility χ_ρ. Remember that for a Fermi system χ_ρ can be written in terms of the quasiparticle parameters as

$$\chi_\rho = N_{QP}(1 - A_0^s) \tag{1}$$

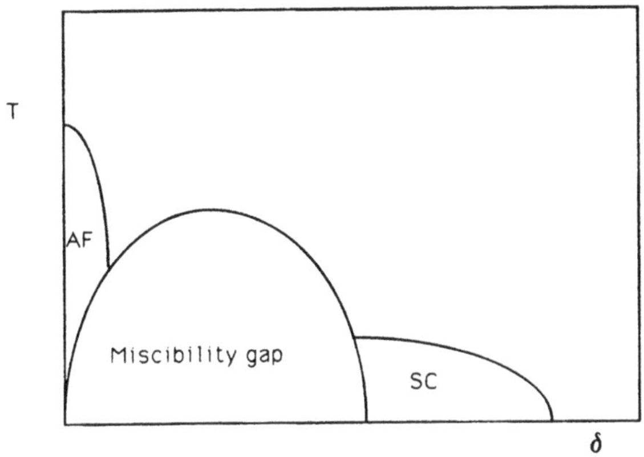

Figure 1: *Schematic phase diagram within the first Scenario*

where N_{QP} is the quasiparticle density of states. A_0^s is the symmetric zero harmonic Landau interaction amplitude and is given by the q and ω zero limit of the quasiparticle interaction at the Fermi surface

$$A_0^s = N_{QP} lim_{q \to 0} lim_{\omega \to 0} \int d\widehat{kk'} V_{kk'}(q, \omega) \tag{2}$$

The boundary of the region of negative compressibility has infinite χ_ρ and is then identified by the condition $A_0^s = -\infty$ i.e. by an infinite effective attraction at zero transferred energy and momentum (with $\omega/q = 0$ and neglecting any irrelevant k,k'-dependence in $V_{k,k'}$).

In a loose sense, the condition for a superconducting instability is less stringent than the condition for phase separation since pairing requires only that some average of $V_{kk'}$ over the Fermi surface in the Cooper channel is negative. For instance the s-wave Cooper coupling is given by

$$A_s^c = N_{QP} \int d\widehat{kk'} V_{k,-k}(q = k - k', \omega = 0) \tag{3}$$

It is evident that not only the q=0 term contributes to A_s^c; however if the negative V region has sufficient weigth in q-space, then A_s^c can be negative before phase separation appears. A realization of that was discussed by Grilli and al. [3] and reported at this Workshop in Di Castro's contribution. The expected phase diagram for this scenario is depicted in Fig. 1, where, assuming the presence of a phase separation between an antiferromagnetic phase with zero doping and a metal at finite doping, superconductivity

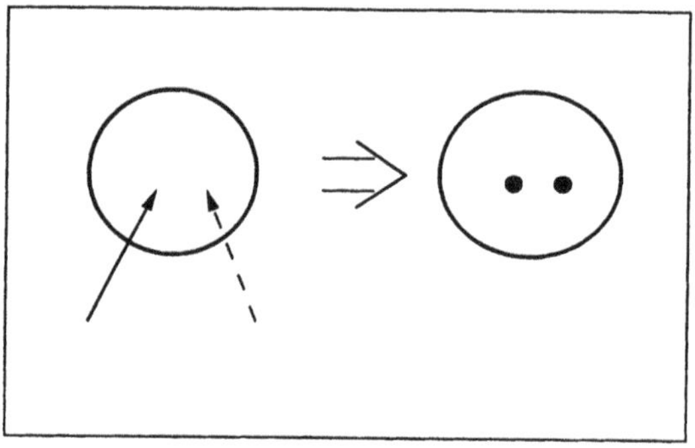

Figure 2: *An additional hole is attracted to form locally the stable hole rich phase*

takes place at the metallic side of the miscibility gap. The superconducting phase is by itself a homogeneous phase and for doping in the miscibility gap the system separates into an antiferromagnetic insulator and a superconducting metal. According to the data of Jorgensen et al [4] and to the experimental results presented during this Workshop, there are evidences that this situation applies to lanthanum systems doped with an excess of oxygen.

2nd Scenario: long range forces play a main role. The framework depicted above presents a rather simple (and natural) relation between phase separation and superconductivity, but neglects completely the presence of long range forces. Long range forces are in fact relevant if there is not ion segregation to compensate for hole (electron) segregation, and lead to what Emery called in his talk a frustrated phase separation. The competition between short range forces (inducing phase separation) and long range forces will result in the formation of (possibly fluctuating) microstructures or droplets whose size can be roughly estimated to be of few lattice constants by comparing the electrostatic energy cost to the energy gain from charge segregation . In the presence of such microstructures a possible mechanism for pairing is illustrated in Fig. 2. As soon as a mobile hole enters a $\delta = 0$ (antiferromagnetically "ordered") region, this region is no more stable; therefore an additional hole is attracted to form locally the hole rich stable phase. In order to work, this mechanism of pairing requires a specific optimal size for the microstructures. Let N_s be the number of copper sites in the microregion and let N_h be the hole number in the antiferromagnetic local phase with $N_h/N_s = 1$. Taking $\delta_{metal} = .2$ as a reasonable value for the doping in the hole rich phase, we should have

$$(N_h + 2)/N_s = 1 + \delta_{metal} = 1.2 \tag{4}$$

which leads to $N_s = 10$ in agreement with the previous rough estimate of the droplet size.

As pointed out by Emery, the interplay between phase separation and long range forces can have important consequences even for the normal properties of the doped systems at low doping. The presence of slowly fluctuating microstructures resembles in fact a glassy phase for the charge degrees of freedom. In these conditions, the density fluctuations act on the carriers as a source of scattering with no lower cutoff in energy and this leads naturally to a marginal Fermi liquid behavior in the metallic regime.

The models. At this point, a few remarks are in order on the underlying microscopic models for phase separation. In this context, the most relevant models on a physical ground are the t-J model (single band or multiband) and the extended three band Hubbard model. They pick up two different features which are possibly relevant in the cuprates, i.e. the presence of strong magnetic effects and the presence of excitonic effects induced by a nearest neighbour copper oxygen repulsion (see Di Castro's talk and references therein). Both models present phase separation, albeit with different mechanisms [1, 5, 6, 3, 7]. In the t-J model phase separation resolves the frustrating competition between the kinetic energy of the holes from doping and the number of broken antiferromagnetic bonds, while in the excitonic model the competition is between the kinetic energy and the oxygen-copper repulsion.

The t-J model is strongly supported by the experimental finding of a phase separation between an antiferromagnet and a metal (a superconductor at low temperature). As we have heard here, the presence of such a phase separation in some cuprates is by now experimentally established. On the other hand, while the neutron data of Jorgensen el al. [4] for strontium doped lanthanum compounds did not find support in this Workshop, the presence of two superconducting phases with two different dopings (as suggested for $La_2CuO_{4+\delta}$ by Grenier) resembles the predictions of the excitonic model.

It is a still open and interesting problem whether the two mechanisms for phase separation cooperate and to what extent this cooperation depends on doping concentration.

3rd Scenario: back to phonons. The scenarios I have discussed up to now both relate superconductivity to phase separation. Let me finally comment on the opposite view that superconductivity is not related to phase separation and that the mechanisms inducing phase separation in the models with short range forces do not play any relevant role in the pairing mechanism. In this context the phonons (possibly in the presence of the superstructures Bianconi talked about) are the "natural" candidates for pairing [9]. Proposals along this line have in fact been discussed during the Workshop.

Such a scenario could be "boring" from a theoretical point of view, in the sense that it remains within the traditional framework of the low temperature superconductivity. However, it would be not "trivial", in the sense that it is not trivial to reconnect the phonon picture with the known properties of the cuprates. Some relevant questions to be answered are in my opinion the following ones.

- The bipolaronic picture seems to be the fate of a strong electron-phonon coupling. If this is the case in the cuprates, the problem is how the bipolaronic limit can be reconciled with the normal state properties which are not "anomalous" (for instance the Fermi surface in agreement with band structure calculations and the not very large effective mass, probably no more than 4m [8]).

-Away from the bipolaronic limit a simplified formulation of the McMillan approach gives

$$T_c \propto e^{-1/(\lambda-\mu^*)} \tag{5}$$

How large should be λ to overcome a μ^* which could be quite large since correlations are in any case strong in these systems? Add to this the difficulty to find simple screening effects for μ^* over distances of order of the coherence length $\xi_0 \approx 10 - 15\mathring{A}$.

-The cuprates in the metallic regime have narrow bands, with ϵ_F of order of .2-.4 eV. Within these values the ratio ϵ_F/ω_D (where ω_D is some typical phonon frequency) is not very large and a naive application of the screening formula for μ^*

$$\mu^* = V_{Coulomb}/(1 + V_{Coulomb} N_{QP} ln \epsilon_F/\omega_D) \tag{6}$$

leads to a very poor screening of the Coulómb repulsion. Moreover the Migdal "theorem" does not apply, i.e. vertex corrections cannot be neglected. Whether these corrections can favour pairing (as suggested by Pietronero [10]) or further reduce T_c is a very intriguing open problem.

References

[1] V. J. Emery, S. A. Kivelson, and H. Q. Lin, Phys. Rev. Lett. 64, 475 (1990).

[2] J. G. Bednorz, and K. A. Müller, Z. Phys. B 64, 189 (1986).

[3] M. Grilli, R. Raimondi, C. Castellani, C. Di Castro, and G. Kotliar, Phys. Rev. Lett. 67, 259 (1991).

[4] J. D. Jorgensen et al., Phys. Rev. B 38, 11337 (1988); M. Hundley et al., Phys. Rev. B 41, 4062 (1990); J. D. Jorgensen et al., *Third International Symposium on Superconductivity*, Sendai, Japan, 6-9 November 1990, Springer-Verlag Tokyo.

[5] C. Castellani, M. Grilli, and G. Kotliar, Phys. Rev. B 43, 8000 (1991).

[6] M. Grilli, C. Castellani, and G. Kotliar, Phys. Rev. B 45, 10805 (1992)

[7] Y. Bang, G. Kotliar, C. Castellani, M. Grilli and R. Raimondi, Phys. Rev. B 43, 13724 (1991).

[8] For a review see e.g. W. E. Pickett, H. Krakauer, R. E. Cohen, and D. J. Singh, Science 255, 46 (1992)

[9] Obviously further scenarios (still involving short range forces away from the phase separation regime) would also deserve consideration. However these topics are not strictly connected with Phase Separation and were not addressed during the Workshop.

[10] L. Pietronero, and S. Strässler, Europhys. Lett. 18, 627 (1992)

www.ingramcontent.com/pod-product-compliance
Lightning Source LLC
Chambersburg PA
CBHW041304210926
51851CB00002B/5
9789810212742